XVII Brazilian Colloquium on Orbital Dynamics (CBDO 2014)

Journal of Physics: Conference Series Volume 641

Aguas de Lindoia, Brazil
1 - 5 December 2014

ISBN: 978-1-5108-1468-4
ISSN: 1742-6588

Printed from e-media with permission by:

Curran Associates, Inc.
57 Morehouse Lane
Red Hook, NY 12571

Some format issues inherent in the e-media version may also appear in this print version.

Printed by Curran Associates, Inc. (2016)

For permission requests, please contact the Institute of Physics
at the address below.

Institute of Physics
Dirac House, Temple Back
Bristol BS1 6BE UK

Phone: 44 1 17 929 7481
Fax: 44 1 17 920 0979

techtracking@iop.org

Additional copies of this publication are available from:

Curran Associates, Inc.
57 Morehouse Lane
Red Hook, NY 12571 USA
Phone: 845-758-0400
Fax: 845-758-2634
Email: curran@proceedings.com
Web: www.proceedings.com

Table of contents

Volume 641

XVII Brazilian Colloquium on Orbital Dynamics, CBDO 2014
1–5 December 2014, Águas de Lindoia, SP, Brazil

Accepted papers received: 1 September 2015
Published online: 7 October 2015

Preface

Papers

Journal of Physics: Conference Series **641** (2015) 011001 doi:10.1088/1742-6596/641/1/011001

Celestial Mechanics: from the bases of the past to the challenges of the future

C F de Melo[1], A F B A Prado[2], E E N Macau[2], O C Winter[3], V M Gomes[3]

[1] Federal University of Minas Gerais – UFMG, Belo Horizonte (MG), Brazil
[2] National Institute for Space Research – INPE, São José dos Campos (SP), Brazil
[3] São Paulo University State – UNESP, Guaratinguetá (SP), Brazil

E-mail: cristianofiorilo@ufmg.br

Abstract: This special issue of Journal of Physics: Conference Series brings a set of 31 papers presented in the Brazilian Colloquium on Orbital Dynamics (CBDO), held on December 1 - 5, 2014, in the city of Águas de Lindoia (SP), Brazil. CBDO is a traditional and important scientific meeting in the areas of Theoretical and Applied Celestial Mechanics. The meeting takes place every two years, when researchers from South America and also guests from other continents present their works and discuss the paths trodden by the space sciences.

1. Introduction

Astronomy is considered the oldest of the sciences because of interest and, especially, the need to understand the celestial phenomena since the beginning of mankind. Celestial mechanics is a branch of astronomy which studies the dynamics of the bodies under the action of gravitational interaction. Its origin was motivated by administrative and often religious aspects of ancient civilizations. The observation of periodicities in the celestial bodies' motion allowed the Babylonians, Egyptians, Greeks, Arabs, Chinese, Incas, Aztecs, Mayas and others to build calendars and to develop agriculture. They were able to forecast phenomena such as eclipses, the seasons and the precession of the equinoxes.

2. Modern Celestial Mechanics

The start of modern celestial mechanics is rooted in the 16th and 17th centuries with works such as those of Nicolaus Copernicus (1473 - 1543), Tycho Brahe (1546 - 1601), Johannes Kepler (1572 - 1630), Galileo Galilei (1564 - 1642) - the first man to use a telescope to see the sky - and Isaac Newton (1642 - 1727). Newton presented the first formulation of gravitational force, which was published in his book entitled *Phylosophiae Naturalis Principia Mathematica*, in 1687. In the following centuries, many scientists added important contributions to celestial mechanics, such as Leonhard Euler (1707 - 1783), Jean le Rond d'Alambert (1717 - 1783), Giuseppe Lagrange (1736 - 1813), Pierre Laplace (1749 - 1827), Adrien-Marie Legendre (1752 - 1833), Hamilton William (1805 - 1865), Henri Poincaré (1854 - 1912), Lyapunov Aleksandr (1857 - 1918), and many others. The works of these scientists have resulted in extraordinary advances in terms of understanding the origin and evolution of the Solar System. They have also paved the way for the dawn of the space age of the 20th century [1].

Journal of Physics: Conference Series **641** (2015) 011001 doi:10.1088/1742-6596/641/1/011001

3. Celestial Mechanics in 20th and 21st centuries

The twentieth century brought new requirements for mankind. After wars, the consolidation of urban societies and population growth, regular activities like communication between people, governments, research institutions, banks and commercial and industrial corporations require high speed and accuracy. In addition, high speed and accuracy are also needed for weather forecasting to reduce the impact of natural disasters, support high-productivity agriculture, navigation, transportation of people and the trade of goods, public safety, natural resources exploitation, etc. In this environment and in order to respond to these new demands, astronautics (or astrodynamics) arose in the second half of the 20th century as a branch of celestial mechanics. The development of rockets allowed the placement and maintenance of thousands of satellites orbiting the Earth which carry out all sorts of missions in response to the requirements of the modern lifestyle. The astronautics of the 20th century also allowed the expansion of knowledge on the Solar System. Probes were sent to all planets, man walked on the Moon, robots explored the surface of Mars and the Hubble space telescope revealed new "old" objects of the Solar System and the Universe [2]. All of this has driven astronomy and celestial mechanics forward.

Now, in the 21st century, about 3,500 satellites are in operation around the Earth, probes and/or robots are exploring the Moon, Mars, Ceres, Saturn, Pluto, asteroids and comets. The Kepler telescope is right now in space finding exoplanets and new planetary systems, and thereby providing new challenges for celestial mechanics. Astronautics uses, and will keep using, the theories of celestial mechanics to send more and more probes and manned spacecrafts to the farthest corners of the Solar System, without forgetting, however, a problem it itself created: space debris. Because of this, one can glimpse years and years of hard work and development for celestial mechanics.

4. Brazilian Colloquium on Orbital Dynamics, CBDO

In order to discuss and develop celestial mechanics in Brazil, a group of Brazilian scientists formed by Sylvio Ferraz-Mello, Rodolpho V. de Moraes, Atair Rios Neto and Wagner Sessin (1946-1997) organized the first Brazilian Colloquium on Orbital Dynamics, CBDO, in 1982. At that time, Brazil was preparing the means to develop, build and launch its first satellites, while consolidating important research groups on celestial mechanics. Since then, the CBDO has taken place every two years, gathering scientists from South America as well as guests from other continents [3].

The main areas the CBDO deals with are: theoretical celestial mechanics, dynamical and planetary astronomy, planetary science and Solar System dynamics, astronautics and dynamical systems and chaos applied to space research.

This special issue of Journal of Physics: Conference Series brings papers presented in the 17th CBDO, held on December 1-5, 2014, in Águas de Lindoia (SP), Brazil. The papers deal with areas of theoretical and applied celestial mechanics such as: planetary science, trajectory optimization, attitude control of spacecrafts, swing-by dynamics, flying formation, space propulsion, space robotic exploration and space debris.

The 31 papers selected have between six and nine pages and bring together the outcomes of research studies developed in Brazilian universities and research institutes. They represent the results of the strengthening of important research groups and the formation of new researchers and engineers who will work in Brazilian space activities in the coming decades. This issue presents papers discussing planetary formation, the optimization of lunar transfer trajectories, low-thrust maneuvers around the Earth and Mars, swing-by strategies, analysis of missions to Europa, a moon of Jupiter, and the triple asteroid (153591) 2001 SN263 - target of the first Brazilian mission to the deep space, called ASTER project [4]. This special issue also contains works regarding the development of continuous propulsion systems, attitude control of spacecraft, CubeSats, space debris, tethered space systems, space robots and chaos applied to space exploration. In the next section, the papers are presented one by one.

Journal of Physics: Conference Series **641** (2015) 011001 doi:10.1088/1742-6596/641/1/011001

5. Introducing the papers

In this section, we present the 31 selected papers of this special issue. We also take this opportunity to thank all the authors for their valuable contributions, and for investing a lot of work and effort into the results of their research work. Special thanks go to reviewers who dedicated their precious time in providing numerous comments and suggestions, criticism and constant and enthusiastic support. The papers are presented one by one in sequence.

The paper "Preliminary Analysis of Optimal Round Trip Lunar Missions", by L. A. Gagg Filho and S. S. Fernandes, studies optimal bi impulsive transfer maneuvers considering the particular problem of sending a spacecraft to the Moon and back to the Earth with minimum fuel consumption [5-7].

The paper "Estimating the trajectory of a space vehicle passing by the Moon using Kalman Filter", by A. F. S. Ferreira, H. K. Kuga, A. F. B. A. Prado and O. C. Winter, uses the Kalman filter to try to estimate parameters of a swing-by maneuver [8] using the Moon as the celestial body for the close approach.

The paper" Multiple island chains in wave-particle interactions", by M. C. de Sousa, I. L. Caldas, A. M. Ozorio de Almeida, F. B. Rizzato and R. Pakter, presents an analysis of isochronous island chains in Poincaré sections near integrable twist perturbation systems [9,10].

The paper "Space Tether Systems – Stability Solutions of a Dumbbell-like System", by D. P. S. dos Santos, S. A. R. F. A. Morant, A. D. Guerman and A. A. Burov, searches for geometries that generate stable solutions for tethers [11] in astrodynamics.

The paper "Particle transport induced by electrostatic wave fluctuations", by K. C. Rosalen, M. Roberto and I. L. Caldas, is a study on the numerical investigation of particle transport driven by electrostatic waves at the plasma edge for a large ratio tokamaks, by considering a kinetic model derived from guiding-center equations of motion [12].

The paper "Nonlinear Least Squares Method for Gyros Bias and Attitude Estimation using Satellite Attitude and Orbit Toolbox for Matlab", by W. R. Silva, H. K. Kuga and M. C. Zanardi, considers the problem of estimating the attitude [13,14] and the gyro bias of a satellite using techniques based on least squares.

The paper "ALR – Laser altimeter for the ASTER deep space mission. Simulated operation above a surface with crater", by A. B. V. de Brum, F. C. da Cruz and A. Hetem Jr., considers details of a laser altimeter that is designed to study the surface of the three bodies of the asteroid 2001SN$_{263}$, which is the target of the ASTER mission [4,15].

The paper "Close Approach Maneuvers around an Oblate Planet", by G. M. C. Oliveira, A. F. B. A. Prado and D. M. Sanchez, studies the effects of the non-spherical form of a celestial body in a swing-by maneuver [8,16], by making numerical simulations of the restricted three-body problem with the additional force due to the J_2 term of the potential of the Earth.

The paper "Three-dimensional Two-Body Tether System – Equilibrium Solutions", by D. P. S. dos Santos and A. Ferreira, searches for equilibrium points in space where a tether [11,17] can be located and have a zero resultant force acting on it.

The paper "A study of optimal low-thrust limited-power transfers between coplanar orbits with small eccentricities", by F. C. Carvalho and S. S. Fernandes, studies the important problem of transferring a spacecraft between coplanar orbits with small eccentricities [7,18] using a propulsion system that is able to deliver a low thrust to the spacecraft.

The paper "Searching for less perturbed elliptical orbits around Europa", by J. C. dos Santos, J. P. S. Carvalho, A. F. B. A. Prado and R. V. de Moraes, uses the integral of the perturbing forces [19,20] to generate maps of orbits that can show their level of perturbation, indicating the ones that are less perturbed to be used by spacecraft.

The paper "Study of the influence of computational parameters on the formation of a giant gaseous planet", by R. A. Moraes and E. Vieira Neto, studies the problem of the formation of giant planets similar to Jupiter and Saturn [21,22] from a practical point of view of the computational parameters used in the simulations.

Journal of Physics: Conference Series **641** (2015) 011001 doi:10.1088/1742-6596/641/1/011001

The paper “Experiment on signal filter combinations for the analysis of information from inertial measurement units in AOCS”, by M. N. Pontuschka, I. M da Fonseca and M. A. A. Melo, introduces an investigation into an FDIR software subsystem that may be part of the attitude and orbit control subsystem, AOCS. An experiment encompassing an accelerometer and a wireless communication system so as to provide input signals to be filtered by the filtering algorithm is also shown [23,24].

The paper "Out-of-Plane Orbital Maneuvers Using Swing-bys with the Moon", by J. B. S. Neto, A. F. B. A. Prado and J. K. S. Formiga, verifies the conditions where a maneuver using a swing-by with the Moon is advantageous with respect to a standard one-impulsive maneuver to change the inclination of the orbit of a spacecraft around the Earth [7,25].

The paper “If there dissipation the particle can gain energy”, by R. E. de Carvalho, describes two different mechanisms to gain energy from the presence of dissipation in a time-dependent non-linear system [26,27].

The paper "Permanent Magnet Hall Thrusters Development and Applications on Future Brazilian Space Missions", by J. L. Ferreira, A. A. Martins, R. Miranda, A. Schelling, L. Souza, E. G. Costa, H. O. Coelho, A. C. B. Serra and F. S. Nathan, studies several aspects of a low thrust engine, that can be used to maneuver a satellite, which is being developed in the Plasma Physics Laboratory of the National University of Brasilia [28].

The paper "Ball’s motion, sliding friction, and internal load distribution in a high-speed ball bearing subjected to a combined radial, thrust, and moment load, applied to the inner rings center of mass: Mathematical model", by M. C. Ricci, develops set of non-linear algebraic equations, which must to be solved using a numerical procedure, for Ball’s motion, sliding friction and internal loading distribution computation for a high-speed ball bearing [29,30].

The paper "Ball’s motion, sliding friction, and internal load distribution in a high-speed ball bearing subjected to a combined radial, thrust, and moment load, applied to the inner ring’s center of mass: Numerical Procedure", by M. C. Ricci, studies the same problem of the previous paper, but from a numerical point of view [31].

The paper "Attitude control system design using a flywheel suspended by gimbals", by, Raphael W. Peres and M. C. Ricci, studies different forms to keep the attitude of a satellite within specified values [32], in order to maintain the satellite in conditions to satisfy the mission requirements.

The paper "Ionosphere influence on success rate of GPS ambiguity resolution in a satellite formation flying" by, L. Baroni, considers the problem of how to deal with the ambiguity resolution of the GPS [33,34] sign in the case of satellite constellations.

The paper "Evasive Maneuvers in Route Collision With Space Debris Cloud", by A. D. C. Jesus, R. T. Sousa and E. Vieira-Neto, shows alternatives to maneuver [35,36] a satellite such that it can avoid collision with a cloud of space debris that is in its trajectory.

The paper "Influence of the external torques in the angle between the spin axis and the Sun direction Spin stabilized Satellite", by G. B. Motta and M. C. Zanardi, studies of the influence of the environmental torques in the angle between the spin axis and the Sun's direction (solar aspect angle) for a spin stabilized satellite [37].

The paper "Magnitude of Solar Radiation Torque and the Earth Shadow Region", by R. E. S. Cabette, M. C. Zanardi and I. Kolesnikov, describes the effects of the passage of a satellite in the shadow of the Earth using the magnitude of the torque generated by the solar radiation pressure [38,39].

The paper "Development of a Helicon Double Layer Thruster" by J. L. Ferreira, F. N. O. Lopes, H. O. Coelho Júnior, A. C. B. S. Serra and E. G. Costa, shows the evolution in the developments of a helicon double layer thruster [40,41], which is a propulsion system with a high potential to be used in space activities.

The paper "The State-of-art in Space Robotics", by I. M. da Fonseca and M. N. Pontuschka, is a survey indicating the latest results in the developments of robots made for space applications and it also includes a discussion of the applications of new concepts such as robonauts, space tugs and robots for future planetary exploration [42,43].

The paper "Study of the decay time of a CubeSat type Satellite considering Perturbations due to the Earth's Oblateness and Atmospheric Drag", by T. P. Brito, C. C. Celestino and R. V. Moraes, introduces a study of the best re-entry time for a CubeSat when it is affected by the Earth's oblateness and atmospheric drag [44].

The paper "Gravitational disturbances generated by the Sun, Phobos and Deimos in orbital maneuvers around Mars with automatic correction of the semi-major axis", by E. M. Rocco, presents analyses into maneuvers of a spacecraft into orbit around Mars that is disturbed by the gravitational attraction of the Sun, Phobos and Deimos. Impulsive optimal maneuvers considering the Lambert's Problem to reduce fuel consumption are also considered [45,46]

The paper "Analysis of the influence of perturbative forces in an artificial lunar satellite", L. D. Gonçalves, E. M. Rocco and R. V. de Moraes, contains numerical models to study the influence of the orbital disturbance forces in trajectories of lunar satellites [47,48] such as the gravitational attraction of the Sun and Earth, lunar albedo and the solar radiation pressure.

The paper "simulations due to the lunar gravitational model uncertainty", by L. D. Gonçalves, E M Rocco, R. V. de Moraes and H. K. Kuga, simulates part of the orbital trajectory of the Lunar Prospector mission in order to analyze the relevance of the Kalman filter to estimate the trajectory. Disturbance due to the lunar potential gravitational (LP100K model) was also considered [49].

The paper "H infinity controller design to a rigid-flexible satellite with two vibration modes", by A. G. de Souza and L. C. G. de Souza, is a study into the rigid-flexible satellite ACS design using the H infinity method [50,51].

The last paper of this special issue is entitled "Navigation and control of an UVA quadrotor in search and surveillance missions", by L. A. Frederico, L. S. Martins-Filho and A. L. da Silva, and it describes the studies into the application of specific random motion, known as Levy flights, to control the trajectory of an Unmanned Aerial Vehicle (UAV) quadrotor designed for surveillance [52,53].

6. Conclusions

Brazil has important groups researching theoretical and applied celestial mechanics. They are distributed in research institutes such as the National Observatory (Observatório Nacional, ON), the National Institute for Space Research (Instituto Nacional de Pesquisas Espaciais, INPE), the Department of Aerospace Science and Technology (Departamento de Ciência e Tecnologia Aeroespacial, DCTA), and the many departments of Mathematics, Physics and Engineering in public universities, such as the Federal Universities of Rio de Janeiro (UFRJ), São Paulo (UNIFESP), Minas Gerais (UFMG), and ABC (UFABC), and the São Paulo State Universities (USP and UNESP).

These research and education institutions work to conduct advanced scientific investigations into the many areas of celestial mechanics, and the papers presented here show the efforts in that direction.

Reference

[1] Macau E E N, de Melo C F, Prado A F B A and Winter O C 2014 Comp. Appl. Math.
[2] de Melo C F and Winter O C 2007 in: Winter O. C. and Prado A. F. B. A. (Organizers), São Paulo, Editora da USP, 37–74
[3] http://www.astro.iag.usp.br/~dinamica/historic.html, accessed in August 31, 2015
[4] Sukhanov A A, Velho H F de C, Macau E E and Winter O C 2010 Cosmic Research **48** 443-450
[5] Sukhanov A A and Prado A F B A 2007 Cosmic Research **45** 417-423
[6] Sukhanov A A and Prado A F B A 2008 Cosmic Research **46** 49-59
[7] da Silva Fernandes S and Marinho C M P 2012 Math. Probl. Eng. **2012**
[8] Prado A F B A and Broucke R A 1995 Appl. Mech. Rev. **48** (4)
[9] Walker G H and Ford J 1969 Phys. Rev. **188** (416)

[10] de Sousa M C, Caldas I L, Ozorio de Almeida A M, Rizzato F B and Pakter R 2013 Phys. Rev. **88** 064901
[11] Troger H, Alpatov A P, Beletsky V V, Dranovskii V I, Khoroshilov V S, Pirozhenko A V and Zakrzhevskii A E 2010. Dynamics of Tethered Space Systems. CRC Press.
[12] Devynck P, Stckel J, Admek J, Duran I, Hron M and Ooost G V 2003 Czech. J. Phys **53** 853
[13] Crassidis J L and Markley F L 2003 *J. Guid. Cont. Dyn.* **26** 536-542
[14] Crassidis J L and Junkins J L2012 Optimal Estimation of Dynamics Systems. CRC Press, New York, N.Y., 2nd ed.
[15] Yano H 2012 Presentation for the NASA Planetary Protection Subcommittee 2012. NASA Headquarters, Washington, D.C., USA
[16] Gomes V M and Prado A F B A 2010 Transactions on Mathematics. **9** 811-820
[17] Burov A A, Kosenko I I, and Guerman A D 2014 Proceed. of 65th International Astronautical Congress, Toronto, Canada
[18] Song Y J, Park S Y, Choi K H and Sim E S 2009 Adv. in Space Research. **43** (9)1391-1406
[19] Lara M and San-Juan J F 2005 *J. Guid. Cont. Dyn.* **28** (2) 291–297
[20] Lara M 2008 *J. Guid. Cont. Dyn.* **31** (1)172–181
[21] Adachi I, Hayashi C and Nakazawa K 1976 Progress of Theoretical Physics **56** 1756
[22] Masset F 2000 Astronomy and Astrophysics **141** (165)
[23] Bartys M, Patton R, Syfert M, Heras S de Las and Quevedo J 2006 Control Engineering Practice **14** (6) 577-596
[24] Patton R J, Uppal F J, Simani S and Polle B 2010 Control Engineering Practice **18** (9) 1093-1109
[25] Torres K S and Prado A F B A 2006 Math. Probl. Eng. **2006** 1-13
[26] Grebogi C, Ott E and Yorke J A 1982 Phys. Rev. Lett. **48** 1507
[27] Egydio de Carvalho R and Abud C V 2011 Phys. Rev. E **84** 036204
[28] Ferreira J L, Moraes B S, Santos J C, Miranda R, Gessini P and Possa G 2013 Proc. of the 33rd International Electric Propulsion Conference Washington D.C. USA
[29] Harris T 2001 Rolling Bearing Analysis, 4th ed
[30] Ricci M C 2010 Proc. 11th Pan-American Congress of Appl. Mech.
[31] Rasolofondraibe L, Pottier B, Marconnet P and Perrin E 2013 IEEE Sens. J. **13** 3067–72
[32] Kaplan M H 1976 Modern Spacecraft Dynamics & Control
[33] Chiaradia A P M, Kuga H K and Prado A F B A 2003 Acta Astronautica **53** 123-133
[34] Liu J, Cannon M E, Alves P, Petovello M G, Lachapelle G, Macgougan G and Degroot L 2003 Strategies GPS Solutions **7** (2) 87-100
[35] Solórzano C R H, Prado A F B A and Sukhanov A A 2010 Math. Probl. Eng. **2010** 1-12
[36] Smirnov N N 2002 (ed.): Space Debris Hazard Evaluation and Mitigation. Taylor Francis
[37] Garcia R V, Zanardi M C and Kuga H K 2009 Math. Probl. Eng. **2009**
[38] Shivastava S K and Modi V J 1993 AIAA Journal **6** (6)
[39] Takeichi N 2010 Cel. Mech. and Dynam. Astron.
[40] Ferreira J L, Ferreira I S, Santos J C, Miranda R, Gessini P, Possa G and Lui T C IEPC-2013-428
[41] Sutherland O, Charles C, Plihon N and Boswell R W 2005 Phy. Rev. Let. **95** 205002
[42] Alexis O and Kostas B R L 2010 10th International Symposium on Artificial Intelligence, Robotics and Automation in Space (i-SAIRAS)
[43] Liam P, David K, David W, Illah N and Trey S 2004 NASA Exploration Team Report. NASA/TM-2004
[44] Colombo C 2014 Proceedings of Key Topics in Orbit Propagation Applied to SSA KePASSA
[45] Lemoine F G, Smith D E; Rowlands D D, Zuber M T, Neumann G A and Chinn D S 2001 J. Geophys. Res. **106** 23359-76
[46] Zuber M T, Lemoine F G, Smith D E, Konopliv A S, Smrekar S E and Asmar S W 2007 J. Geophys. Res. **112** E05S07

[47] Carvalho J P S, Moraes R V and Prado A F B A 2010 Cel. Mech. and Dyn. Astron. **108** 371-388
[48] Carvalho J P S, Moraes R V and Prado A F B A 2011 Math. Probl. Eng. **2011** 1-19
[49] Konopliv A S, Asmar S W, Carranza E, Sjogren W L and Yuan D N 2001 Icarus **150** 1-18
[50] Souza L C G and Souza A G 2014 Shock and Vibration **2014** 1-8
[51] Skogestad S and Postlethwaite I 2001 Multivariable Feedback Control Analysis and Design. 2 ed. Chichester: John Wiley Sons
[52] Viswanathan G M, Afanasyev V, Buldyrev S V, Murphy E J, Prince P A and Stanley H E 1996 Nature **381** 413–415
[53] Mellinger D, Michael N and Kumar V 2010 The International Journal of Robotics Research p. 11

Peer review statement

All papers published in this volume of *Journal of Physics: Conference Series* have been peer reviewed through processes administered by the proceedings Editors. Reviews were conducted by expert referees to the professional and scientific standards expected of a proceedings journal published by IOP Publishing.

Journal of Physics: Conference Series **641** (2015) 012001 doi:10.1088/1742-6596/641/1/012001

Preliminary Analysis of Optimal Round Trip Lunar Missions

L A Gagg Filho and S da Silva Fernandes

Departamento de Matemática, InstitutoTecnológico de Aeronáutica, 12228-900, São José dos Campos, SP, Brasil.

E-mail: luizarthur.gagg@gmail.com

Abstract. A study of optimal bi-impulsive trajectories of round trip lunar missions is presented in this paper. The optimization criterion is the total velocity increment. The dynamical model utilized to describe the motion of the space vehicle is a full lunar patched-conic approximation, which embraces the lunar patched-conic of the outgoing trip and the lunar patched-conic of the return mission. Each one of these parts is considered separately to solve an optimization problem of two degrees of freedom. The Sequential Gradient Restoration Algorithm (SGRA) is employed to achieve the optimal solutions, which show a good agreement with the ones provided by literature, and, proved to be consistent with the image trajectories theorem.

1. Introduction

The problem of transferring a space vehicle from one orbit to another orbit has been growing in importance in last decades. The solution of this problem is of great concern to the space science. Also commercial applications can be found such as: the maintenance of telecommunications satellites, GPS constellations satellites, geostationary satellites and others [7]. However, the science development is the greatest benefit of this field because the trajectories design is one of many keys necessary to perform a space exploration and solve fundamentals issues of science, including the origin of life on Earth.

In the majority of the examples above the actuators of the space vehicle are assumed impulsive, which means that they produce an instantaneous velocity increment to put the vehicle in the desired trajectory. The determination of the velocity increments of the space vehicle depends upon the dynamics constraints utilized in the mathematical modeling of the system. These dynamics constraints can be divided in four class according to the literature [7]: two body model, perturbed two body model, three body model and N-body model. The two body model is the simplest one and it is more studied among all because it provides a good approximation for more complex problems and it holds analytical solutions. The so-called Hohmann transfer appears from this model [8]. The four-body model was studied by several researches [3] [10] [2]. Da Silva Fernandes and Marinho [2] have employed the gradient algorithm coupled with the Newton-Raphson method to obtain low energy Earth-Moon transfer with the Sun's perturbation.

To accomplish the optimization and to obtain the desirable trajectory it is necessary to provide an initial guess for the solution. For lunar or interplanetary mission, the well-known patched-conic approximation can be used to obtain the initial guess. It can be also applied for preliminary mission analysis [1].

The focus of the present work is to calculate and analyze optimal round trip lunar missions based on the patched-conic approximation, which has a more detailed geometry. The solution obtained by this model can be used to feed the optimization algorithm in more complex models.

Journal of Physics: Conference Series **641** (2015) 012001 doi:10.1088/1742-6596/641/1/012001

2. Objectives

The main purpose of this work is to study a full lunar patched-conic approximation considering a round trip mission and utilize this model to optimize the fuel consumption of the space vehicle by the application of the Sequential Gradient Restoration Algorithm (SGRA).

3. Formulation of patched-conic approximation

The aim of this section is to explain the round trip lunar patched-conic approximation and the optimization problem related to it. To achieve this purpose the outgoing trip is explained first. It consists of transferring a space vehicle from a circular low Earth orbit (LEO) to a circular low Moon orbit (LMO) minimizing the fuel consumption, represented by the total velocity increment. Next, the return flight is formulated similarly to the outgoing trip, but transferring the space vehicle from a LMO to a LEO. The velocity increments are assumed to be instantaneous and tangential to the corresponding orbits.

3.1. Outgoing Trip

The patched-conic approximation is based on the two-body dynamics, in which the motion of the space vehicle is described by a conic, and it uses the concept of sphere of influence. It is a simple strategy of patching two conics to describe the entire trajectory. In the lunar patched-conic approximation an ellipse and a hyperbole are used. The ellipse describes the motion of the space vehicle when the Earth gravitational field is considered (geocentric phase), and, the hyperbole describes the motion when the Moon gravitational field is considered (selenocentric phase). To obtain the complete trajectory of the relative motion of the vehicle, the two conics are connected together at the edge of the sphere of influence of the Moon. Similar to the Hohmann transfer, the patched-conic approximation involves two impulses with the total increment of velocity representing the fuel consumption [4]. A mathematical formulation of the patched-conic approximation for the outgoing trip can be found in [1]. The following hypotheses are assumed:

1) The Earth planet is fixed in space;
2) The Moon orbit around the Earth is circular;
3) The flight of the space vehicle lies in the orbital plane of the Moon;
4) The gravitational field of the Earth and the Moon are central and obey the inverse square law;
5) The transfer trajectory has two distinct phases: the geocentric phase, which it starts immediately after the first velocity increment; and, the selenocentric phase, which begins when the vehicle reaches the sphere of influence of the Moon;
6) The two impulses model is considered. Each velocity increment is applied tangentially to the initial orbit (LEO) and the final (LMO) orbit.

The patched-conic approximation for the outgoing trip is completely determined by four variables: the radial distance r_0, the velocity v_0, the flight path angle φ_0 of the space vehicle at the point of insertion in the geocentric trajectory; and, the phase angle $\lambda_1$of the vehicle with the Earth relatively to the Moon at the moment when the space vehicle reaches the sphere of influence of the Moon. From the hypothesis 6), $\varphi_0 = 0$.

3.2. Return Trip

The mathematical formulation of the return trip is basically the same as the outgoing mission, but with the selenocentric trajectory as the starting path. The hypotheses remain the same; however, the fifth hypothesis can be better stated as it follows:

5) The transfer trajectory has two distinct phases: the selenocentric phase, which starts immediately after the first velocity increment; and, the geocentric phase, which begins when the vehicle reaches the edge of the sphere of influence of the Moon.

Journal of Physics: Conference Series **641** (2015) 012001 doi:10.1088/1742-6596/641/1/012001

The return trip based on the patched-conic approximation is completely solved when the set of variables $(r_0, v_0, \varphi_0, \lambda_1)$ is known. Unlike the outgoing trip, these variables are related with the beginning of the return mission so that the radial distance r_0, the velocity v_0 and the flight path angle φ_0 are related at the point of insertion in the selenocentric trajectory. λ_1 is the same phase angle but with a distinct configuration. Again, according to the hypothesis 6), the flight path angle φ_0 is 0.

3.3. Optimization Problem based on Patched-Conic Approximation

The patched-conic approximations for the outgoing and return trips, as mentioned before, are determined through a boundary value problem, in which the iteration variable is v_0. The variables r_0, φ_0 and λ_1 must be previously specified. In order to solve this problem one can use the Hohmann transfer as initial guess of v_0. If the set (r_0, φ_0) is prescribed, then there is just one solution for each value of λ_1 if the solution exists. In this way, an optimization problem can be enunciated in order to determine some value of λ_1 which minimizes the total fuel consumption, represented by the total velocity increment. The complete optimization problem is stated as below:

"Determine the set of variables (λ_1, v_0) which minimizes the function

$$F(\lambda_1, v_0): \Delta v_{Total} = \Delta v_{LEO} + \Delta v_{LMO}$$

subjected to the constraint

$$g(\lambda_1, v_0): r_f - r_p = 0$$

where Δv_{LEO} is the velocity increment at the circular low Earth orbit (LEO) and Δv_{LMO} is the velocity increment at the circular low Moon orbit (LMO). Note that $r_p = r_{pM}$ for the outgoing trip, and, $r_p = r_{pE}$ for the return trip, where r_{pM} and r_{pE} are, respectively, the distance at the periapsis of the selenocentric trajectory in the outgoing trip, and, the distance at the periapsis of the geocentric trajectory in the return trip. In both cases r_f is the prescribed radius"

The Sequential Gradient Restoration Algorithm (SGRA) is employed to solve this two degrees of freedom optimization problem. This algorithm has two distinct phases [6]: in the gradient phase, a displacement is applied in order to reduce the value of the function, but only satisfying the constraints at first order. In the restoration phase, a displacement is applied in order to restore the constraints, which is originally not completely satisfied due to the displacement of the gradient phase. To initialize this algorithm, an initial guess that satisfies the constraint must be provided [5]. Therefore, the boundary value problem must be solved in order to obtain this initial guess and to ensure the algorithm's convergence. The boundary value problem consists of determining v_0 to a specified set (r_0, φ_0) such that it satisfies the final condition $r_p = r_f$ to a certain λ_1.

4. Results

This section shows the main results obtained to the problem of minimizing the fuel consumption of a space vehicle in a round trip mission from Earth to Moon. The outgoing and the return trips are considered separately by patched-conic approximations. The path of the space vehicle in the LMO can be clockwise or counterclockwise. All the trajectories calculated are direct ascent. The LEO altitude is specified at 463 km, which corresponds to the altitude of the International Space Station, and the LMO altitude is set at 100 km. The table 1 illustrates a list with the values of others parameters necessary to specify the entire problem.

In the lunar patched-conic approximation, the derivatives of the expressions involved in the boundary value problems and in the optimization problem can be obtained analytically and numerically (centered differences). These two approaches are implemented in this work and good agreement of the results has been noticed. Before the optimization problem, the boundary value problem has been solved to verify the behavior of the curve of the total velocity increment with respect to λ_1 as it is illustrated in figure 1. The results of the outgoing trip are analyzed first.

Table 1. Parameters specifications

ω_M(Moon's angular velocity)	$2.649 \times 10^{-6}\ rad/s$
v_M(Moon's velocity)	$1.018\ \text{km}/s$
R_E (Earth radius)	6378.2 km
R_M(Moon radius)	1738 km
μ_E (Earth's gravitational parameter)	$398600\ \text{km}^3/s^2$
μ_M(Moon's gravitational parameter)	$4902.83\ \text{km}^3/s^2$
D (Earth-Moon mean distance)	384400 km
R_s (Moon's sphere of influence radius)	66300 km

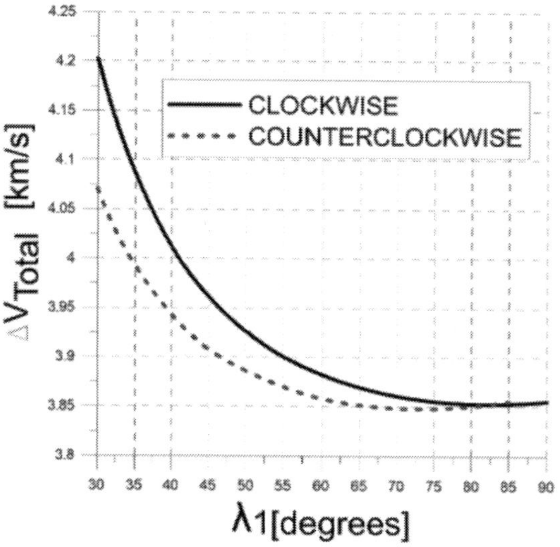

Figure 1. Outgoing trip. LEO = 463 km, LMO = 100 km.

The aspect of the curve shown in figure 1 remains the same for the others values of LMO and LEO altitudes. Note from this figure that it is possible to find a value to λ_1 which minimizes Δv_{Total} in the both cases: clockwise and counterclockwise arrival. So, the SGRA is employed to perform this task. The tolerance in the Newton-Raphson algorithm is 10^{-8}. In the optimization problem it is used a tolerance of 10^{-12} with numerical derivatives and a tolerance of 10^{-17} with analytical derivatives. The tolerance of the restoration phase in the SGRA is set in 10^{-15}.

The table 2 compares the optimized results of the patched-conic approximation obtained in this work with some results provided by the literature. The symbol '[4]' in the tables refers to the results given by [1]; and the expression 'Miele' corresponds to the results given by [6]. According to this table, a good agreement is noticed between the results obtained in the present work with the results provided by literature, mostly with the results given by [1]. When the same results is compared with the ones found by [6], a slight discrepancy occurs among the initial phase angles γ_0. Such variable is very sensitive to the algorithm precision, so its value can change with the algorithm tolerance.

Journal of Physics: Conference Series **641** (2015) 012001 doi:10.1088/1742-6596/641/1/012001

The return trip optimization problem based on the patched-conic approximation is basically the same as the outgoing mission. Furthermore, the image trajectory theorem is easily verified with the round trip trajectory in a rotating reference frame with origin coincident with the center of Earth, as shown in figures 2-5.

All the trajectories obtained in this paper are direct ascent maneuvers and feasible, i.e., there is no collision with the Moon. Figure 1 shows there is only one minimum in the total velocity increment curve. However, others sets of minimum can be found if the time of flight is considerably increased and more revolutions are performed by the vehicle around the Earth. Topputo [9] develops a systematic analysis of these minimum in a four-body model.

Table 2. Altitude LEO: 463 km. Altitude LMO: 100 km. Couterclockwise arrival. Outgoing trip.

Altitude LMO [km]	Model	Δv_{Total} [km/s]	Δv_{LEO} [km/s]	Δv_{LMO} [km/s]	*Time of flight* [days]	γ_0 [degrees]	Feasibility
100	*Patched-conic* [4]	3.8482	3.0655	0.7827	4.794	-115.723	Yes
	Patched-conic	3.8482	3.0655	0.7827	4.797	-115.691	Yes
	PCR3BP simplified [4]	3.8758	3.0649	0.8109	4.564	-116.800	Yes
	PCR3BP classic [4]	3.8777	3.0658	0.8119	4.573	-116.410	Yes
	Miele	3.8760	3.0650	0.8110	4.370	-118.980	—

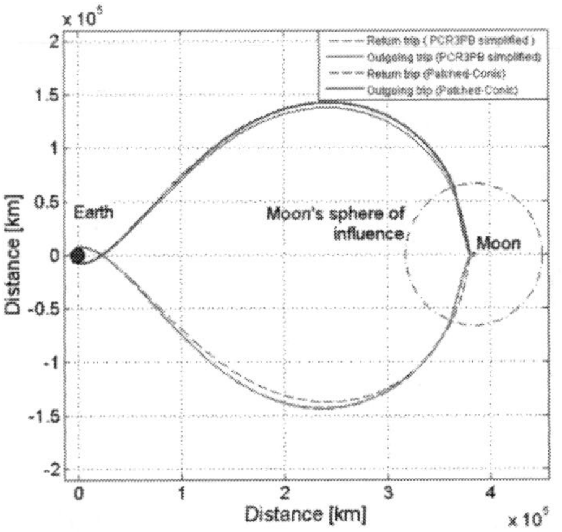

Figure 2. Round trip in the rotating reference frame. LEO = 463 km, LMO = 100 km. Counterclockwise.

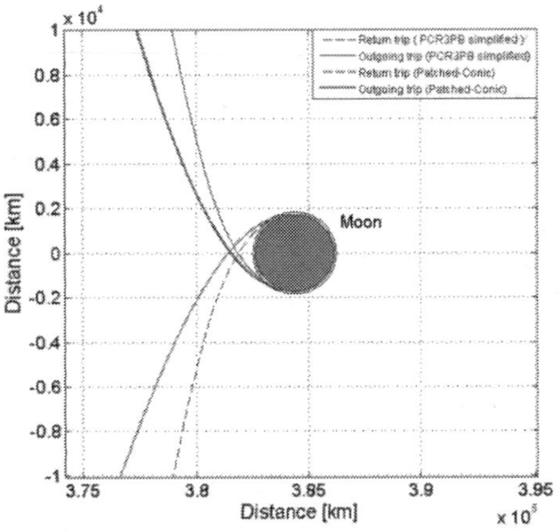

Figure 3. Zoom at LMO. Counterclockwise case.

Note that the flight of time obtained for the optimal trajectories is larger than that of Apollo missions (about 3 days), however, these last trajectories are not an optimal solution with minimum fuel consumption [6].

Journal of Physics: Conference Series **641** (2015) 012001 doi:10.1088/1742-6596/641/1/012001

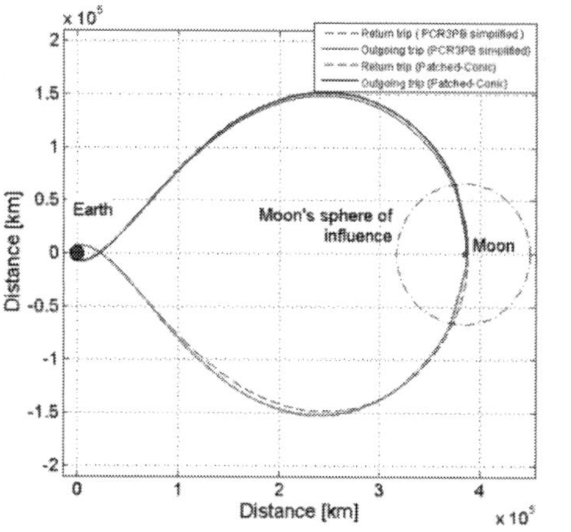

Figure 4. Round trip in the rotating reference frame. LEO = 463 km, LMO = 100 km. Clockwise.

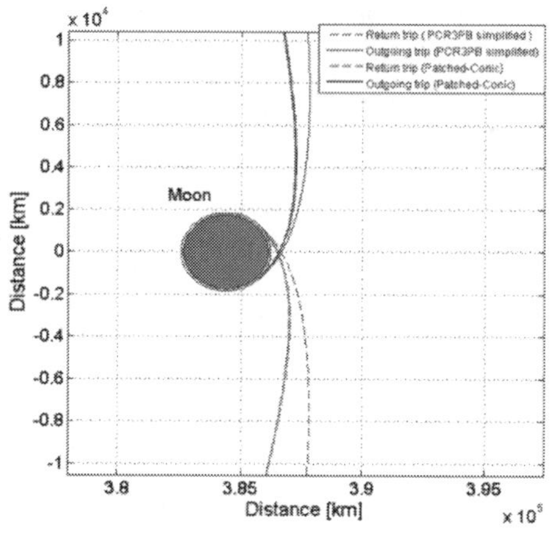

Figure 5. Zoom at LMO. Clockwise case.

5. Conclusion

In this paper, a full lunar patched-conic approximation is presented. The round trip journey of the space vehicle is obtained through optimal impulsive trajectories based on these patched-conic approximations. The Sequential Gradient Restoration Algorithm is utilized in the optimization problem, which is conducted separately for the outgoing and the return trip. The results demonstrate to be consistent with those provided by the three body problem and found in the literature. The image trajectories theorem is proven due to the symmetry between the outgoing trajectory path with the return trajectory path in the rotating reference frame. Therefore, the patched-conic approximations can be applied for preliminary mission analysis as well as used as initial guess for more complex models.

6. Acknowledgment

The research was supported by Fapesp under contracts 2012/25308-5 and 2012/21023-6 and CNPq under contract 304913/2013-8.

7. References

[1] Da Silva Fernandes S and Marinho C M P 2012 Optimal two-impulse trajectories with moderate flight time for earth-moon missions *Mathematical Problems in Engineering* vol **2012.** 34 pages. Article ID 971983. DOI: 10.1155/2012/971983.

[2] Da Silva Fernandes S and Marinho C M P. 2013 Sun perturbations on optimal trajectories for earth-moon flight. *Proc. Int. Congress of Mechanical Engineering COBEM 2013 (Ribeirão Preto)*. vol **2013** (Ribeirão Preto: Brazil/ABCM). ISSN 2176-5480.

[3] Koon W S, et al. 2001 Low energy transfer to the Moon. In *Dynamics of Natural and Artificial Celestial Bodies* vol **2001** (Springer Netherlands). pp 63-73. DOI: 10.1007/978-94-017-1327-6_8.

[4] Marec J-P 1979 *Optimal space trajectories* vol **1** (New York: Elsevier)

[5] Miele A, Huang H Y, and Heideman J C 1969 Sequential gradient-restoration algorithm for the minimization of constrained functions: ordinary and conjugate gradient versions *Journal of Optimization Theory and Applications* vol **4** n 4 (Kluwer Academic Publishers-Plenum Publishers) pp 213-243. DOI: 10.1007/BF00927947

[6] Miele A and Mancuso S 2001 Optimal trajectories for Earth–Moon–Earth flight *Acta Astronautica* vol **49** n 2 (ELSEVIER) pp 59-71. DOI: 10.1016/S0094-5765(01)00007-8

[7] Prado A F B A and Rios-Neto A T A I R 1993 Um estudo bibliográfico sobre o problema de transferências de órbitas *Revista Brasileira de Ciências Mecânicas* vol **15** n 1 pp 65-78
[8] Szebehely V G and Mark H 1998 *Adventures in celestial mechanics* 2 ed (John Wiley & Sons)
[9] Topputo F 2013 On optimal two-impulse Earth–Moon transfers in a four-body model *Celestial Mechanics and Dynamical Astronomy* vol **117** n 3 (Springer Netherlands) pp 279-313. DOI: 10.1007/s10569-013-9513-8
[10] Topputo F and Belbruno E 2013 Optimization of Low-Energy Transfers. In *Modeling and Optimization in Space Engineering 2013* vol **73** (New York:Springer) pp 389-404. DOI: 10.1007/978-1-4614-4469-5_16

Estimating the trajectory of a space vehicle passing by the Moon using Kalman Filter

A F S Ferreira[1], H K Kuga[1], A F B A Prado[1] and O C Winter[2]

[1] Instituto Nacional de Pesquisas Espaciais, INPE, São José dos Campos, Brazil
[2] Faculdade de Engenharia - Universidade Estadual Paulista, FEG-UNESP, Guaratinguetá, Brazil

E-mail: aleferrazsilva@hotmail.com

Abstract. The trajectory of a spacecraft going to the periapsis of its orbit around the smaller body of a binary system is estimated from the observed data using a Kalman Filter. Simulations of the trajectories will be made to generate the data of the spacecraft position. When observed from the Earth, using a fixed star as a reference, we have the spacecraft position as an angle function ϕ. This is the angle between spacecraft position vector and the star position vector. The result shows that Kalman Filter can be used successfully for trajectory estimation, if adequate accuracy is available for the angle observations.

1. Introduction

When a satellite passes next to a celestial body and uses the gravity of this body to gain or lose energy, there is a maneuver called Swing-by [1,2]. There are different approaches to use this maneuver. It can be propelled or not, can vary the activation point of the propellant and can be applied to circular or elliptical orbits. There are several works available in the literature about this problem [3-7].

In this paper we will estimate the spacecraft trajectory going to the periapsis of its orbit around the smaller body of a binary system. We use observed data and used a Kalman Filter [8,9]. We will estimate only the trajectory part before the passage by the periapsis, because there are multiple options after that point, e. g., the spacecraft can be captured, crash with the body or continue its trajectory.

We consider the orbital plane of the binary system as a reference plane, and the system origin is choose as the center of mass of the two main bodies.

To have the observational data, simulations are made. These simulations were performed by integrating a trajectory around the smaller body of the binary, using the model given by the restricted three body problem. When observed from Earth, using a fixed star as a reference, we have the spacecraft position as a function of an angle, given by the scalar product of the spacecraft position vector ($\mathbf{r}_s$) and the star position vector ($\mathbf{r}_e$).

The data describing the position for each instant of the trajectory, inserted with random errors, will be the observational data to be used by the Kalman Filter, to estimate the correct trajectory.

Therefore, the goal of this work is to obtain an algorithm that estimates the correct trajectory of a spacecraft, around a binary system, from this type of observed data.

2. Methods

The first step was to simulate the observations. We considered the radius of the periapsis of the trajectory of the spacecraft equals to 10 radius of the smaller body of the binary. The canonical system of units was used. Numerical integration were made to obtain the results.

The observations are the angle (ϕ) that describes the spacecraft's position at each instant of time, observing the Earth, having a fixed star as a reference. The ϕ value was calculated by the scalar product of the spacecraft position vector ($\mathbf{r_s}$) (originated on the Earth) and the star position vector ($\mathbf{r_e}$). Being $\mathbf{r_{cm}} = (r_{cm_x}, r_{cm_y}, r_{cm_z})$ the binary system center of mass position vector (origin on the Earth), $\mathbf{r_e} = (r_{e_x}, r_{e_y}, r_{e_z})$ the star position vector, $\mathbf{r_s} = \mathbf{r_{cm}} + \mathbf{r}$ and $\mathbf{r} = (x, y, z)$ the spacecraft position vector with origin at the binary system center of mass. Since, the spacecraft moves in fixed plane $(z = 0)$. Then ϕ is given by:

$$\phi = a\sin\left(\frac{r_{e_x}(r_{cm_x} + x) + r_{e_y}(r_{cm_y} + y) + r_{e_z}(r_{cm_z})}{\left|\vec{r_e}\right|\left|\vec{r_s}\right|}\right) \tag{1}$$

$$\left|\mathbf{r_e}\right| = \sqrt{r_{ex}^2 + r_{ey}^2 + r_{ez}^2} \tag{2}$$

$$\left|\mathbf{r_s}\right| = \sqrt{(r_{cmx} + x)^2 + (r_{cmy} + y)^2 + (r_{cmz})^2} \tag{3}$$

2.1 Kalman Filter

Kalman filter is a mathematical method developed by Rudolf Kalman, in 1960, to estimate states that tend to be closer to actual values, from measurements along the time. There are related different versions of the filter, and in this work we used the extended Kalman filter, that treats nonlinear models. This version generates reference trajectories that are updated with every processing measurements. Given the dynamical system:

$$\dot{X} = f(X) + \omega = \begin{bmatrix} x_3 \\ x_4 \\ \frac{-(1-\mu)(x_1+\mu)}{((x_1+\mu)^2+x_2{}^2)^{3/2}} - \frac{-\mu(x_1-1+\mu)}{((x_1-1+\mu)^2+x_2{}^2)^{3/2}} \\ \frac{-(1-\mu)x_2}{((x_1+\mu)^2+x_2{}^2)^{3/2}} - \frac{\mu x_2}{((x_1-1+\mu)^2+x_2{}^2)^{3/2}} \end{bmatrix} + \omega, \tag{4}$$

the spacecraft state is $X = [x\ y\ \dot{x}\ \dot{y}] = [x_1\ x_2\ x_3\ x_4]$, the equations of motion are given by the restricted three body problem and ω is noise of the process that has normal distribution with zero mean and covariance Q ($\omega = N(0, Q)$).

The measurement model is given by:

$$g = h(X) + v \tag{5}$$

being y the measurements vector, $h(X) = \phi(X)$, where $h(X)$ relates the actual states with the observed states and v is the measurements noise, assumed to be a white noise with zero mean and covariance R ($v = N(0, R)$).

The Kalman filter is divided in time-update and measurement-update cycles. The time-update process propagates the state and the covariance from the time t_{k-1} to t_k, using the equations:

Journal of Physics: Conference Series **641** (2015) 012002 doi:10.1088/1742-6596/641/1/012002

$$\dot{\overline{X}} = f(\overline{X}) \tag{6}$$

$$\overline{P}_k = \varphi_{k,k-1}\hat{P}_{k-1}\varphi_{k,k-1}^t + \Gamma_k Q_k \Gamma_k^t \tag{7}$$

where $\overline{X}$ is the propagated state vector, $\overline{P}$ is the propagated covariance matrix, $\hat{P}$ is the former updated covariance matrix and φ is the transition matrix, given by $e^{F\Delta t}$, being F the jacobian matrix of f with respect to X ($F = \frac{\partial f}{\partial X}$) and $\Delta t = t_k - t_{k-1}$ is the time interval. The measurement-update process, at this stage, corrects the state and the covariance for the instant t_k in accordance with the measurement g_{t_k}, according to the equations:

$$K = \overline{P}H^t(H\overline{P}H^t + R)^{-1} \tag{8}$$

$$\hat{P} = (I - KH)\overline{P} \tag{9}$$

$$\hat{X} = \overline{X} + K[g - h(\overline{X})] \tag{10}$$

where H is the jacobian matrix of $h(X)$ with respect to X ($H = \frac{\partial h}{\partial X}$), K is the Kalman gain, I is the identity matrix and $\hat{X}$ is the estimated state vector.

3. Results

For the results that follow we adopt the initial covariance matrix:

$$\overline{P}_0 = \begin{bmatrix} 0.1 & 0 & 0 & 0 \\ 0 & 0.1 & 0 & 0 \\ 0 & 0 & 0.1 & 0 \\ 0 & 0 & 0 & 0.1 \end{bmatrix},$$

the matrix Q of the process noise is given by

$$Q = \begin{bmatrix} 10^{-5} & 0 & 0 & 0 \\ 0 & 10^{-5} & 0 & 0 \\ 0 & 0 & 10^{-4} & 0 \\ 0 & 0 & 0 & 10^{-4} \end{bmatrix},$$

and $R = 10^{-10}$.

The following data will show the estimated trajectory. Figure 1 shows the measurement residues of the time-update and measurement-update state as a time function. The residue is given by the difference between the measurements and the function that relates the actual states with the observed states, i.e., $g - h(\overline{X})$ for the residue of the propagated state vector (time-update process) and $g - h(\hat{X})$ for residue of the estimated state vector (measurement-update process). The data are in canonical units.

Journal of Physics: Conference Series **641** (2015) 012002 doi:10.1088/1742-6596/641/1/012002

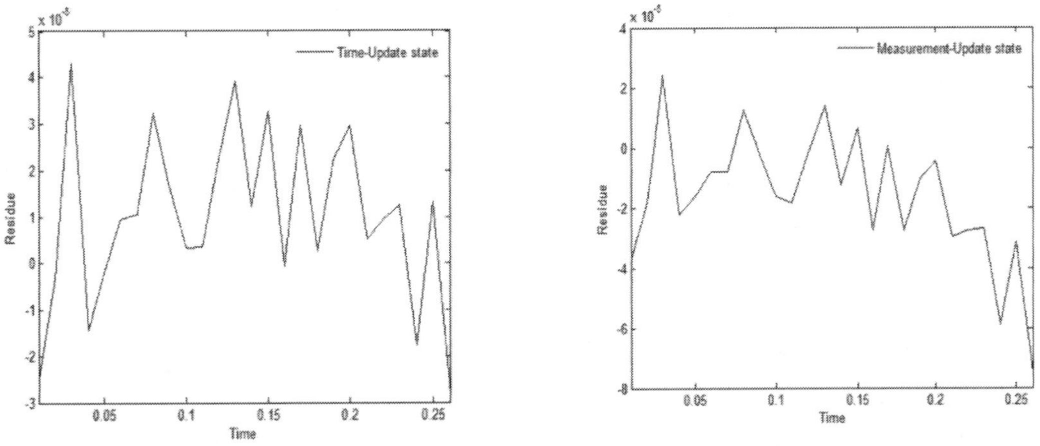

(a) (b)

Figure 1. Measurement residues of the (a) time-update and (b) measurement-update state as a time function.

Figure 2 and 3 show the standard deviation and error for position and velocity, respectively. The error is the difference between the actual state and the estimated state.

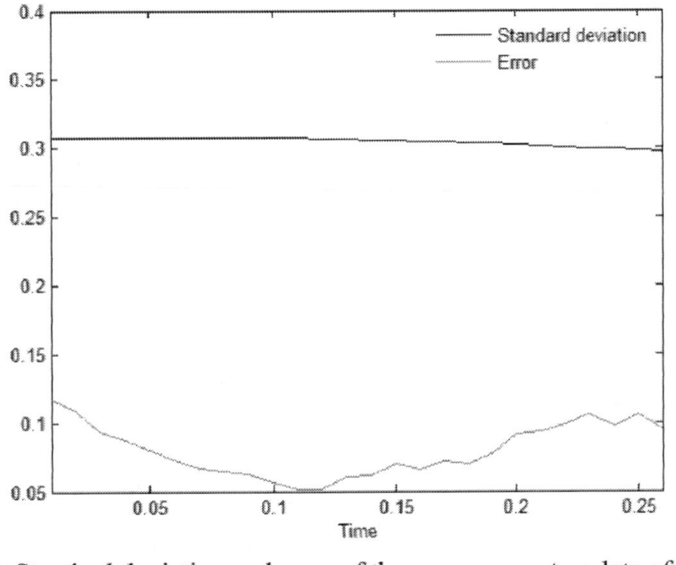

Figure 2. Standard deviation and error of the measurement-update of position.

Journal of Physics: Conference Series **641** (2015) 012002 doi:10.1088/1742-6596/641/1/012002

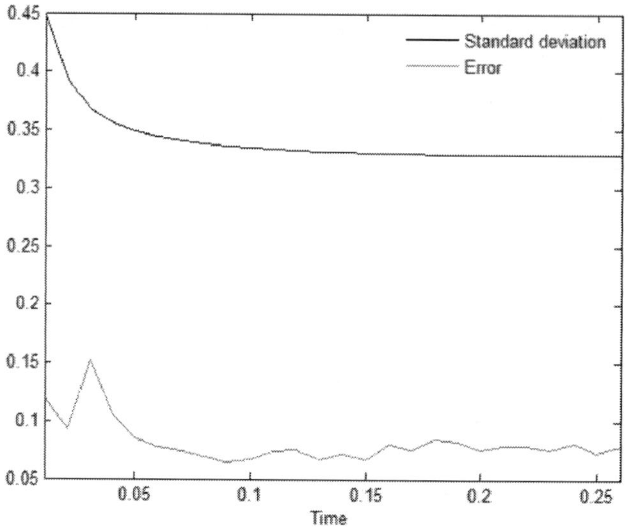

Figure 3. Standard deviation and error of the measurement-update of velocity.

A decrease in the value of the standard deviation means that the estimated value approach the actual value. Figure 4 shows the angle ϕ (in rad), for the actual, estimated and measured trajectory, as a function of time.

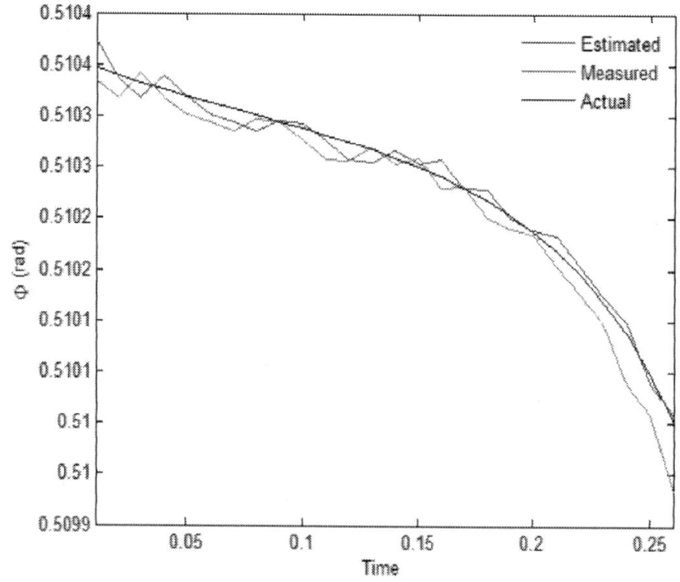

Figure 4. Position of the spacecraft given by the angle ϕ (in rad).

4. Conclusions

The spacecraft trajectory that makes a Swing-By maneuver around a binary system, estimated by the Kalman filter, was presented. The maneuver was estimated until the orbit periapsis, because there are multiple options for the sequence from there, like the capture of the spacecraft, collision or escape.

Measurements were obtained from observations made from Earth, given by the angle between the spacecraft's position and a fixed star.

It is shown that such a technique can be used successfully for trajectory estimation, if adequate accuracy is available for the angle observations. Besides, the Kalman filter was important to reduce the error in estimating the correct trajectory.

Acknowledgements

The authors wish to express their appreciation for the support provided by grants #473387/2012-3, 473164/2013-2 and 304700/2009-6 from the National Council for Scientific and Technological Development (CNPq); grants # 2011/08171-3, 2011/13101-4, 2014/06688-7 and 2012/21023-6 from São Paulo Research Foundation (FAPESP) and the financial support from the National Council for the Improvement of Higher Education (CAPES).

References

[1] Prado A F B A and Broucke R A 1995 A Classification of Swing-By Trajectories Using the Moon *Appl. Mech. Rev.* **48** 4

[2] Prado A F B A 1996 Powered Swing-By *J. of Guid., Control and Dyn.* **19** 5 pp. 1142-47

[3] Broucke R A 1988 The Celestial mechanics of gravity assist In:AIAA/AAS Astrodynamics Conference, Minneapolis, MN, (AIAA paper 88-4220)

[4] Mcconaghy T T, Debban T J, Petropoulos A E and Longuski J M 2003 Design and optimization of low-thrust gravity trajectories with gravity assist *J. Spac. and Rock.* **40** 380-387

[5] Silva A F, Prado A F B A and Winter, O C 2013a Optimal impulsive control in a powered Swing-By In: AIAA Guidance, Navigation, and Control (GNC) Conference

[6] Silva A F, Prado A F B A and Winter O C 2013b Powered Swing-By Maneuvers around the Moon *J. Phy.: Conf. Ser.* **465** 012001

[7] Strange N J and Longuski J M 2002 Graphical method for gravity-assist trajectory design *J. Spac. and Rock.* **39** 9-16

[8] Kuga H K 2005 Practical notions of estimation techniques (in Portuguese) Class Notes on Optimization of Dynamical Systems - Part II

[9] Maybeck P S 1979 Stochastic models, estimation and control. New York: Academic Press

Multiple island chains in wave-particle interactions

M C de Sousa[1], I L Caldas[1], A M Ozorio de Almeida[2], F B Rizzato[3] and R Pakter[3]

[1] Instituto de Física, Universidade de São Paulo, São Paulo, Brazil
[2] Centro Brasileiro de Pesquisas Físicas, Rio de Janeiro, Brazil
[3] Instituto de Física, Universidade Federal do Rio Grande do Sul, Porto Alegre, Brazil

E-mail: meirielenso@gmail.com

Abstract. We analyze the isochronous island chains that appear in the Poincaré sections of near integrable twist systems. When the system presents just one resonant perturbation with a winding number, the number of chains is constant and it is completely determined by the perturbation. However, for systems that are perturbed by an infinite number of resonant perturbations with the same winding number, the number of isochronous chains depends on the superposition of the perturbations and it is a function of the parameters. Considering a system that describes wave-particle interaction, we show that the number of island chains increases without limit when the wave period or wave number are increased.

1. Introduction

The Poincaré-Birkhoff Fixed Point Theorem states that the resonances of a near integrable system present an even number of periodic points [1-6]. Half of these points are stable and half of them are unstable. The stable points are surrounded by resonant islands and the set of islands that appear around a single periodic trajectory forms a chain [1-4, 6, 7]. When the resonance presents more than one stable periodic trajectory, the island chains surrounding these trajectories are called isochronous [5, 8].

The winding number, also known as rotation number, is not a monotonic function for nontwist systems [5, 6, 9]. In this case, it is common to observe more than one isochronous chain with different values for the action variables in the Poincaré sections, as can be seen in Refs. [6-10]. For twist systems, the winding number is a monotonic function of the action variables [5, 6, 9], which makes it impossible for the system to present isochronous chains in different regions of the Poincaré sections. Although the Poincaré-Birkhoff Fixed Point Theorem does not make any claim on the number of isochronous chains [1-3], we generally observe in the literature twist systems that present only one chain [2]. However, Refs. [11-13] show that it is possible for twist systems to present more than one isochronous chain with the same value for the action variables.

In this paper, we consider two near integrable twist systems that may present more than one isochronous chain in their Poincaré sections. We begin by analyzing a Hamiltonian that describes a nonlinear two-oscillator system, as discussed by Walker and Ford in Ref. [11]. This Hamiltonian presents two perturbative terms with different winding numbers. By changing the perturbative terms, we observe that the number of isochronous chains is altered, although the winding numbers remain the same. On the other hand, the number of chains does not vary with the parameters of the system.

In Ref. [13], we studied a Hamiltonian describing wave-particle interaction. When the wave is given as a series of periodic pulses, the system presents an infinite number of resonant perturbations with the same winding number, and each of these perturbations may generate islands in the same region of phase space. This superposition makes the number of island chains vary as a function of the wave parameters. For sufficiently high values of the parameters, all the resonances of the system present more than one isochronous chain.

2. Two-oscillator system

A (m,n) primary resonance is characterized by the winding number $\Omega = n/m$, where m and n are two positive relative primes. For a two dimensional system described by the action-angle variables $(\boldsymbol{J},\boldsymbol{\varphi})$, the Poincaré-Birkhoff Fixed Point Theorem states that a (m,n) resonance presents L isochronous chains with m islands each in the Poincaré section $J_1 \times \varphi_1$, and L isochronous chains with n islands each in the Poincaré section $J_2 \times \varphi_2$. However, the theorem does not make any claim on the value of L.

We point out that the winding number is not sufficient to determine all the features of a resonance because there is an infinite number of perturbations characterized by the same winding number, i.e., all the (lm, ln) perturbations, with l a not null integer, are characterized by $\Omega = n/m$. Therefore, to determine the number of isochronous chains in the Poincaré sections, it is necessary to know which (lm, ln) perturbations act on the system.

In Ref [11], Walker and Ford analyzed a Hamiltonian describing a nonlinear two-oscillator system:

$$\begin{aligned} H &= H_0(J_1,J_2) + H_1(J_1,J_2,\varphi_1,\varphi_2), \\ H &= J_1 + J_2 - J_1^2 - 3J_1J_2 + J_2^2 + \alpha J_1 J_2 \cos(r_1\varphi_1 - s_1\varphi_2) + \beta J_1 J_2^{3/2}\cos(r_2\varphi_1 - s_2\varphi_2), \end{aligned} \tag{1}$$

where r_i and s_i are not null integers with the same signal [13] that can be written as $(r_i, s_i) = (l_i m_i, l_i n_i)$. Walker and Ford used the polar coordinates

$$q_j = \sqrt{2J_j}\cos\varphi_j \quad \text{and} \quad p_j = -\sqrt{2J_j}\sin\varphi_j$$

to represent the Poincaré sections of the system, and they considered $(r_1, s_1) = (2,2)$ and $(r_2, s_2) = (2,3)$. They verified that the (1,1) resonance generated by the (2,2) perturbation presents 2 island chains in the Poincaré section $p_2 \times q_2$, while the (2,3) resonance generated by the (2,3) perturbation presents just one chain.

In figure 1, we show three Poincaré sections built for the same values of the parameters but considering different resonant perturbations. In each panel of figure 1, the $(m_2, n_2) = (2,3)$ resonance (inner resonant islands) is generated by a different perturbation. In panel (a), $(r_2, s_2) = (2,3)$ and the $(m_2, n_2) = (2,3)$ resonance presents one isochronous chain with three islands, i.e., the three inner islands in the Poincaré section are generated by a single initial condition. In panel (b), we consider $(r_2, s_2) = (4,6)$, and the $(m_2, n_2) = (2,3)$ resonance exhibits two chains with three islands each. It means that we need two different initial conditions to generate the six inner islands. Moreover, we

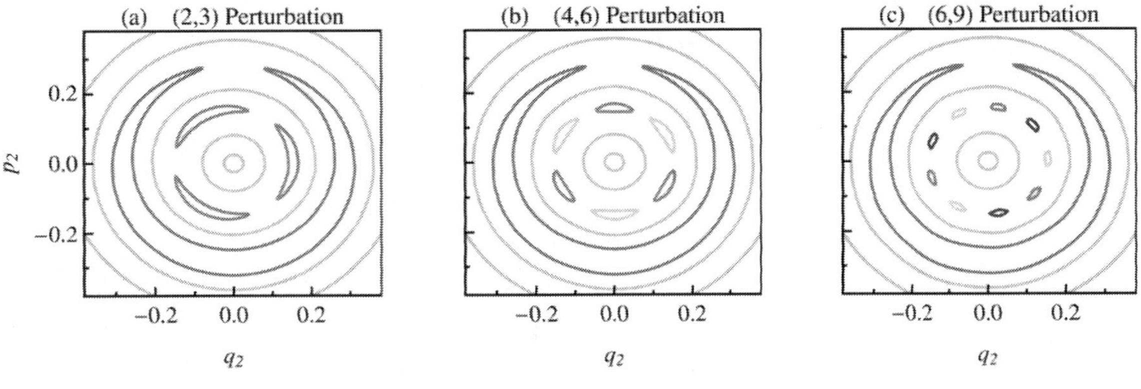

Figure 1. Poincaré section $p_2 \times q_2$ showing the $(m_2, n_2) = (2,3)$ resonance (inner islands) generated by different resonant perturbations: (a) $(r_2, s_2) = (2,3)$; (b) $(r_2, s_2) = (4,6)$; and (c) $(r_2, s_2) = (6,9)$.

Journal of Physics: Conference Series **641** (2015) 012003 doi:10.1088/1742-6596/641/1/012003

observe that the islands of the two chains alternate in the Poincaré section. In panel (c), the $(r_2, s_2) = (6,9)$ perturbation generates three island chains with three islands each.

In all the panels of figure 1, $(r_1, s_1) = (1,1)$ and this perturbation generates the outer resonant island in the pictures. For $(r_1, s_1) = (1,1)$, the $(m_1, n_1) = (1,1)$ resonance presents only one isochronous chain with one island. When we change the resonant perturbation generating the $(m_1, n_1) = (1,1)$ resonance, the number of chains varies similarly to the behavior observed for the $(m_2, n_2) = (2,3)$ resonance.

While the winding number $\Omega = n_i / m_i$ defines the number of islands per chain (as stated by the Poincaré-Birkhoff theorem), figure 1 shows that the number L_i of chains is defined by $|l_i|$. Therefore, to determine the number of isochronous chains it is necessary to know not only the winding number characterizing the resonance, but also which $(l_i m_i,\ l_i n_i)$ resonant perturbation acts on the system to generate the islands.

3. Wave-particle interaction

In Ref. [13], we analyzed the islands of a system describing wave-particle interaction. One particle moves under the action of a uniform magnetic field and a stationary electrostatic wave given as a series of periodic pulses. The dimensionless Hamiltonian that describes the dynamics transverse to the magnetic field is given by [14, 15]

$$H(I,\theta,t) = \sqrt{1+2I} + \frac{\varepsilon}{2}\cos(k\sqrt{2I}\sin\theta)\sum_{l=-\infty}^{+\infty}\delta(t-lT), \tag{2}$$

where (I,θ) are action-angle variables, T is the wave period, k is the wave number and $\varepsilon/2$ is the wave amplitude. The Hamiltonian (2) can be expanded in a Fourier-Bessel series as [13]

$$H = \sqrt{1+2I} + \frac{\varepsilon}{2T}\sum_{r=-\infty}^{+\infty}\sum_{s=-\infty}^{+\infty} J_r(k\sqrt{2I})\cos\left(r\theta - \frac{2\pi st}{T}\right), \tag{3}$$

where $J_r(k\sqrt{2I})$ are Bessel functions of order r.

In the resonant islands of the system, the wave transfers energy to the particle and it is regularly accelerated [14-16]. References [14, 16] calculate the parameters interval for which the acceleration can be maximum. However, to achieve the condition of maximum acceleration, it is also necessary to properly adjust the initial conditions of the particle to make it follow the best trajectory in phase space. For this purpose, one should know the position of the resonances and the number of isochronous chains in phase space.

Reference [13] shows that Hamiltonian (3) presents an infinite number of resonant perturbations with the same winding number. Actually, all the (r, s) perturbations that can be written as $(r, s) = (l_i m,\ l_i n)$, with m and n fixed and different values of l_i, are characterized by the winding number $\Omega = n / m$. For example, the perturbations (1,1), (2,2), (-1,-1), (-3,-3), etc., they all have $\Omega = 1/1$.

All the resonant terms that present the same winding number may generate islands in the same region of phase space. This superposition of terms acting on the same region makes the number of isochronous chains vary as a function of the wave period and wave number [13], as can be seen in figure 2. This figure shows the phase space near the (2,3) resonance for different values of the wave number. In panel (a), $k = 1$ and the (2,3) resonance presents just one island chain, i.e., the two islands are generated by the same initial condition. Panel (b) shows two isochronous chains with two islands each for $k = 3$. The islands of one chain are centered at $\theta = 0,\ \pi$, while the islands of the other chain are centered at $\theta = \pi/2,\ 3\pi/2$. The islands of the two distinct chains can also be recognized by their different size and shape. Similarly to figure 1, the islands of the different chains in figure 2(b) alternate in phase space.

Figure 3 shows the parameter spaces for the (2,3) and (5,2) resonances. The parameter spaces represent the number of isochronous chains as a function of T and k, as indicated in the pictures. For the (2,3) resonance, the number of chains increases monotonically with the parameters of the waves. For the (m, n) resonances with $m > 4$, the number of chains is not a monotonic function of the

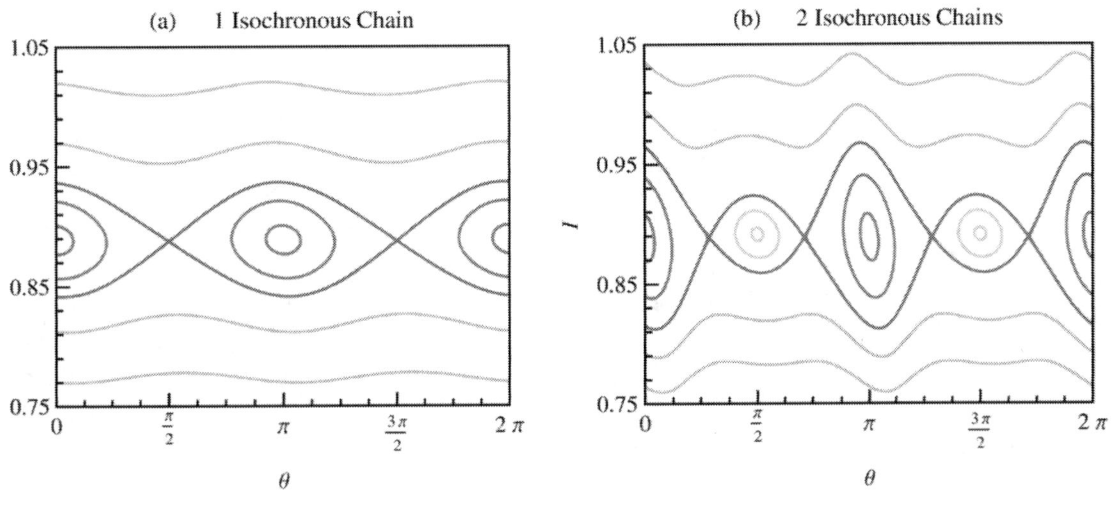

Figure 2. Phase space near the (2,3) resonance for $T = 5\pi$, $\varepsilon = 0.01$, and (a) $k = 1$; (b) $k = 3$.

parameters, as can be seen in figure 3(b) for the (5,2) resonance. As we increase the value of m, the pattern becomes more complicated.

In Ref. [13], we discussed how the number of isochronous chains are related to the resonant perturbations in Hamiltonian (3). For a (m, n) resonance with $m \leq 4$, the L chains observed in phase space are generated by the (r, s) = $(l_i m,\ l_i n)$ perturbations for which $|l_i| \leq L$. It happens because the amplitudes of the $(l_i m,\ l_i n)$ perturbations with $|l_i| > L$ are generally much smaller and they do not contribute to the generation of islands.

When $m > 4$, the coupling of the $(l_i m,\ l_i n)$ resonant terms becomes more complicated and the previous statement is not true anymore [13]. For the (5,2) resonance in figure 3(b), for example, we observe two isochronous chains when $T \to 4\pi/5$ or $k \to 0$. These two chains are generated by the (r, s) = $(5l_i,\ 2l_i)$ perturbations with $|l_i| \leq 2$. Increasing the values of T or k, figure 3(b) presents a very thin region corresponding to four island chains, which are generated by the perturbations with $|l_i| \leq 4$. If we continue to increase the values of T or k, the (5,2) resonance exhibits a second region in the parameter space that corresponds to two chains. However, this second region is generated by the coupling of the $(5l_i,\ 2l_i)$ perturbations for which $|l_i| \leq 4$ and not $|l_i| \leq 2$ as observed for the first region representing two chains.

Although the L chains may be generated by the $(l_i m,\ l_i n)$ perturbations with $|l_i| > L$ when $m > 4$, the number of resonant terms that actually contribute to the generation of islands in phase space is always finite [13]. We also showed in Ref. [13] that the borders separating two regions with a different number of chains in the parameter space respect the same kind of power law for all the resonances. This power law is related to the position of the resonance in phase space with respect to the action variable.

Due to a symmetry in Hamiltonian (3), all the resonances of the system present an even number of islands in phase space [13]. Therefore, all the (m, n) resonances with m odd present an even number of isochronous chains, as can be seen in figure 3(b). On the other hand, when the number m of islands per chain is even, the number of chains may be even or odd, as shown in figure 3(a). This symmetry in (3) implies that most of the resonances of the system always presents more than one chain in phase space. Furthermore, when the values of the wave period and wave number are sufficiently high, all the resonances have more than one isochronous chain. Such behavior is not commonly observed in the literature that generally presents twist systems with single chains [2].

4. Conclusions

The Poincaré-Birkhoff Fixed Point Theorem deals with the winding number that characterizes a

Journal of Physics: Conference Series **641** (2015) 012003 doi:10.1088/1742-6596/641/1/012003

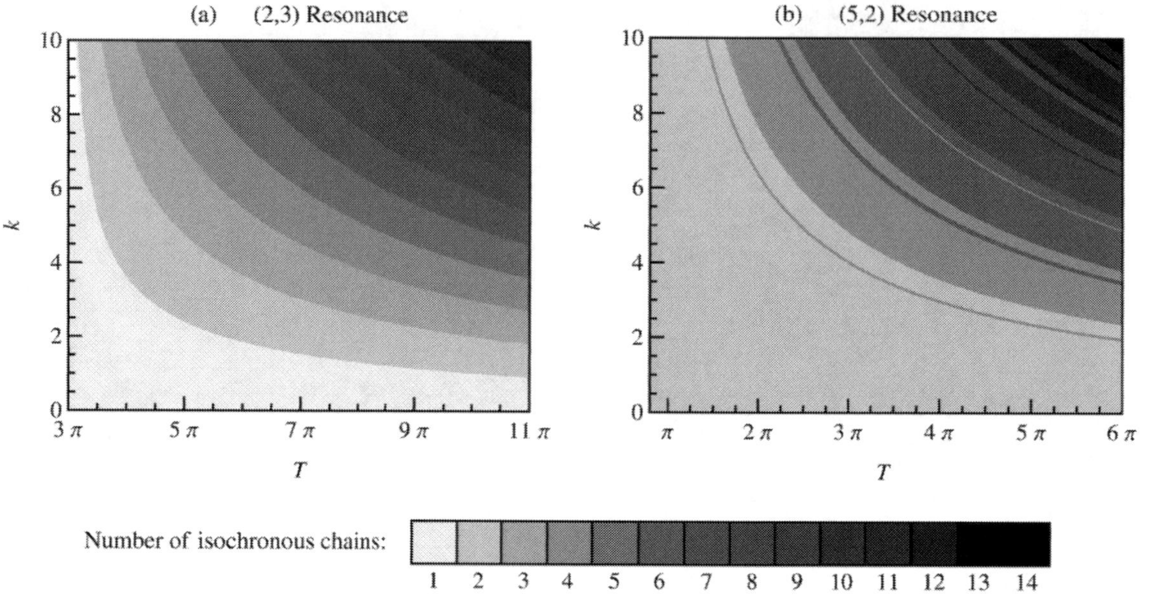

Figure 3. Number of isochronous chains as a function of T and k for the (a) (2,3); and (b) (5,2) resonances.

resonance. In this paper, we showed that the winding number is not enough to determine all the features of a resonance because an infinite number of resonant perturbations have the same winding number. In particular, the number of isochronous chains present in phase space depends on the perturbations that act on the system and their superposition.

When the system has just one resonant perturbation with the same winding number, the number of chains is completely determined by this perturbation and it is a constant for the system. For systems that present an infinite number of perturbations with the same winding number, all these perturbations may generate islands in the same region of phase space. Such a superposition alters the number of isochronous chains according to the amplitudes of the perturbations and the coupling between them. For a system describing wave-particle interaction as discussed in this paper, we observed that the number of chains increases without limit as we increase the wave period or wave number. For sufficiently high values of these parameters, all the resonances of the system exhibit more than one chain in phase space.

We point out that the results presented in this paper are not a particular feature of the analyzed systems. They are valid for near integrable twist systems that are perturbed by either one or infinite resonant terms with the same winding number. For systems that are perturbed by a finite number of resonant terms with the same winding number, we also observe that the number of chains varies as a function of the parameters of the system. However, in this case, the number of chains is limited to a finite set and it does not increase without bounds.

Acknowledgments

We acknowledge financial support from São Paulo Research Foundation (FAPESP, Brazil) under Grants No. 2011/20794-6 and No. 2011/19296-1, CNPq (Brazil), CAPES (Brazil), and US-AFOSR under Grant No. FA9550-12-1-0438.

References

[1] Birkhoff G D 1927 *Acta Math.* **50** 359
[2] Lichtenberg A J and Lieberman M A 1992 *Regular and chaotic dynamics* 2nd ed. (New York: Springer)

Journal of Physics: Conference Series **641** (2015) 012003 doi:10.1088/1742-6596/641/1/012003

[3] Berry M V 1978 *AIP Conf. Proc.* **46** 16
Reichl L E 2004 *The transition to chaos: Conservative classical systems and quantum manifestations* 2nd ed. (New York: Springer)
[4] Ozorio de Almeida A M 1988 *Hamiltonian systems: Chaos and quantization* (Cambridge: Cambridge University Press)
[5] Voyatzis G, Meletlidou E and Ichtiaroglou S 2002 *Chaos, Solitons and Fractals* **14** 1179
[6] Petrisor E, Misguich J H and Constantinescu D 2003 *Chaos, Solitons and Fractals* **18** 1085
[7] Van Der Weele J P, Valkering T P, Capel H W and Post T 1988 *Physica* A **153** 283
[8] Egydio de Carvalho R and Ozorio de Almeida A M 1992 *Phys. Lett.* A **162** 457
[9] del-Castillo-Negrete D, Greene J M and Morrison P J 1996 *Physica* D **91** 1
[10] Oda G A and Caldas I L 1995 *Chaos, Solitons and Fractals* **5** 15
Portela J S E, Caldas I L and Viana R L 2008 *The European Physical Journal Special Topics* **165** 195
[11] Walker G H and Ford J 1969 *Phys. Rev.* **188** 416
[12] Van Der Weele J P, Capel H W, Valkering T P and Post T 1988 *Physica* A **147** 499
[13] de Sousa M C, Caldas I L, Ozorio de Almeida A M, Rizzato F B and Pakter R 2013 *Phys. Rev.* E **88** 064901
[14] de Sousa M C, Steffens F M, Pakter R and Rizzato F B 2010 *Phys. Rev.* E **82** 026402
[15] de Sousa M C, Caldas I L, Rizzato F B, Pakter R and Steffens F M 2012 *Phys. Rev.* E **86** 016217
[16] Pakter R and Corso G 1995 *Phys. Plasmas* **2** 4312

Journal of Physics: Conference Series **641** (2015) 012004 doi:10.1088/1742-6596/641/1/012004

Stability solutions of a dumbbell-like system in an elliptical orbit

Denilson Paulo Souza dos Santos[1], **Simão António da Rocha e Brito de Aguiã Morant**[2], **Anna D. Guerman**[3], **Alexander A. Burov**[4]

[1−2−3] Centre for Mechanical and Aerospace Science and Technologies (C-MAST) – UBI – Portugal.
[1] Division of Space Mechanics and Control – INPE (National Institute for Space Research) – São José dos Campos – SP, Brasil.
[4] Dorodnotsyn Computing Center of the Russian Academy of Sciencies, Russia.

E-mail: denilson.paulo@gmail.com

Abstract. A dumbbell-like system is analyzed, considering two mass points connected with a massless and rigid tether with length variations, and the center of mass described by Keplerian orbits. This kind of system, in a certain type of configuration, is a simple conceptualization of the space elevator. The system motion is obtained with the Lagrangian Formulation in a Central Gravitational Field, and the perturbations of motion are neglected. Laws of control are considered for the angle of systems rotation around the center of mass. Those include uniform rotations or no rotation at all. Stability conditions were obtained for the first case, analyzing its neighborhood and using Floquet Theory. The results shown there are regions of eccentricities were stability is found. Lastly, a dynamic numerical simulator was created, where the implementation of the results could be tested. The dynamic behavior of the system showed regular and chaotic properties.

1. Introduction

Researchers have been studying, since 1960th, the possibilities to control attitude motion of space systems with variable mass distributions. Perturbations of motion can occur for several such atmospheric drag, solar pressure, system deployment, n-body perturbations else which will not be discussed here. It will be analyzed stability conditions inside solutions in Keplerian orbits for a model for two-body tether systems in a Central Gravitational Field (Newtonian). The motions equation will be written by Lagrangian formulation, explaining the Kinetic and Potential Energy, considering the possibility to control the tether length or the angle of rotation around the center of mass. Several authors studied attitude dynamics of TSS of different configurations. Two-body and three-body systems have been mostly studied, but even tethered satellite constellations were analyzed. In [1] the dynamics of three-body TSS were examined, with the cm (center of mass) following a circular orbit. A. Burov, I. Kosenko and A. Guerman, [2] – [5], studied the dynamics of a moon-anchored tether with a material point at its end, for variable tether length. The bi-dimensional problem can be addressed as a simple model for the lunar elevator. In this problem, some particular solutions of the motion equation are derived, choosing a specific control law for the tether length and finding radial and oblique configurations for the system geometry, while defining regions of stability and instability for each case.

Journal of Physics: Conference Series **641** (2015) 012004 doi:10.1088/1742-6596/641/1/012004

2. Mathematical model of a two point tethered satellite system (TSS)

The main objective of this paper is to analyze equilibrium conditions for a two masses point TSS in an elliptical orbit, where the local vertical of the system must be align with an inert axis fixed on Earth. The model is a dumbbell-like system, with the two body connected with a rigid and massless tether. The systems center of mass is described by a Keplerian orbit fixed on an inert axis in primary bodys center [6].

Figure 1. Model of the satellite system connected with tethers.

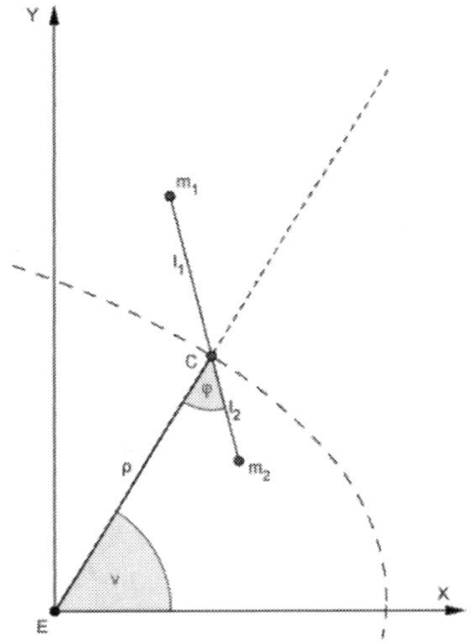

Figure 2. Dumbbell-like system composed by a massless tether and two mass points.

The simplifying scheme can be seen in Figure 2. Although the simplification is rough, since the moment of inertia of a $3D$ system may not be aligned with the local vertical, it allows a fundamental mission conceptualization. In this model, the position of the two masses point will be described in order to the position of the center of mass (C), and the systems motion in an elliptic orbit. The angles ν and φ represent the true anomaly and the angle between the tether and ρ, respectively. ρ is the distance from E to C, l_1 and l_2 are the tether lengths from the mass points m_1 and m_2 to C, respectively.

2.1. Positions of the system

This allows the model to be described by planar Keplerian motion:

$$\rho = \frac{p}{1 + e\cos(\nu)} \tag{1}$$

where p is the focal parameter, e the eccentricity an ν the true anomaly. The system coordinates are:

$$\begin{cases} x_0 = \rho\cos(\nu) \\ y_0 = \rho\sin(\nu) \\ x_1 = x_0 + l_1\cos(\nu + \varphi) \\ y_1 = y_0 + l_1\sin(\nu + \varphi) \\ x_2 = x_0 - l_2\cos(\nu + \varphi) \\ y_2 = y_0 - l_2\sin(\nu + \varphi) \end{cases} \tag{2}$$

Journal of Physics: Conference Series **641** (2015) 012004 doi:10.1088/1742-6596/641/1/012004

where, x_0 and y_0 being the coordinates of the position of the center of mass, $(\vec{r}_0)$; x_i and y_i being the coordinates of the position of the point mass i, $(\vec{r}_i)$.

$$\begin{cases} \vec{r}_0 = (x_0, y_0) \\ \vec{r}_1 = (x_1, y_1) \\ \vec{r}_2 = (x_2, y_2) \end{cases} \tag{3}$$

2.2. Lagrange Equations of Motion
Lagrange equations of motion are known as differential equations of system motion in generalized coordinates. In fact, these Lagrange equations provide a simple way to obtain the system positions in function of time. It was decided to choose the Lagrangian formulation because it allows us to define the global positions of the system through Potential and Kinetic Energy and in function of generalized coordinates, unlike the needs to define all the forces involved, as in Newtonian formulation. The general formula of the Lagrange Equation is:

$$\frac{d}{dt}\left(\frac{\partial L}{\partial \dot{q}_i}\right) - \frac{dL}{dq_i} = Q_i \tag{4}$$

where $L = T - V$, q_i is the generalized coordinate and Q_i is the generalized force actuating in the system. For the following analysis, the generalized coordinates are φ and l and the system is only under the gravity-gradient. Defining φ or l allows to control the system, as can be seen in [2]. It has been chosen to define φ and consecutively obtain the l behavior.

$$l\left(3\mu_0 sin(2\varphi)(1 + ecos(\nu))^3 + 2p^3\left(\ddot{\nu} + \ddot{\varphi}\right)\right) + 4p^3\dot{l}\left(\dot{\nu} + \dot{\varphi}\right) = 0 \tag{5}$$

All these variables are time-related. For a more advantageous formulation, consider a timeless relation, based only on general positions and given by the new independent variable true anomaly ν [6]. Substituting all these variables it can be rearranged in:

$$(1 + e\cos(\nu))\varphi'' + 2\left(\frac{l'}{l}(1 + e\cos(\nu)) - e\sin(\nu)\right)(\varphi' + 1) + 3\cos(\varphi)\sin(\varphi) = 0 \tag{6}$$

a well-known equation ([2] – [6]).

3. φ Control laws
At this stage, the Lagrange Equations of Motion were derived and control laws can now be considered for φ or l. Opting for one is sufficient and convenient to know the system motion. The chose to define over one or other parameter is arbitrary and, therefore, were chosen φ control laws. The tether length (l) is small compared to the orbit parameter (p). Two control laws for the behavior of φ are presented: fixed angle and uniform rotations. Each case will be analyzed independently.

3.1. Fixed Angle
The fixed angle $(\varphi = \varphi_0)$ for non-rotational motion about the center of mass, allows the satellite to point its face (or antenna) through Earth, not rotating locally. This has practical and real-life meaning, as GPS satellites, for instance, need to be pointing every time through Earth. This control law (Eq. 6)can assume the following expression:

$$\frac{l'(\nu)}{l(\nu)} = \frac{e\sin(\nu)}{1 + e\cos(\nu)} - \frac{3\sin(2\varphi_0)}{4(1 + e\cos(\nu))} \tag{7}$$

Chosen different φ_0, the logarithmic behaviours of the tether are: (Figures 3 - 10).

Journal of Physics: Conference Series **641** (2015) 012004 doi:10.1088/1742-6596/641/1/012004

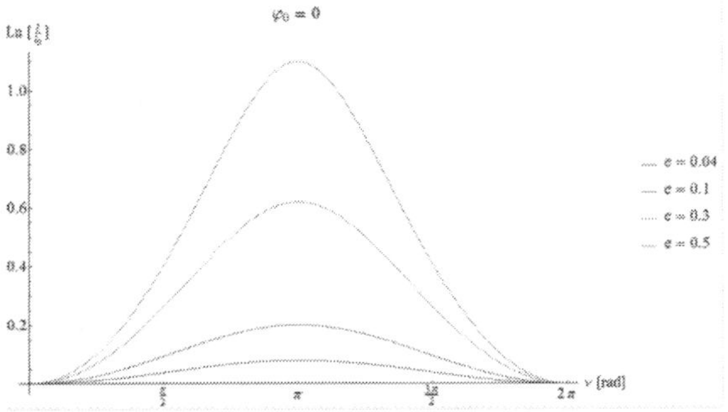

Figure 3. Tether logarithmic ratio for different eccentricities for $\varphi_0 = 0$.

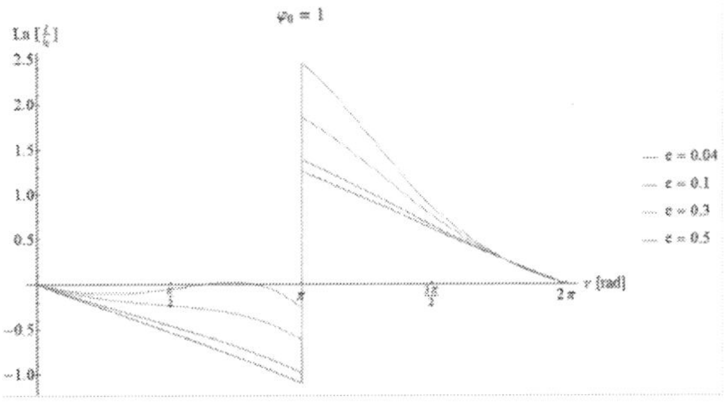

Figure 4. Tether logarithmic ratio for different eccentricities for $\varphi_0 = 1$.

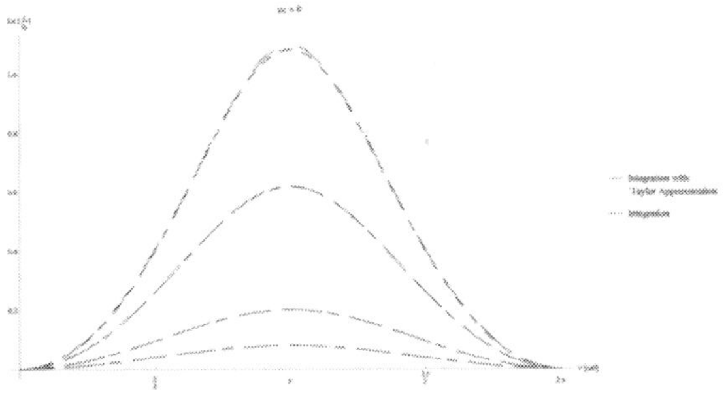

Figure 5. Tether logarithmic ratio approximation comparison.

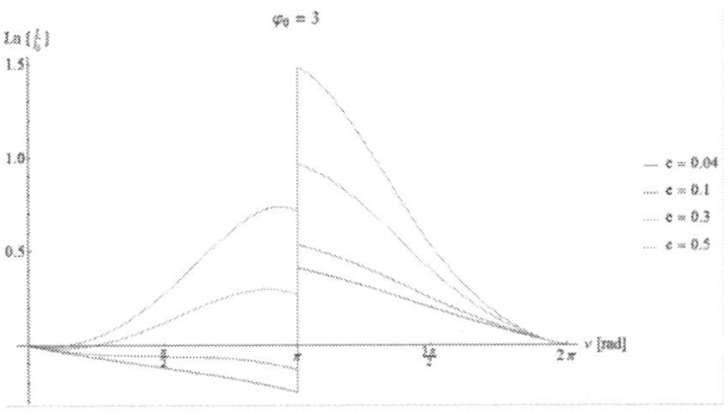

Figure 6. $\varphi_0 = 3$.

There are uniform solutions to values $\varphi_0 = \frac{k\pi}{2}$ (k is integer). For an analytical solution, an approximation of 4th order using Taylor Series was also done allowing the integration of Eq. 7 showed in Figure 5, analyzed for various cases with different eccentricities.

3.2. Uniform Rotation

For uniform rotations, the control law can assume the following equation ($\varphi = \omega\nu + \varphi_0$). This relation means that the actual angle of rotation depends on the initial condition φ_0, the variation of ν and a constant ω. Equation 6 is rewritten

$$\frac{l'(\nu)}{l(\nu)} = \frac{e\sin(\nu)}{1+e\cos(\nu)} - \frac{3\sin\left(2\left(\omega\nu+\varphi_0\right)\right)}{4(\omega+1)(e\cos(\nu)+1)} \tag{8}$$

The Eq. 8 allows to know the tether length for each ν. A numerical integration was done, where ω and φ_0 were substituted previously. Also, for $\omega = -1$ nothing can be analyzed since the Eq. 8 has singularities around that point.

The monodromy matrix is used to analyze stability conditions around small oscillations ($\delta\varphi$) of φ. Replacing ($\varphi = \omega\nu + \delta\varphi$) in Eq. (6) allows one obtain the nonlinear equation of perturbed motion, and linearized equation constricts this analysis to small variations of φ.

$$(1+e\cos(\nu))\delta\varphi'' + \frac{3}{2}\left(2\cos(2\omega\nu)\delta\varphi - \frac{\sin(2\omega\nu)\delta\varphi'}{\omega+1}\right) = 0 \tag{9}$$

The Eq. 9 is the equation for the variations. It is now possible to analyze the stability using the Floquet theory, the linearized equation constricts this analysis to small variations of φ. Applying the Floquet theory, find the monodromy matrix for this system.

Journal of Physics: Conference Series **641** (2015) 012004 doi:10.1088/1742-6596/641/1/012004

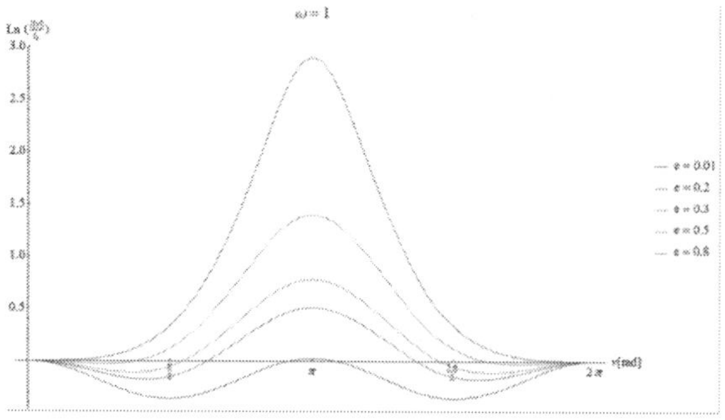

Figure 7. The control law for the tether length for $\varphi_0 = 0$ and $\omega = 1$.

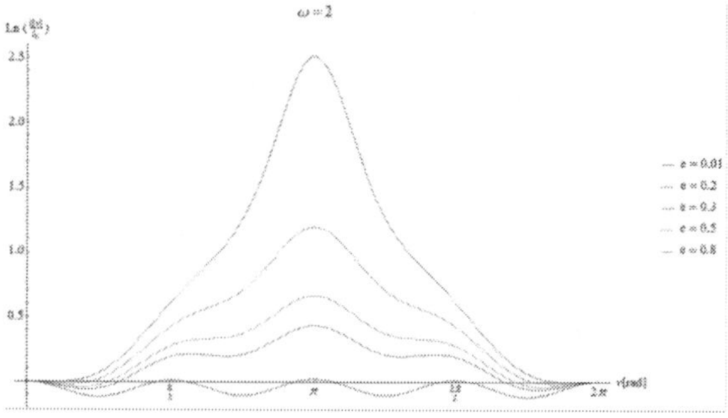

Figure 8. The control law for the tether length for $\varphi_0 = 0$ and $\omega = 2$.

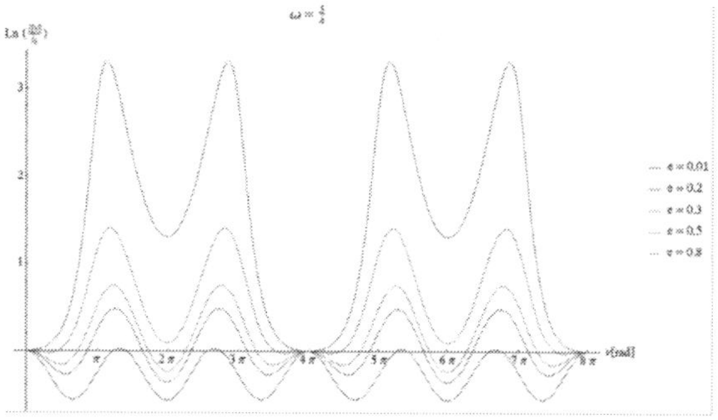

Figure 9. The control law for the tether length for $\varphi_0 = 0$ and $\omega = \frac{3}{4}$.

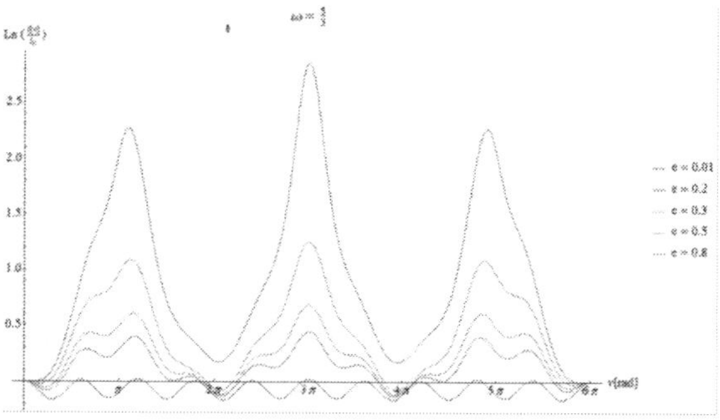

Figure 10. The control law for the tether length for $\varphi_0 = 0$ and $\omega = \frac{5}{3}$.

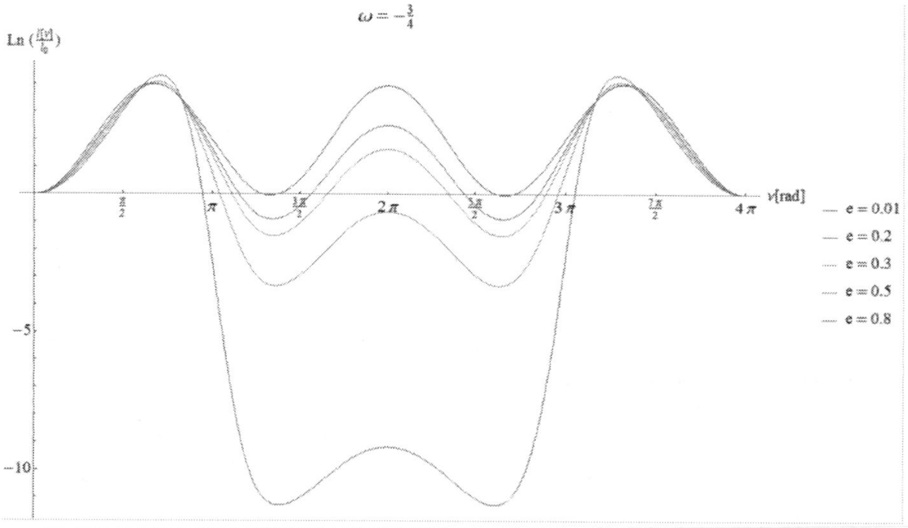

Figure 11. The control law for the tether length for $\varphi_0 = 0$ and $\omega = -\frac{3}{4}$.

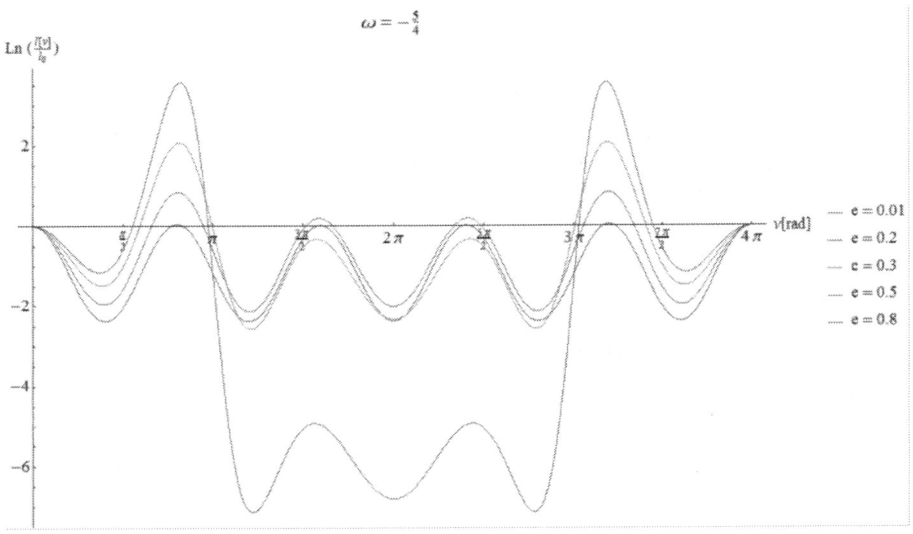

Figure 12. The control law for the tether length for $\varphi_0 = 0$ and $\omega = -\frac{5}{4}$.

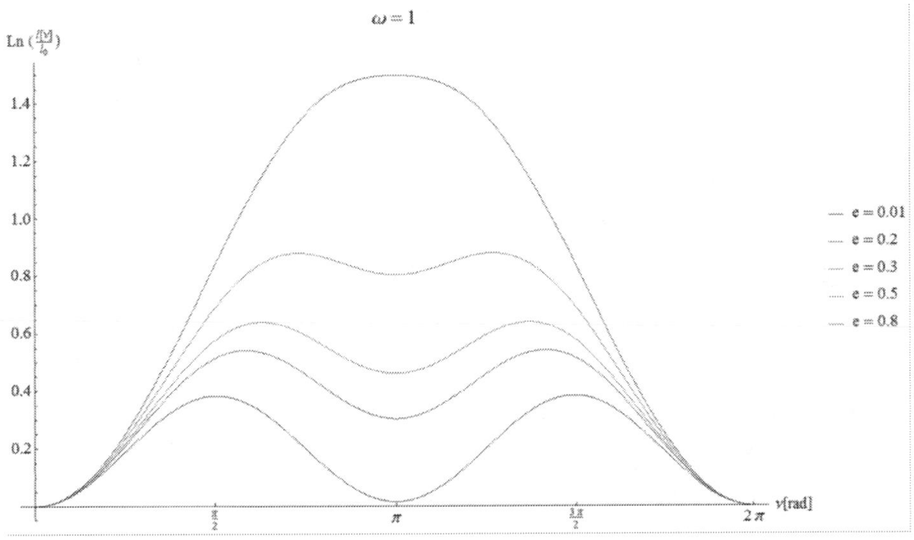

Figure 13. The control law for the tether length for $\varphi_0 = \frac{\pi}{2}$ and $\omega = 1$.

4. Conclusion
Laws of control are considered for the angle of systems rotation around the center of mass. The necessary conditions of stability for uniform rotations were analyzed using the Floquet Theory analyzing the parameters ω and the eccentricity of the orbit, generating the control laws for the tether length. Stability conditions give the values of e and ω where the system is stable. As reported before, this means that small perturbations wont change the behavior of the system. Stability is found for several values of eccentricity, as shown in stability curves for the parameter ω. The uniform rotations of φ proved to have smooth tether variations, for every simulation done. For the fixed angle analysis, there was only one value where φ behave properly, which is $\varphi_0 = 0$. At last, the simulations should always be considered in an engineering study. They allow to fully understand the dynamics of the system, and serve as a guide to future work. The

results obtained in the simulator were always expected.

Acknowledgments

This work was accomplished with the support of CAPES – Brazil, under Contract BEX 10689-13-3, INPE – National Institute for Space Research – Brazil and Centre for Mechanical and Aerospace Science and Technologies (C-MAST) – UBI Portugal.

References

[1] Misra, A. K., Z. Amier, and V. J. Modi. "Attitude dynamics of three-body tethered systems." Acta Astronautica (1988):1059-1068.
[2] A.A. Burov and H. Troger, The relative equilibria of an orbital pendulum suspended on a tether, Journal of Applied Mathematics and Mechanics, Vol. 64, Issue 5, 2000b, pp. 723-728.
[3] A. Burov, O. I. Kononov, and A. D. Guerman, Relative equilibria of a Moon - tethered spacecraft,Advances in the Astronautical Sciences, v. 136, 2011a, pp. 2553-2562.
[4] A. Burov and I.I. Kosenko, Plane oscillations of a body with variable mass distribution in an elliptic orbit, Proc. of ENOC 2011, July 24-29, 2011b, Rome, Italy.
[5] A.A. Burov, I.I. Kosenko, and A. D. Guerman, Dynamics of a moon-anchored tether with variable length. Advances in the Astronautical Sciences, 2012, Vol. 142, pp.3495-3507.
[6] Simão Antônio da Rocha e Brito de Aguiã Morant. *Space Tether Systems – Stability Solutions of a Dumbbell-like System*, Dissertation Master Engineering Aeronautical - UBI - Portugal 2013.
[7] Belestsky, Vladimir V.; Levin, Evgenii M., 1993. Dynamics of Space Tethers Systems. San Diego - California: Advances in the Astronautical Sciences, Vol. 83 - American Astronautical Society ISSN 0-87703-370-6.
[8] M.P. Cartmell, D.J. McKenzie, A review of space tether research, Progress in Aerospace Sciences, Volume 44, Issue 1, January 2008, Pages 1-21, ISSN 0376-0421, http://dx.doi.org/10.1016/j.paerosci.2007.08.002.

Nonlinear Least Squares Method for Gyros Bias and Attitude Estimation Using Satellite Attitude and Orbit Toolbox for Matlab

W R Silva[1], H K Kuga[1] and M C Zanardi[2]

[1]Space Mechanics and Control Division, National Institute for Space Research (INPE), Av dos Astronautas, 1758, Jd. da Granja, CEP:12227-010, São José dos Campos, S.P., Brazil.
[2]Federal University of ABC (UFABC), Av. dos Estados, 5001, Bangu, CEP:09210-580, Santo André, S.P., Brazil.

E-mail: reis.william@gmail.com

Abstract. The knowledge of the attitude determination is essential to the safety and control of the satellite and payload, and this involves approaches of nonlinear estimation techniques. Here one focuses on determining the attitude and the gyros drift of a real satellite CBERS-2 (China Brazil Earth Resources Satellite) using simulated measurements provided by propagator PROPAT Satellite Attitude and Orbit Toolbox for Matlab. The method used for the estimation was the Nonlinear Least Squares Estimation (NLSE). The attitude dynamical model is described by nonlinear equations involving the Euler angles. The attitude sensors available are two DSS (Digital Sun Sensor), two IRES (Infra-Red Earth Sensor), and one triad of mechanical gyros. The two IRES give direct measurements of roll and pitch angles with a certain level of error. The two DSS are nonlinear functions of roll, pitch, and yaw attitude angles. Gyros are very important sensors, as they provide direct incremental angles or angular velocities. However gyros present several sources of error of which the drift is the most troublesome. Results show that one can reach accuracies in attitude determination within the prescribed requirements, besides providing estimates of the gyro drifts which can be further used to enhance the gyro error model.

1. Introduction

Attitude estimation is a process of determining the orientation of a satellite with respect to an inertial reference system.There are a large number of estimation methods, thus, it is necessary to evaluate the processing time and the desired precision in project requirements.The state estimation methods of attitude, that was used in this work, correct successive estimates of attitude parameters, different from the deterministic attitude method in which the same number of observations as variables is used to obtain one or more discrete attitude solutions[9].

In this work, the attitude is represented by Euler angles, due to its easy geometrical interpretation. The state estimation is performed by the Nonlinear Least Squares Estimation using simulated measurements provided by propagator PROPAT [1]. This method is capable of estimating nonlinear systems states from data obtained from different sensors of attitude. It was taken simulated data supplied by one triad of mechanical gyros, two IRES (Infra-Red Earth Sensor) and two DSS (Digital Sun Sensors). These sensors configurations are similar to those found in real satellite CBERS-2 (China Brazil Earth Resources Satellite).

Journal of Physics: Conference Series **641** (2015) 012005 doi:10.1088/1742-6596/641/1/012005

2. Nonlinear Least Squares Estimation

This method, otherwise known as Gaussian Least Squares Differential Correction, was originally developed by Gauss and employed to determine planetary orbits (during the 1800s) from telescope measurements of the line of sight angles to the planets [6].

The Nonlinear Least Squares Estimation assumes that the dynamic model of the state vector is perfect. Thus the nonlinear dynamic system is gven by [8, 3]:

$$\begin{aligned} \dot{\boldsymbol{x}} &= f(\boldsymbol{x}) \\ \boldsymbol{y}_k &= h_k(\boldsymbol{x}_k) + \boldsymbol{v}_k \end{aligned} \tag{1}$$

where $f(\boldsymbol{x})$ represents the nonlinear vector function of state $\boldsymbol{x}$, $\boldsymbol{y}_k$ represents the vector of observation of the sensors, $h(\boldsymbol{x}_k)$ is the function associated with the model of sensor observations, $\boldsymbol{x}_k$ represents the state vector at instant t_k and $\boldsymbol{v}_k = N(0, \boldsymbol{R}_k)$ represents the vector associated with the noise of the observations at this point and the weighting matrix $\boldsymbol{R}_k$ are symmetric positive definite.

The linearized dynamic equation by the expansion of $f(\boldsymbol{x})$ in Taylor series with truncation on linear term is given by:

$$f(\boldsymbol{x}) = f(\bar{\boldsymbol{x}}) + \left.\frac{\partial f}{\partial \boldsymbol{x}}\right|_{\bar{\boldsymbol{x}}} (\boldsymbol{x} - \bar{\boldsymbol{x}}) \tag{2}$$

By setting the following deviations

$$\begin{aligned} \delta\boldsymbol{x} &\equiv \boldsymbol{x} - \bar{\boldsymbol{x}} \\ \delta\dot{\boldsymbol{x}} &\equiv \dot{\boldsymbol{x}} - \dot{\bar{\boldsymbol{x}}} \equiv \dot{\boldsymbol{x}} - f(\bar{\boldsymbol{x}}) \end{aligned} \tag{3}$$

the following expression arises

$$\delta\dot{\boldsymbol{x}} = \boldsymbol{F}\delta\boldsymbol{x} \tag{4}$$

where $\boldsymbol{F}$ is the Jacobian matrix given by $\boldsymbol{F} = \left.\frac{\partial f}{\partial \boldsymbol{x}}\right|_{\bar{\boldsymbol{x}}}$, and of course this linearization is valid only while $\delta\boldsymbol{x}$ is small.

The observation equation can also be linearized by the expansion of h_k into Taylor series with truncation on linear term:

$$\boldsymbol{y}_k = h_k(\bar{\boldsymbol{x}}) + \left.\frac{\partial h_k}{\partial \boldsymbol{x}}\right|_{\bar{\boldsymbol{x}}} (\boldsymbol{x} - \bar{\boldsymbol{x}}) + \boldsymbol{v}_k \tag{5}$$

The residues are defined by $\delta\boldsymbol{y}_k = \boldsymbol{y}_k - h_k(\bar{\boldsymbol{x}})$, such that the observation equation is given by the linearized equation:

$$\delta\boldsymbol{y}_k = \boldsymbol{H}_k\delta\boldsymbol{x}_k + \boldsymbol{v}_k \tag{6}$$

where $\boldsymbol{H}_k$ is given by $\boldsymbol{H}_k = \left.\frac{\partial h_k}{\partial \boldsymbol{x}}\right|_{\bar{\boldsymbol{x}}}$.

The a-priori information to the state and covariance in instant t_0 given by $\hat{\boldsymbol{x}}_0 = \hat{\boldsymbol{x}}_0(t_0)$ and $\hat{\boldsymbol{P}}_0 = \hat{\boldsymbol{P}}_0(t_0)$. The method of nonlinear least squares must be implemented in an iterative way, which refines the variations instead of states [3, 8]. By setting the following deviations:

$$\begin{aligned} \delta\bar{\boldsymbol{x}}_k &\equiv \hat{\boldsymbol{x}}_k - \hat{\boldsymbol{x}}_0 \\ \delta\hat{\boldsymbol{x}}_k &\equiv \hat{\boldsymbol{x}}_k - \hat{\boldsymbol{x}}_{k-1} \end{aligned} \tag{7}$$

Finally, the equations that implement the estimation algorithm for nonlinear least squares method are [3, 8]:

$$\hat{\boldsymbol{P}}_k = \left(\hat{\boldsymbol{P}}_0^{-1} + \boldsymbol{H}^T\boldsymbol{R}^{-1}\boldsymbol{H}\right)^{-1} \tag{8}$$

Journal of Physics: Conference Series **641** (2015) 012005 doi:10.1088/1742-6596/641/1/012005

$$\delta\hat{\boldsymbol{x}}_k = \hat{\boldsymbol{P}}_k \left(\hat{\boldsymbol{P}}_0^{-1} \delta\bar{\boldsymbol{x}}_{k-1} + \boldsymbol{H}^T \boldsymbol{R}^{-1} \delta\boldsymbol{y}_k\right)^{-1} \tag{9}$$

Normally the iterations continue until convergence. Basically, the most widely used criterion to terminate the algorithm is to verify when the deviation $\delta\hat{\boldsymbol{x}}_k$ becomes small enough, in this work we use six iterations of least squares method for state estimation.

In this way the final solution to the States will be

$$\hat{\boldsymbol{x}}_k = \hat{\boldsymbol{x}}_{k-1} + \delta\hat{\boldsymbol{x}}_k \tag{10}$$

with covariance $\hat{\boldsymbol{P}}_k$ given by Eq. (8).

3. Attitude Representation by Euler Angles

In the case of CBERS-2, attitude stabilization is done in three axes namely geo-targeted, and can be described in relation to the orbital system. The rotation sequence adopted in this work was the 3-2-1, see Figure 1, and the following sequences of rotations are used [4, 5]

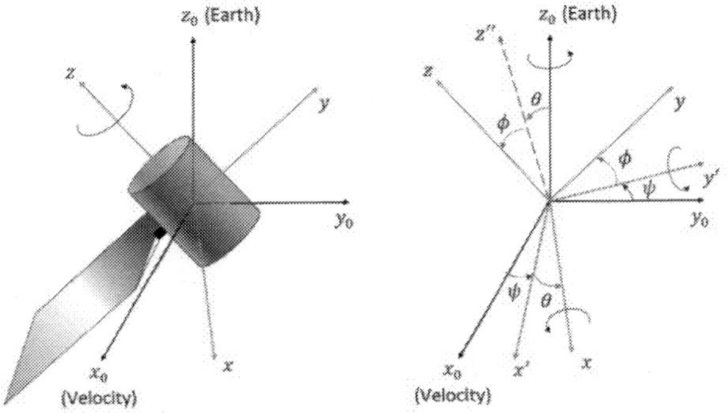

Figure 1. Illustration to represent the orbital local system $(\mathbf{x}_o, \mathbf{y}_o, \mathbf{z}_o)$ and the attitude system $(\mathbf{x}, \mathbf{y}, \mathbf{z})$ and the relationship between them by Euler angles (ϕ, θ, ψ)

- 1st rotation of an angle ψ (*yaw* angle) around the $\mathbf{z}_o$ axis.
- 2nd rotation of an angle θ (*pitch* angle) around an intermediate $\mathbf{y}'$ axis.
- 3rd rotation of an angle ϕ (*roll* angle) around the $\mathbf{x}$ axis.

Thus, the rotational matrix that relates the orbital local system $(\mathbf{x}_o, \mathbf{y}_o, \mathbf{z}_o)$ and the attitude system $(\mathbf{x}, \mathbf{y}, \mathbf{z})$ is given by [7]:

$$R = \begin{bmatrix} \cos\theta\cos\psi & \cos\theta\sin\psi & -\sin\theta \\ \sin\phi\sin\theta\cos\psi - \sin\psi\cos\phi & \sin\phi\sin\theta\sin\psi + \cos\phi\cos\psi & \sin\phi\cos\theta \\ \cos\phi\sin\theta\cos\psi + \sin\phi\sin\psi & \cos\phi\sin\theta\sin\psi - \sin\phi\cos\psi & \cos\phi\cos\theta \end{bmatrix} \tag{11}$$

From (11), can be obtained the kinematic equation of the problem used in the estimation process.

4. Computer Simulation by PROPAT and Results

The nonlinear system that represents the process and measurements equations of the problem is given by [9, 5, 7]:

Journal of Physics: Conference Series **641** (2015) 012005 doi:10.1088/1742-6596/641/1/012005

$$\begin{bmatrix} \dot{\phi} \\ \dot{\theta} \\ \dot{\psi} \\ \dot{\varepsilon_x} \\ \dot{\varepsilon_y} \\ \dot{\varepsilon_z} \end{bmatrix} = \begin{bmatrix} \hat{\omega}_x + \hat{\omega}_y \sin\phi \tan\theta + \hat{\omega}_z \cos\phi \tan\theta + \omega_0 (\cos\phi \sin\psi + \sin\theta \sin\psi \tan\theta) \\ \hat{\omega}_y \cos\phi - \hat{\omega}_z \sin\phi + \omega_0 \cos\psi \\ \left(\dfrac{\hat{\omega}_y \sin\phi + \hat{\omega}_z \cos\phi + \omega_0 \sin\theta \sin\psi}{\cos\theta} \right) \\ 0 \\ 0 \\ 0 \end{bmatrix} + \boldsymbol{w} \quad (12)$$

$$\boldsymbol{y}_k = \begin{bmatrix} \arctan\left(\dfrac{-(S_{0y} - \psi S_{0x} + \phi S_{0z})}{(S_{0x} + \psi S_{0y} - \theta S_{0z})\cos 60^\circ + (S_{0z} - \phi S_{0y} - \theta S_{0z})\cos 150^\circ} \right) \\ 24^\circ + \arctan\left(\dfrac{S_{0x} + \psi S_{0y} - \theta S_{0z}}{S_{0z} - \phi S_{0y} - \theta S_{0z}} \right) \\ \phi \\ \theta \end{bmatrix} + \boldsymbol{v}_k \qquad (13)$$

The state vector is composed by the attitude angles (ϕ, θ, ψ) and by gyros bias (ε_x, ε_y, ε_z), present in the angular velocity of the satellite represented on the satellite system $\hat{\omega}_x$, $\hat{\omega}_y$ and $\hat{\omega}_z$, respectively; the term ω_0 is the orbital angular velocity. The elements S_{0x}, S_{0y} e S_{0z} are the components of the solar vector in orbital coordinate system [4]. The matrices $\boldsymbol{w} = \begin{bmatrix} w_\phi & w_\theta & w_\psi & w_{\varepsilon_x} & w_{\varepsilon_y} & w_{\varepsilon_z} \end{bmatrix}^T$ and $\boldsymbol{v}_k = \begin{bmatrix} v_{\alpha_\psi} & v_{\alpha_\theta} & v_{\phi_H} & v_{\theta_H} \end{bmatrix}^T$ are the process and measurements noises, respectively. The orbit and attitude simulation were made by propagator PROPAT [1] and the Nonlinear Least Squares Method is coded in MatLab software with a sampling rate of $0.5s$ for $10min$ of observation.

Before analyzing the accuracy of the filter in question, it is important to analyze their convergence done through residual frequency configuration for the 6th iteration of the Nonlinear Least Squares Method, that presenting aspects of a gaussian. See Figure 2

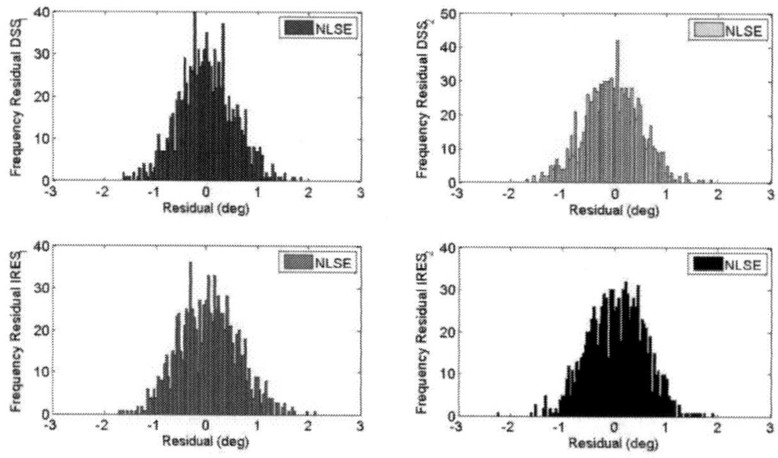

Figure 2. Frequency Residuals for the two Digital Sun Sensor (DSS) and for the two Infrared Earth Sensor (IRES) on board the CBERS-2 satellite

It is said that a Filter is converging when the residual is close to zero average and it happens with the results presented in Table 1. The average results for the residuals of the DSS_1, DSS_2,

Journal of Physics: Conference Series **641** (2015) 012005
doi:10.1088/1742-6596/641/1/012005

$IRES_1$ and $IRES_2$ get better when higher is the number of iterations for the Nonlinear Least Squares Method.

Table 1. Mean and standard deviation statistics of the DSS and IRES Residuals

	1st iteration	3rd iteration	6th iteration
DSS_1 Res.(deg)	5.2115 ± 0.9852	0.0223 ± 0.6063	0.0162 ± 0.5967
DSS_2 Res.(deg)	6.7599 ± 0.9236	0.0282 ± 0.7122	-0.0035 ± 0.6803
$IRES_1$ Res.(deg)	2.9663 ± 0.5875	0.0006 ± 0.5846	-0.0519 ± 0.5715
$IRES_2$ Res.(deg)	5.4776 ± 0.6950	-0.0073 ± 0.6781	-0.0331 ± 0.6802

The high values in the 6th iteration for the $IRES_1$ and $IRES_2$ are a small oscillation in the average result, the larger the number of iterations, more, the residue approaches to zero.

Fig. 3 and 4 present the attitude and gyros bias estimated.

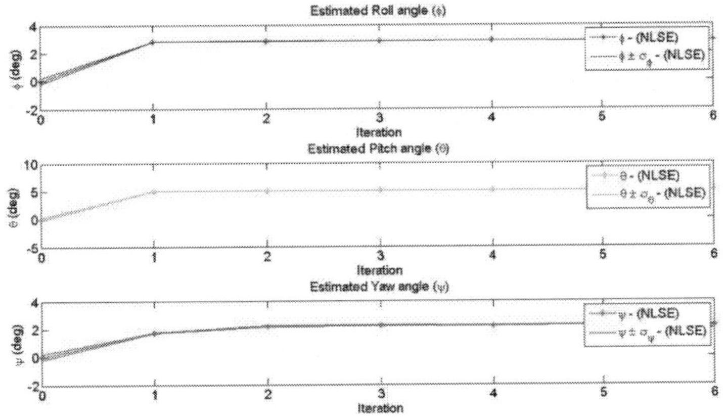

Figure 3. Estimated *roll*, *pitch* and *yaw* angles respectively

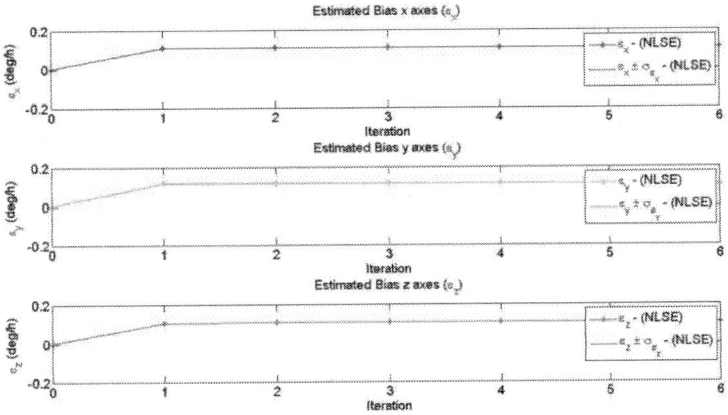

Figure 4. Estimated bias gyros around the **x**, **y** and **z** axes respectively

It can be seen that the attitude and gyros bias estimation converge quickly after the third iteration of the NLSE.

5. Conclusions

The main objective of this study was to estimate the attitude of a CBERS-2 like satellite, using simulated data provided by PROPAT Satellite Attitude and Orbit Toolbox for Matlab considering the sensors available on board this satellite. To verify the consistency of the estimator, the attitude and gyros bias was estimated by Nonlinear Least Squares Estimation (NLSE).

The usage of the simulated data from on board attitude sensors poses difficulties like mismodelling, mismatch of sizes, misalignments, unforeseen systematic errors and post-launch calibration errors. However, it is observed that the attitude and gyros bias estimated by the NLSE are in close agreement with the results obtained in a previous work [5] which used the Extended Kalman Filter for estimation.

However, checking the robustness of the estimation method, it was noted that the greater the deterioration of the initial conditions, the NLSE took longer to achieve convergence compared to [5]. In this case, the linearization performed by NLSE was made distant from the initial conditions of the real problem with the intention to validate the convergence of the estimation method. Thus, the NLSE produces satisfactory results for attitude estimation.

The contribution of the NLSE, derived here, assumed that the process noise do not exist and the measureaments noise is zero-mean and white (uncorrelated with itself from one time step to the next), and that the variance of the measurements noise is known. The method applied in this research is an alternative to the criterion of minimum variance estimation. The approach requires an assumption about the sources of uncertainty in the statistics of the problem, assuming the dynamic model of the state vector is perfect, that is, without noise.

Finally, it can be concluded that the algorithm of the NLSE converges, for initial conditions close to the actual values, providing a kinematic attitude solution besides estimating biases (gyro drifts).

6. Acknowledgment

The authors would like to thank the financial support received by CAPES, FAPESP 2012/ 21023-6, CNPQ 303119/ 2010-1, and the partial support from project SIA-DCTA-INPE under contract FINEP 0.1.06.1177.03

References

[1] Carrara, V. PROPAT Satellite Attitude and Orbit Toolbox for Matlab, *Avaliable in:* http://www2.dem.inpe.br/val/projetos/propat/default.htm, 1994. Access in: July 22th, 2014.

[2] Crassidis, J. L.; Markley, F. L. Unscented filtering for spacecraft attitude estimation, *Journal of Guidance, Control and Dynamics*, vol.26, (536 - 542), 2003.

[3] Crassidis, J. L.; Junkins, J. L. Optimal Estimation of Dynamics Systems, *CRC Press*, New York, N.Y., 2nd ed., 2012.

[4] Fuming; H., Kuga, H. K. CBERS simulator mathematical models, *CBTT Project, CBTT/ 2000/ MM/ 001*, 1999 (1999) 1-12.

[5] Garcia, R. V.; Kuga, H. K.; Zanardi, M. C. Using Extended Kalman Filter and Least Squares Method for Spacecraft Attitude Estimation, *Mathematics in Engineering, Scince and Aerospace*, vol.2, (445 - 453), 2011.

[6] Gauss, K. F. Theory of the Motion of the Heavenly Bodies Moving about the Sun in Conics Sections, A Translation of Theoria Motus, *Dover Publications*, New York , N.Y., 1963.

[7] Silva, W. R.; Kuga, H. K.; Zanardi, M. C. Application of the Extended H_∞ Filter for Attitude Determination and Gyro Calibration, *24th AAS/AAIA Space Flight Mechanics Meeting*, (2014) 1-15.

[8] Simon, D. Optimal State Estimation: Kalman, H_∞, and Nonlinear Approaches, Wiley-Interscience, New York, N.Y., 2006.

[9] Wertz, J. R. Spacecraft attitude determination and control, D. Reidel, Dordrecht, Holanda, 1978.

Journal of Physics: Conference Series **641** (2015) 012006 doi:10.1088/1742-6596/641/1/012006

Particle transport induced by electrostatic wave fluctuations

K C Rosalem[1], M Roberto[1] and I L Caldas[2]
[1] Departamento de Física, Instituto Tecnológico de Aeronáutica, 12228-900, São José dos Campos, SP, Brazil
[2] Instituto de Física, Universidade de São Paulo, 05315-970, São Paulo, SP, Brazil

E-mail: kaue@ita.br

Abstract. Particle transport driven by electrostatic waves at the plasma edge is numerically investigated, for large aspect ratio tokamaks, by considering a kinetic model derived from guiding-center equations of motion. Initially, the transport is estimated for trajectories obtained from differential equations for a wave spectrum generated by a dominant spatial mode and three time modes. Then, in case of infinite time modes, the differential equations of motion are used to introduce a symplectic map that allows to analyze the particle transport. The particle transport barriers are observed for spatial localized dominant perturbation and infinite modes. In presence of infinite spatial modes, periodic islands arise in between chaotic trajectories at the plasma edge.

1. Introduction

Several experiments in tokamaks indicate a broad frequency spectrum of plasma drift waves for each wave number [1-3] and the drift waves show substantial amplitudes that imply chaotic particle trajectories [4-8]. Thus, the drift waves are essential features to properly describe the observed driven turbulent transport. Accordingly, this transport is mainly caused by the particle $\boldsymbol{E} \times \boldsymbol{B}$ drift due to the equilibrium and the perturbing electric fields [9,10] and such transport is much affected by the equilibrium field radial profiles [11-13].

In this work, we integrated numerically particle trajectories to investigate the relation between transport and electrostatic wave fluctuations at the plasma edge for large aspect ratio tokamaks. Initially, we assume different amplitude levels in the wave spectra for three specific time modes. The particle trajectories are obtained by integrating differential equations. Complementary, to investigate infinite mode perturbations, these equations of motion are rewritten as a symplectic map in which all modes have the same amplitude.

Early works on electrostatic wave fluctuations derive a drift wave maps around the radial position associated to the minimum of safety factor profiles [4-6]. However, the confined particles can leave the neighborhoods of this minimum point for long time integration. Avoiding this radial limitation, we show how the transport is affected by changing the shear of the radial electric and poloidal magnetic fields [14]. To describe drift waves in tokamak plasmas, the electrostatic spatial modes have to be defined by a dominant perturbations or treat as a generalized modes. Besides, we also study how the particle orbits are modified when the local perturbations are replaced by infinite spatial modes.

In section 2, we introduced the drift-kinetic model that leads to the differential equations for the particle trajectories in the vicinity of a given magnetic and electric field profiles. In section 3, we

[1] To whom any correspondence should be addressed.

Journal of Physics: Conference Series **641** (2015) 012006 doi:10.1088/1742-6596/641/1/012006

compared the Poincaré maps obtained by integrating the drift-kinetic model for specific resonances with the symplectic map for infinite spatial modes. The observed particle transport is summarized and concluded in section 4.

2. Drift-kinetic model

This section provides the model for which the particle trajectories can be calculated in presence of fluctuating electrostatic potential, describing drift waves propagating in the poloidal and toroidal directions according to the spatial modes. For that, we consider the particle trajectories along the magnetic field lines and the displacement due to the drift velocity in the guiding-center equation of motion,

$$\frac{d\boldsymbol{x}}{dt} = v_{\parallel}\frac{\boldsymbol{B}}{B} + \frac{\boldsymbol{E}\times\boldsymbol{B}}{B^2}, \tag{1}$$

where the components of $\boldsymbol{x} = (r, \vartheta, \varphi)$ are defined as the local polar coordinates. Thus, the plasma configuration corresponds to a layer of large aspect ratio tokamak.

The electric field is composed by a fluctuating component $\tilde{\boldsymbol{E}} = -\boldsymbol{\nabla}\tilde{\phi}$ plus an equilibrium radial profile $\boldsymbol{E}_r$, and the magnetic curvature is introduced by the safety factor profile for $B \approx B_\varphi \gg B_\vartheta$. In order to consider a fluctuating potential in this model we use the wave spectrum,

$$\tilde{\phi}(\boldsymbol{x}, t) = \sum_{m,l,n} \phi_{m,l,n} \, cos(m\vartheta - l\varphi - n\omega_0 t)\,, \tag{2}$$

where $\phi_{m,l,n}$ is the mode amplitude, ω_0 the lowest angular frequency with substantial amplitude in the drift wave spectrum, m and l define the spatial modes and n the time mode in this coordinate system. In the numerical simulations, we assume a dominant M/L spatial mode and three harmonics n time modes.

We introduce action and angle variables as $I = (r/a)^2$ and $\Psi = M\vartheta - L\varphi$, respectively [5]. Taking these assumptions we obtain the differential equations for the particle trajectories,

$$\frac{dI}{dt} = \frac{2M}{a^2B}\sum_{M,L,n}^{N} \phi_{m,l,n} \, sin(\Psi - n\omega_0 t)\,, \tag{3}$$

$$\frac{d\Psi}{dt} = \frac{v_{\parallel}}{Rq(I)}[M - Lq(I)] - \frac{ME_r(I)}{aB\sqrt{I}}\,. \tag{4}$$

To solve this system, the equilibrium fields and the parallel velocity along the magnetic field lines have to be defined for a set tokamak parameters.

3. Transport barrier and diffusion

We investigate numerically the particle transport properties by integrating the equations of motion. Simulations are carried out using the TCABR tokamak parameters, assuming a dominant spatial mode $M/L = 16/4$, particle parallel velocity constant $v_{\parallel} = 1.5\,km\,s^{-1}$ along the trajectories [15] and lowest angular frequency $\omega_0 = 6\times10^4\,rad\,s^{-1}$ [3]. For this tokamak configuration we use $R = 61\,cm$, $a = 18\,cm$ and $B = 1.1\,T$. We choose the non-monotonic electric field and monotonic magnetic field profiles discussed in [14].

In order to analyze changes in the radial transport due to the amplitude modes ϕ_{MLn}, we compare the wave spectra for the cases (a) and (b) in figure 1. These perturbing spectra are composed by three time modes n at the plasma edge, whose resonance radial positions can be determined by taking the condition $d/dt\,(\Psi - n\omega_0 t) \cong 0$. In figure 2 we observe the resonance conditions around $n = 3$ with double location for the radial positions.

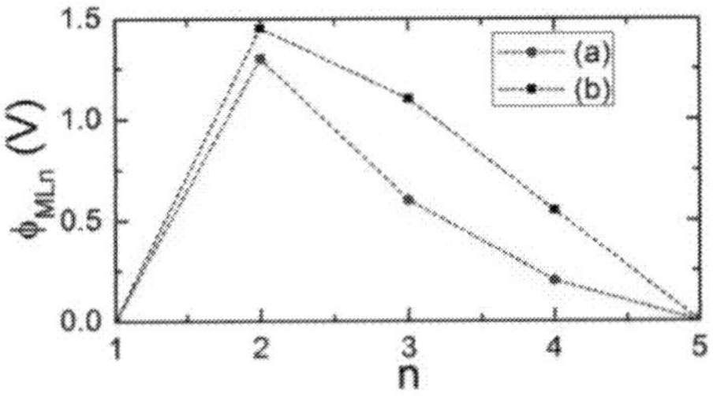

Figure 1. Eletrostatic wave spectra (a) and (b) for a dominant $M/L = 16/4$ spatial mode.

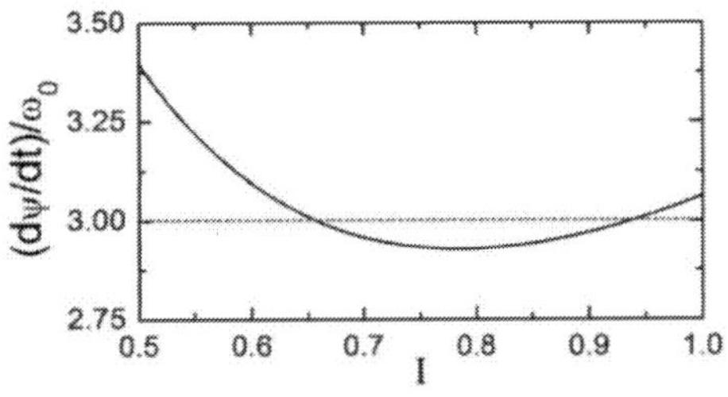

Figure 2. Resonance conditions around $n = 3$ and assuming the lowest angular frequency as $\omega_0 = 6 \times 10^4\ rad\ s^{-1}$ for the cases (a) and (b) in figure 1.

In figure 3(a) we present the Poincaré map obtained by integrating equations (3) and (4) for various initial conditions and assuming the wave spectrum (a) shown in figure 1. The resonance $n = 3$ corresponds to the twin islands separated by the shearless curve depicted in red, which is located by the extremum value of the rotational number profile [14]. In figure 3(b) we assumed the wave spectrum (b) of figure 1, and all the other parameters are the same as those used in the previous map. The onset of meanders [16,17] are induced around of the shearless curve by increasing the wave mode amplitudes. Furthermore, this mode increasing also gives rise to noticeable chaotic trajectories around the inner radial perturbation created by $n = 4$, as seen in figure 3(b). The invariant shearless barrier persists for the modified spectrum.

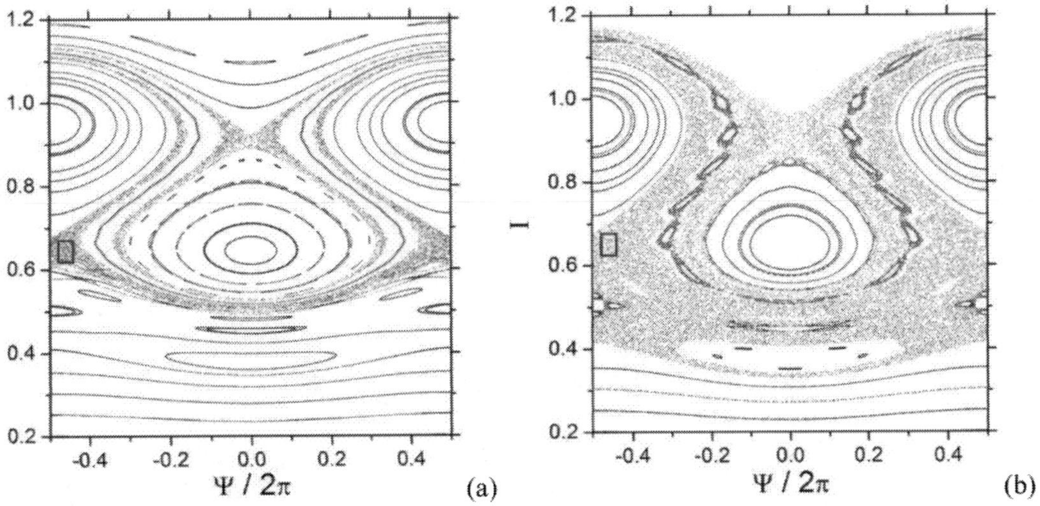

Figure 3. Poincaré maps for the wave spectra (a) and (b) in figure 1. The potential mode amplitudes in case of (a) are $\phi_{MLn=2} = 1.30\ V$, $\phi_{MLn=3} = 0.60\ V$, and $\phi_{MLn=4} = 0.20\ V$, for (b) are $\phi_{MLn=2} = 1.45\ V$, $\phi_{MLn=3} = 1.10\ V$, and $\phi_{MLn=4} = 0.55\ V$. The shearless curves can be identified by the invariant in red. Initial conditions for running diffusion coefficients are shown in the blue grids.

Figure 4 shows the time dependence of the running diffusion coefficient [4] calculated for the particle radial positions for cases (a) and (b) of wave spectra in figure 1. For these calculations, we considered the blue grid in the Poincaré maps of figure 3. One observes the increase of particle transport as increasing the amplitude modes by modifying the wave spectrum.

Journal of Physics: Conference Series **641** (2015) 012006 doi:10.1088/1742-6596/641/1/012006

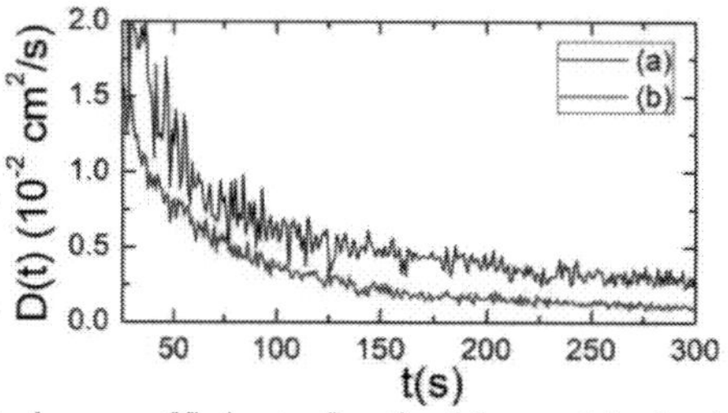

Figure 4. Running diffusion coefficients for the blue grids depicted in figure 3.

Despite this model relates a specific electrostatic amplitude for each time mode, it is limited by the wave spectrum characterized by finite n modes. To simulate how a continuum wave spectrum affects the particle transport, a symplectic map is derived for infinite time modes. In this case, one can approach the oscillations sum by considering the delta function and impulsive iterations at time $t = 2\pi n/\omega_0$. Thus, one can rewrite the equations (3) and (4) as a map given by

$$I_{N+1} = I_N + \frac{4\pi M}{\omega_0 a^2 B}\phi_{ML}\, sin(\chi)\,, \tag{5}$$

$$\chi_{N+1} = \chi_N + R_1(I_{N+1}) + R_2(I_{N+1})\,, \tag{6}$$

$$R_1(I) = \frac{v_\parallel}{\omega_0 R q(I)}[M - Lq(I)]\,, \tag{7}$$

$$R_2(I) = -\frac{M E_r(I)}{\omega_0 a B\sqrt{I}}\,, \tag{8}$$

where the angle variable is defined as $\chi = \Psi/2\pi$.

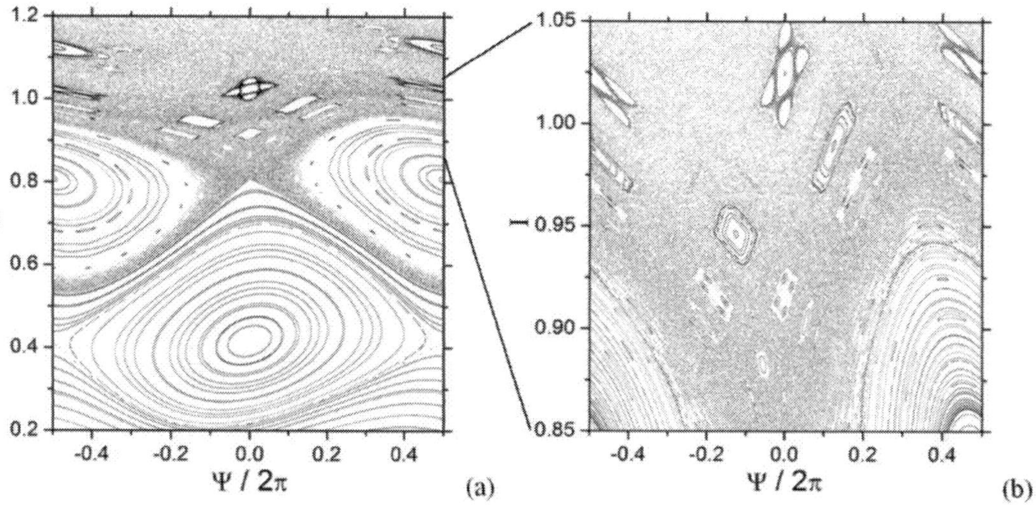

Figure 5. Symplectic map (b) for potential mode amplitude $\phi_{ML} = 0.65\,V$ and (b) the same map amplified at the plasma edge.

In figure 5 we present the symplectic map by using the equations (5) and (6) for various initial conditions and considering the same dominant spatial mode as in figure 3. The symplectic map shows the same structures for particle transport predictions, as seen in figure 5(a). Since the map has infinity time modes at the plasma edge, we expect to identify many islands immersed in a chaotic sea, as it is clearly shown by the amplification in figure 5(b).

Journal of Physics: Conference Series **641** (2015) 012006 doi:10.1088/1742-6596/641/1/012006

4. Conclusions

We have investigated how the particle transport increases with the mode amplitude of the fluctuating electric potential and how this transport changes with the radial field profiles. To analyze this dependence, the running coefficient has been computed and compared for modifications in the drift wave spectrum. We observed in the Poincaré maps robust barriers provided by the electric and magnetic configurations that persist for these spectra. Through the resonance condition, we identified these barriers in between the twin islands region. From the equations of motion we introduced a symplectic map to consider infinite time modes of perturbations. For this map method, we observed many islands where the particles could be trapped along of their trajectories at the plasma edge.

Acknowledgments

This work was supported by São Paulo Research Foundation (FAPESP, Brazil) under Grant No 2013/0340-6 and 2011/19296-1, CNPq and CAPES.

References

[1] Devynck P, Stckel J, Admek J, Duran I, Hron M and Ooost G V 2003 *Czech. J. Phys* **53** 853
[2] Oost G V, Admek J, Antoni V, Balan P, Boedo JA, Devynck P, Duran I, Eliseev L, Gunn J P, Hron M, Ionita C, Jachmich S, Kirnev G S, Martines E, Melnikov A, Schrittwieser R, Silva C, Stöckel J, Tendler M, Varandas C, Schoor M V, Vershkov V, and Weynants R R 2003 *Plasma Phys. Contr. Fusion* **45** 621
[3] Nascimento I C, Kuznetsov Yu K, Guimarães-Filho Z O, Chamaa-Neto I El, Usuriaga O, Fonseca A M M, Galvão R M O, Caldas I L, Severo J H F, Semenov I B, Ribeiro C, Heller M V P, Bellintani V, Elizondo J I, Sanada E 2007 *Nucl. Fusion* **47** 1570
[4] Park H-B, Heo E-G, Horton W, and D-I Choi 1997 *Phys. Plasmas* **4** 3273
[5] Horton W, Park H-B, Kwon J-M, Strozzi D, Morrison P J, and Choi D-I 1998 *Phys. Plasmas* **5** 3910
[6] Kwon J-M, Horton W, Zhu P, Morrison P J, Park H-B, and Choi D-I 2000 *Phys. Plasmas* **7** 1169
[7] Miskane F, Garbet X, Dezairi A, and Saifaoui D 2000 *Phys. Plasmas* **7** 4197
[8] Marcus F A, Caldas I L, Guimarães-Filho Z O, Morrison P J, Horton W, Kuznetsov Y K, and Nascimento I C 2008 *Phys. Plasmas* **15** 112304
[9] Horton W 1999 *Rev. Mod. Phys.* **71** 735
[10] Del-Castillo-Negrete D and Martinelli J J 2012 *Commun. in Nonlinear Sci. and Numer. Simul.* **17** 2031
[11] Ritz C P, Bravenec R V, Schoch P M, Bengtson R D, Boedo J A, Forster J C, Gentle K W, He Y, Hickok R L, Kim Y J, Lin H, Phillips P E, Rhodes T L, Rowan W L, Valanju P M, and Wootton A J. 1989 *Phys. Rev. Lett.* **62** 1844
[12] Hidalgo C, Pedrosa M A, and Gonçalves B 2002 *New J. Phys.* **4** 51
[13] Gürcan Ö D, Garbet X, Hennequin P, Diamond P H, Casati A, and Falchetto G L 2009 *Phys. Rev. Lett.* **102** 255002
[14] Rosalem C K, Roberto M, Caldas I L 2014 *Nucl. Fusion* **54** 064001
[15] Severo J H F, Nascimento I C, Tsypin V S, and Galvão R M O 2003 *Nucl. Fusion* **43** 1047
[16] Caldas I L, Viana R L, Szezech J D r, Portela J S E, Fonseca J, Roberto M, Martins C G L, Silva E J 2012 *Commun. Nonlinear Sci. Numer. Simul.* **17** 2021
[17] Szezech J D Jr, Caldas I L, Lopes S R, Morrison P J, Viana R L 2012 *Phys. Rev. E* **86** 036206

Journal of Physics: Conference Series **641** (2015) 012007 doi:10.1088/1742-6596/641/1/012007

ALR – Laser altimeter for the ASTER deep space mission. Simulated operation above a surface with crater

A G V de Brum[1], F C da Cruz[2] and A Hetem Jr[1]

[1] Universidade Federal do ABC – UFABC, Santo André, SP, Brazil
[2] Universidade de Campinas – UNICAMP/IFGW, Campinas, SP, Brazil

E-mail: antonio.brum@ufabc.edu.br

Abstract. To assist in the investigation of the triple asteroid system 2001-SN263, the deep space mission ASTER will carry onboard a laser altimeter. The instrument was named ALR and its development is now in progress. In order to help in the instrument design, with a view to the creation of software to control the instrument, a package of computer programs was produced to simulate the operation of a pulsed laser altimeter with operating principle based on the measurement of the time of flight of the travelling pulse. This software Simulator was called *ALR_Sim*, and the results obtained with its use represent what should be expected as return signal when laser pulses are fired toward a target, reflect on it and return to be detected by the instrument. The program was successfully tested with regard to some of the most common situations expected. It constitutes now the main workbench dedicated to the creation and testing of control software to embark in the ALR. In addition, the Simulator constitutes also an important tool to assist the creation of software to be used on Earth, in the processing and analysis of the data received from the instrument. This work presents the results obtained in the special case which involves the modeling of a surface with crater, along with the simulation of the instrument operation above this type of terrain. This study points out that the comparison of the wave form obtained as return signal after reflection of the laser pulse on the surface of the crater with the expected return signal in the case of a flat and homogeneous surface is a useful method that can be applied for terrain details extraction.

1. Introduction

In the last years, many space missions dedicated to the investigation of small bodies of the solar system occurred, while others are still in progress or being planned. Among these, missions NEAR-Shoemaker (1996-2001), Hayabusa (2003-2010), Rosetta (in progress; scheduled to end in December 2015), Hayabusa 2 [1] (scheduled to launch in 2015).

For the most part, these and other missions that have not been mentioned carry on board at least one laser altimeter (laser rangefinder) as an instrument of scientific research (shape, topography, mass, etc.) and aid to navigation, in the phase of greater proximity to the target. As an example, the LIDAR, used in the Hayabusa mission [2, 3].

In general terms, these devices are used for determining the distance to the target, within their reach, which is done from the firing of laser pulses towards the target and the detection of signals that return after reflection on the target surface. As an instrument of scientific research, the shape of the pulse that returns after reflection on the surface contains information about the roughness (topography) of the lighted area. This information collected over wide areas of the target surface is used in the construction of a 3D model of the terrain.

Inserted in this sector of contemporary space exploration, the mission ASTER was planned in cooperation with Russia to be the first Brazilian deep space mission. This mission intends to investigate the triple asteroid 2001-SN263 and has its launch date previously planned for 2017 [4, 5].

Among the instruments that will take part in this mission, the laser altimeter, named ALR (ASTER Laser Rangefinder), is being designed to meet the mission objectives concerning the investigation of the shapes, surfaces topographies and masses of the asteroids, together with the calculation of their masses, in addition to supporting the proximity navigation. Reference [6] Brum et al (2011) contains more detailed information about the instrument design.

The ALR is planned to be a solid state pulsed laser of Nd: YAG, diode pumped and Q switched. The frequency of the light emitted is 1064 nm. The modeling and simulation of the operation of a laser altimeter with such features was the subject of a previous article [7]. These studies, still in progress, aim at the creation of the various subroutines that will compose the control software of the apparatus, to be embedded in the Signal Processing and Control (electronic) Unit of the instrument (SPCU). The simulation includes the instrument, its operation and the environment in which it will supposedly operate. The fundamentals involved in the operation of a laser altimeter and the investigation of surfaces with its use are deeply discussed in [8].

With use of the MATLAB environment [9], a software Simulator was created to serve as a laboratory useful in the understanding, implementation and testing of all the concepts involved in the detection of distances using instruments similar to the ALR. The *ALR_Sim* Simulator constitutes the main result of this effort and brings together all the programs previously created. For more details on the program, see [10].

This article describes the modeling performed and the simulations conducted with use of the *ALR_Sim*, in order to obtain and evaluate what should be the expected results of the instrument operation on a surface with crater, and for an emitted pulse with exponential characteristic, which more closely approximates the real pulse emitted by a Q-switched solid-state laser.

2. Modeling and simulation

The complete simulation steps were described in details in the cited papers. The Simulator package covers all major expected conditions involving the operation of the instrument, which are:

- the physical features of the instrument (parameters, specifications and transmitter-detector characteristics);
- the features of the environment where the instrument will operate;
- the physical features of the target surface.

2.1. Emitting pulse modeling

The emitted pulse model that best represents the Q-switched laser emission is given by the exponential distribution. This curve has emitted power rise led by a quadratic term, followed by exponential emission decay [11].

$$P_3 = Bt^2 \exp\left(-\frac{t-\tau}{\omega}\right), \tag{1}$$

In Equation (1), B is an adjusting factor for the peak value of the emitted power, τ and ω are adjusted curve parameters to obtain $P_{max} = 1$ kW, $\omega_p = 10$ ns (pulse width at half maximum power).

The pulse modeling for the general case, i.e., when one chooses P_{max}, ω_p, t_{max} (time to reach maximum power) and t_f (the larger time span to complete the decay), keeping the exponential feature, demands the determination of proper values for B, τ and ω for any case, which represents a complex matter, once these variables are interdependent. To achieve this goal, in addition to $P_3(t)$, described previously, a second curve, $P_4(t)$, was created using a smoothed cubic spline applied to the points in Table 1, created to represent (model) the characteristics of the expected emitted pulse. These points were chosen to model perfectly the emitted pulse curve, according to the available knowledge of it today. Any modifications or advances in this knowledge can be inserted here. The function "CSAPS" (smoothed cubic spline) of MATLAB was used for this purpose

Table 1. Points used to model the shape of the exponential emitted pulse.

t (s)	t_i	t_1	t_{max1}	t_{max2}	t_2	t_3	t_f
P (W)	0	$P_{max}/2$	P_{max}	P_{max}	$P_{max}/2$	0	0

In Table 1,

t_i - is the starting moment of emission of the laser pulse (defined by the user);

t_1 - is the moment (calculated) when the pulse reaches half of its peak power, $P_{max}/2$, in the rising part of the curve, considering exponential emission; t_1 is given by

$$t_1 = t_i - \omega_P \left[1 + exp\left(\frac{\omega_P}{2\omega}\right)\right]\left[1 - exp\left(\frac{\omega_P}{\omega}\right)\right]^{-1};$$

t_2 - is the moment when the pulse reaches half of its peak power, $P_{max}/2$, in the descent part of the exponential curve;

t_{max1} and t_{max2} – subdivisions of t_{max}, instant of time corresponding to the maximum power value; created to assure P_{max} as the maximum value of the adjusted curve, with $t_{max1} = t_{max} - 0.1dt$ and $t_{max2} = t_{max} + 0.1dt$;

t_3 and t_f – used to model de curve decay (more abrupt or less), $t_3 = t_2 + \omega_p$; and $t_f = t_3 + \omega_p$.

2.2. *Surface with crater modeling*

The modeling of the surface with crater followed the profile of the lunar impact crater Eudoxus [12]. A sample of 45 points of half of the crater profile was taken resulting in the set of points (X_{crat}, Z_{crat}). These, in turn, were used to adjust a curve of type "Gauss 5" smoothed (internal function of MATLAB). The smoothed curve adjusted to the points of the crater, z = f (x), $0 \leq x \leq 40$ km, was used to determine the surface of revolution around the +z-axis. The crater surface points were obtained from a grid of equally-spaced points in the xy-plane with x and y taken between -40 and +40 metric units (km, m, etc.). For each of the internal points within the grid, the value of z=f(x, y) was calculated.

The model of the "N reflective point sources of the incident light", created for this simulation [7, 9], requires the identification of a set of points on the surface to be taken as sources of the light that returns to the detector. The determination of these points was made by taking the intersection points between the straight lines connecting the instrument transmitter, located, in the adopted frame, at the point A = (0, 0, R), to each point $P_i = (x_i, y_i, 0)$, of the flat footprint (defined and calculated in [7]), when these lines are prolonged until they reach the surface z = f (x, y), representing the crater surface (below the flat surface). Each intersection determines a point $P_i' = (x_i', y_i', z_i')$ of the crater surface, $z_{crat} = f(x,y)$.

The solution of the system created from the equations of the crater and of the line *r*, from A to each point P_i of the flat footprint, offers the set of intersection points P_i', related to each point P_i. To calculate each of these points, the numeric solver of systems of equations available in MATLAB was used. The crater determined through these points is shown in Figure 1. In this case, the value of N (1396 points) is a consequence of the chosen (input) value for L, the side dimension of one element of area on the flat footprint.

The equations of the line ***r*** and of the crater are given, respectively, by (2) and (3),

$$\frac{x}{x_i} = \frac{y}{y_i} = -\frac{(z-R)}{R} \tag{2}$$

$$Z_{crat} = E_1 + E_2, \tag{3}$$

where E_1 and E_2 given by

$$E_1 = a_1 exp\left[-\left(\frac{g-b_1}{c_1}\right)^2\right] + a_2 exp\left[-\left(\frac{g-b_2}{c_2}\right)^2\right] + a_3 exp\left[-\left(\frac{g-b_3}{c_3}\right)^2\right],$$

$$E_2 = a_4 exp\left[-\left(\frac{g-b_4}{c_4}\right)^2\right] + a_5 exp\left[-\left(\frac{g-b_5}{c_5}\right)^2\right], \text{ with } g = (x^2 + y^2)^{\frac{1}{2}}.$$

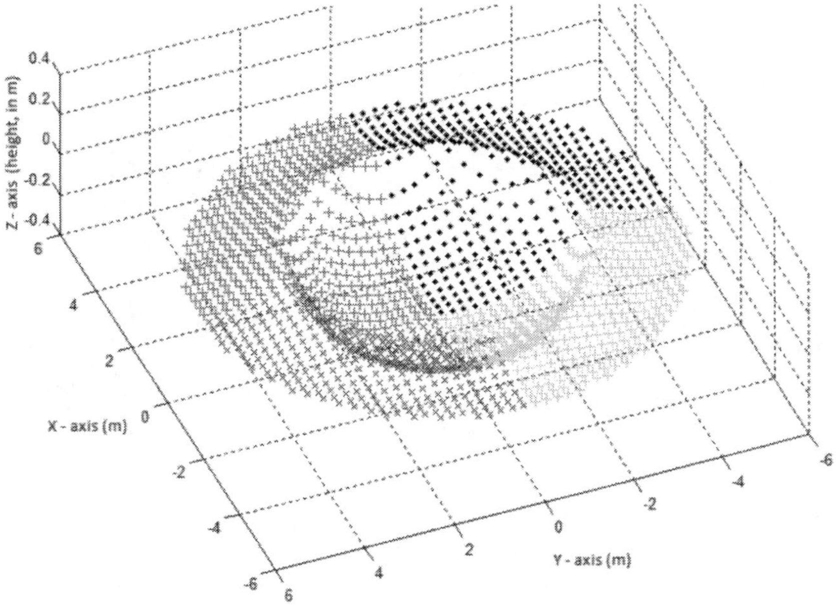

Figure 1. N=1396 point sources that reflect the light incident on the surface of the crater.

Once identified the points P_i', that are considered as emitting point sources of the light incident on the crater surface, one has also to calculate the cosine of the angle between the incident light and the normal to the surface at each point P_i'. This value is necessary for the purpose of calculating the intensity of the diffuse light that returns to the detector. Traditional methods in analytic geometry are used for this purpose. Thus,

$$\cos(\varphi) = \hat{n}_{vet} \cdot \hat{r}, \qquad (6)$$

where $\hat{n}_{vet} = \vec{n}/|\vec{n}|$, $\vec{n} = \left(-\frac{\partial z}{\partial x}, -\frac{\partial z}{\partial y}, 1\right)$, $z = z_{crat}$ and $\hat{r} = \vec{r}/|\vec{r}|$.

Identified the points P_i', i=1,...,N, and the angles between the normal to the surface at each point P_i' and the direction that returns to the detector, the contribution of each source of light from the surface of the crater can be calculated by the same procedure described before for other types of surfaces.

3. Results

An improved version of the *ALR_Sim* was created to include the functions related to the surface with crater. In the simulations performed, the following input data was used:

- Emitted pulse: exponential adjusted with 1,000 Watts of peak intensity and 10 ns in width (at half power);

- Emitter/detector altitude: 15,000 m;
- Approximate dimensions of target surface in the crater: 6 m diameter by 0.3 m depth;
- L=25 cm; side dimension of each element *dA* of the flat footprint area.

For the purpose of performing a comparative analysis, the results obtained with the same parameters in the case of a flat surface are presented in Figure 2.

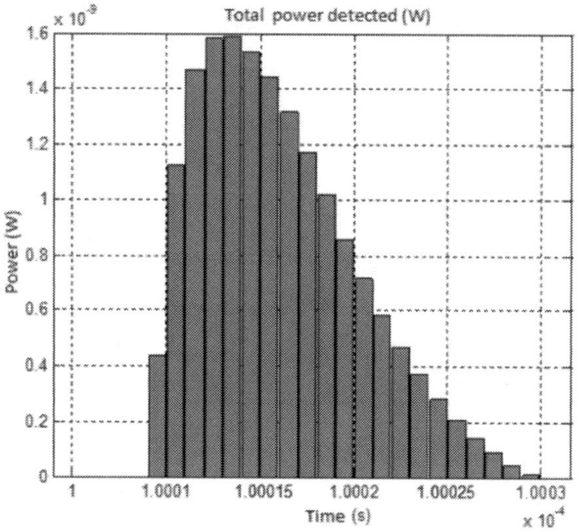

Figure 2(a). Response curve obtained from the return of the emitted light that reaches the flat footprint, distant 15 km from the transmitter, and returns to the receiver.

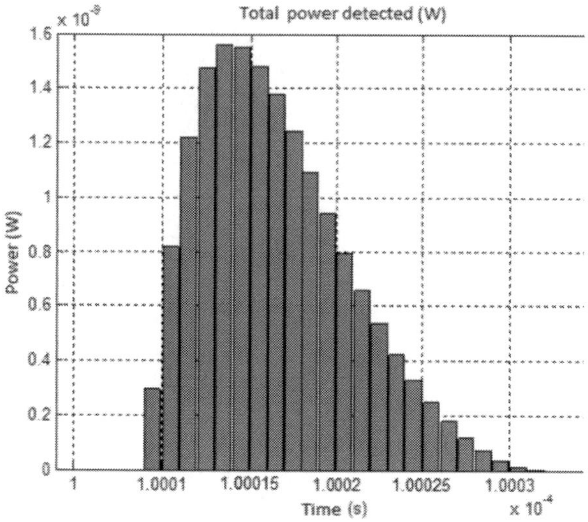

Figure 2(b). Response curve obtained in the case when the surface of the crater is illuminated centrally, in the same conditions of Figure 2(a).

It is observed here that the expected return signal, after reflection of the emitted pulse on a flat and homogeneous surface (constant reflectivity), should mirror the emitted pulse. Thus, the number of subdivisions in the sampling of the emitted pulse equals the number of subdivisions of the return signal (in Figure 2 (a), this means 21 subdivisions = 21 intervals 1 ns wide), the shape of the return wave follows the shape of the emitted wave, the intensity depends only on the distance and on the medium (reflectivity and absorption), the shortest distance between source and target is given by the time interval between the pulse emission starting moment, and the moment of arrival of the first return signal ($d_{min}=c*(t_{1r} - t_{1e})/2$), and the longest distance between source and target is given by the interval of time between the last sampled moment of pulse emission and the last moment of arrival of return signal ($d_{max}=c*(t_{fr} - t_{of})/2$). The difference between these two values gives the greatest difference of level of the lighted terrain. Of course, this account applied to the flat surface results null.

With regard to the surface with crater, the same emitted pulse is used to generate the same flat footprint. The rays of light coming from the emitter get through this plan and reach the surface of the crater, where they are identified as points of this surface and are considered as point sources of (part of) the incident light, as shown in Figure 1. Figure 2(b) shows the waveform obtained as return signal in this case. Comparing the two waveforms (Figure 2), their similarity is clear. Differences encountered are due to the crater features that can be, this way, extracted through a comparative analysis. Important items here to be analyzed are the differences in the power intensities which cause a "flattening" in the shape of the returning wave, and also the spread of this wave.

4. Conclusions and comments

The *ALR_Sim* Simulator was improved to include surfaces with crater and used to simulate the response of the operation of the instrument over one such surface. The models involved are presented

Journal of Physics: Conference Series **641** (2015) 012007 doi:10.1088/1742-6596/641/1/012007

and the results show that this type of surface causes flattening and spreading in the waveform received, when the return signal is compared with the expected for a flat and homogeneous surface.

The comparative analysis used here is indicated as a useful method to the extraction of surface features and can be used also to understand and extract the details of any type of surface.

References

[1] Yano H 2012 Planetary protection of Hayabusa-2 mission, a sample return from 1999 JU3, C-type NEO. Presentation for the NASA Planetary Protection Subcommittee 2012. May 1-2, 2012 NASA Headquarters, Washington, D.C., USA

[2] Barnouin-Jha O S *et al* 2008 Small-scale topography of 25143 Itokawa from the Hayabusa laser altimeter. *Icarus* **198**, pp. 102-124. doi:10.1016/j.icarus.2008.05.026

[3] Tsuno K, Okumura E, Katsuyama Y, Mizuno T, Hassimoto T Nakayama M, Yuasa H 2006 Lidar on board asteroid explorer Hayabusa. *Proc. 6th Internat. Conf. on Space Optics (ESTEC, Noordwijk, The Netherlands, 27-30 June 2006) (ESA SP-621, June 2006)*

[4] Sukhanov A A, Velho H F C, Macau E E N, and Winter O C 2010 The Aster Project: Flight to a Near-Earth Asteroid. *Cosmic Research*. **48**, pp. 433-450

[5] Macau E E N, Winter O C, Velho, H F C, Sukhanov A A, Brum A G V de, Ferreira J L, Hetem Jr A, Sandonato G M and Sfair R 2011 The Aster Mission: Exploring for the First Time a Triple System Asteroid. *In: Proc. of the 62nd Int Astronautical Congress (Cape Town, South Africa, 2011)*

[6] Brum A G V de, Hetem Jr A, Rego I da S, Francisco C P F, Fenilli A, Madeira F, Cruz F C da, and Assafin M 2011 Preliminary Development Plan of the ALR, the Laser Rangefinder for the ASTER Deep Space Mission to the 2001-SN263 Asteroid. *Journal of Aerospace Technology and Management, JATM* **3**, p. 331-338. doi: 10.5028/jatm.2011.03033611.

[7] Brum A G V de, Rodrigues A P 2013 Topographic profile of a target with use of laser pulses. A survey directed to the Brazilian deep space mission ASTER. *Journal of Physics: Conf. Ser.* **465** 01 2003. doi:10.1088/1742-6596/465/1/012003.

[8] Shan J and Toth C K 2009 *Topographic Laser Ranging and Scanning: Principles and Processing* (Boca Raton: CRC Press)

[9] MATLAB Release 2009a, The MathWorks, Inc., Natick, Massachusetts, United States.

[10] Brum A G V de, Cruz, F C da, Hetem Jr A and Rodrigues A P 2014 ALR - A laser altimeter for the first Brazilian deep space mission. Modeling and simulation of the instrument and its operation. *Journal: Computational & Applied Mathematics* (Basel: Springer). doi: 10.1007/s40314-014-0145-8.

[11] Steinvall O 2000 Effects of target shape and reflection on laser radar cross sections. *Applied Optics*, **39**(24):4381–4391.

[12] Tolentino V 2014. Vaz Tolentino Lunar observatory. Retrieved from http://www.vaztolentino.com.br/imagens/6667-As-crateras-ARISTOTELES-e-EUDOXUS

Journal of Physics: Conference Series **641** (2015) 012008 doi:10.1088/1742-6596/641/1/012008

Close approach maneuvers around an oblate planet

G M C Oliveira[1,2], A F B A Prado[2] and D M Sanchez[2]

[1]Federal Center for Technological Education of Minas Gerais – CEFET/MG, Belo Horizonte, MG, 30421-169, Brazil
[2]National Institute for Space Research – INPE, São José dos Campos, SP, 12227-010, Brazil

E-mail: magela@contagem.cefetmg.br

Abstract. There are many applications of the close approach maneuvers in astronautics, and several missions used this technique in the last decades. In the present work, those close approach maneuvers are revisited, but now considering that the spacecraft passes around an oblate planet. This fact changes the distribution of mass of the planet, increasing the mass in the region of the equator, so increasing the gravitational forces in the equatorial plane. Since the present study is limited to planar trajectories, there is an increase in the variation of energy given by the maneuver. The planet Jupiter is used as the body for the close approach, but the value of J_2 is varied in a large range to simulate situations of other celestial bodies that have larger oblateness, but the same mass ratio. This is particularly true in recent discovered exoplanets, and this first study can help the study of the dynamics around those bodies.

1. Introduction

Several papers in the literature describe the close approach maneuvers, like the ones shown in references [1] to [20]. Using the model given by the "patched conics" to explain the close approach maneuvers, it is assumed that the mission can be divided in three stages, all of them studied under Keplerian motion. It is also considered that the system consists of three bodies: M_1, a massive body in the center of the Cartesian system; M_2, a smaller body in a Keplerian orbit around M_1; and M_3, a spacecraft that is traveling in an orbit around M_1 when it passes close to M_2. This passage changes the orbit of M_3, which leads to a variation of eccentricity, semi-major axis, velocity, energy and angular momentum of M_3. This phenomenon is also called a swing-by maneuver. The advantages of this maneuver are the savings of time and fuel for a particular mission. According to these assumptions, the orbits of M_1 and M_2 remain unchanged. For the motion of the spacecraft, the model of two bodies is used in the initial phase, for the spacecraft-central body system. In the second stage, it is assumed that the spacecraft passes through the secondary body in an open trajectory. The third step is again assumed to be a system formed by two bodies, spacecraft-central body. The goal of this maneuver is to modify the Keplerian trajectories spacecraft-central body from the first to the third stage. Figure 1 illustrates this sequence. In this figure, A and B are the initial and final points of the trajectory during the close approach, δ is the angle of curvature of the spacecraft, Ψ is the angle between the periapsis line and the line connecting M_1 and M_2, $\mathbf{V_2}$ is the velocity of M_2 with respect to M_1, $\mathbf{V}_\infty^-$ and $\mathbf{V}_\infty^+$ are the velocities of M_3 with respect to M_2, when approaching and leaving M_2, respectively, and r_p is the periapsis distance of the close approach.

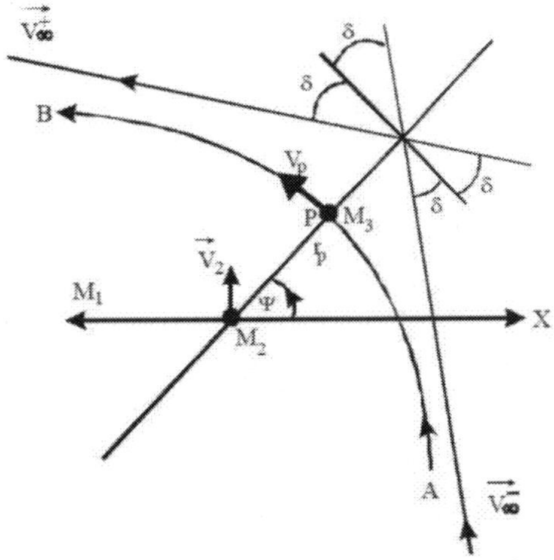

Figure 1. The swing-by maneuver. Adapted from [13].

2. Equations of Motion

To get more accurate results, it is important to improve the mathematical model used to describe this problem. The study is limited to planar motion, because this region is where the effects of the oblateness of the planet is maximized. The planar circular restricted three-body problem is an excellent mathematical model to use in this type of study, because it allows both primaries to act all the time in the motion of the spacecraft. Based on this model, it is possible to get the spacecraft equations of motion, as follows [21].

This problem is studied using the canonical system of units. The unit of mass will be the total mass of the system, where the mass of Sun (M_1) is added to the mass of the planet (M_2) and the spacecraft (M_3) is assumed to have negligible mass. In this system, the admensional mass of M_2 is given by the mass ratio:

$$\mu = \frac{M_2}{M_1+M_2} \tag{1}$$

The mass of M_1 is given by:

$$(1-\mu) \tag{2}$$

Using a fixed reference system, where the origin is located at the centroid of the two massive bodies, it is possible to develop the equations of motion. The x-axis is the line connecting M_1 and M_2 and the vertical axis is the line perpendicular to the x-axis. In this system, M_1 and M_2 have positions that are given by:

$$\begin{aligned} \bar{x}_1 &= -\mu\, r\cos v \\ \bar{y}_1 &= -\mu\, r\sin v \\ \bar{x}_2 &= (1-\mu)\, r\cos v \\ \bar{y}_2 &= (1-\mu)\, r\sin v \end{aligned} \tag{3}$$

Where r is the distance between the two primaries, given by:

$$r = \frac{1-e^2}{1+e\cos v} \tag{4}$$

and v is the true anomaly of M_2. In this system, the equations of motion of the spacecraft are given by:

$$\ddot{\bar{x}} = \frac{-(1-\mu)(\bar{x}-\bar{x}_1)}{r_1{}^3} - \frac{\mu(\bar{x}-\bar{x}_2)}{r_2{}^3} \tag{5}$$

Journal of Physics: Conference Series **641** (2015) 012008 doi:10.1088/1742-6596/641/1/012008

$$\ddot{\bar{y}} = \frac{-(1-\mu)(\bar{y}-\bar{y}_1)}{r_1^3} - \frac{\mu(\bar{y}-\bar{y}_2)}{r_2^3} \quad (6)$$

Where r_1 and r_2 are the distances from the spacecraft to M_1 and M_2, respectively, given by:

$$r_1^2 = (\bar{x}-\bar{x}_1)^2 + (\bar{y}-\bar{y}_1)^2 \quad (7)$$

$$r_2^2 = (\bar{x}-\bar{x}_2)^2 + (\bar{y}-\bar{y}_2)^2 \quad (8)$$

Taking into account the disturbance due to the oblateness (J_2) of the secondary body M_2, there is a new term in the potential energy, given by:

$$U_{J2} = -\mu R^2 J_2 \left(-\frac{1}{2}\frac{1}{r^3}\right) \quad (9)$$

Where R is the equatorial radius of the planet. Therefore, the equation of motion of the spacecraft in this fixed reference system has the additional term:

$$\ddot{r} = -\frac{3}{2}\mu R^2 J_2 \frac{1}{r^5} \quad (10)$$

That can be expressed in terms of x-y components by equations 11 and 12:

$$\ddot{x} = -\frac{3}{2}\mu R^2 J_2 \frac{(\bar{x}-\bar{x}_2)}{r_2^5} \quad (11)$$

$$\ddot{y} = -\frac{3}{2}\mu R^2 J_2 \frac{(\bar{y}-\bar{y}_2)}{r_2^5} \quad (12)$$

The main problem will be to simulate and classify trajectories passing close to Jupiter (M_2). For this task we assume the presence of three bodies: Sun (M_1), Jupiter (M_2) and the spacecraft (M_3). Thus, the problem is to study the motion of the spacecraft near the encounter with the planet Jupiter. The energy (E) and the angular momentum (C) are calculated before and after the close encounter to detect the change of the orbit due to this maneuver. The simulations are made by numerical integration of the equations (5) and (6), adding the effect of the oblateness shown in equations (10), (11) and (12). When the spacecraft reaches the points A and B, far away from Jupiter, the energy and angular momentum before and after the passage, respectively, are calculated assuming a two-body dynamics Sun-spacecraft. The orbits are classified in four categories [13]: direct elliptical (negative energy and positive angular momentum), retrograde elliptical (negative energy and angular momentum), direct hyperbolic (positive energy and angular momentum), and retrograde hyperbolic (positive energy and negative angular momentum).

3. Results

The results consist of graphs showing the variation of the orbit of the spacecraft due to the encounter with Jupiter considering the oblateness of this planet. The horizontal axis represents the angle of approach and the vertical axis represents the velocity of the spacecraft at the periapsis. They are related to the orbits in accordance with the rules shown in table 1:

Table 1. Rules for the establishment of letters to orbits. Adapted from [13].

After: Before:	Direct Elliptical	Retrograde Elliptical	Direct Hyperbolic	Retrograde Hyperbolic
Direct Elliptical	A	E	I	M
Retrograde Elliptical	B	F	J	N
Direct Hyperbolic	C	G	K	O
Retrograde Hyperbolic	D	H	L	P

The letter Z is black and indicates that the spacecraft did not leave Jupiter, so it remains orbiting around the planet. The three independent variables that describe the maneuver are: 1) r_p , the distance from the periapsis, kept constant at 1.3 radius of Jupiter; 2) Vp, the velocity at the periapsis; 3) Ψ, the

Journal of Physics: Conference Series **641** (2015) 012008 doi:10.1088/1742-6596/641/1/012008

approach angle. The graphs were obtained for the spacecraft trajectories considering several values for the oblateness of Jupiter: 0.0, 0.0147 (the real value), 0.0735 (five times the real value), 0.1470 (ten times the real value). The angle Ψ goes from 180° to 360°, because there is a symmetry in the system and the range from 0° to 180° is a mirror image of the range shown here, just replacing escapes by captures. The velocities where chosen in the range from 4.1 to 4.7 canonical units, because this is the regions where captures and escapes occur. The regions below and above these values are dominated by orbits that stay around Jupiter or has negligible effects, respectively. The graphs below show the results. Figures 2 to 5 show that the general forms of the figures are not changed by the oblateness of the planet, but some particular regions are strongly modified by the J_2 term. Note, for example, the bottom right corner of the figures 2 to 5. Other examples are the middle and the bottom left corner of these figures.

To emphasize those modifications new graphs were made, showing the bottom right corner in more details. It is noted that trajectories of type K (that represents orbits that are direct hyperbola before and after the close approach) are transformed in I (that represents orbits that are direct ellipse before the close approach and direct hyperbola after that). The physical reason for this fact is the extra force generated by the largest concentration of mass near the equator of the planet. Figures 6 to 9 show a detailed view for Ψ between 330° and 360° and Vp between 4.1 and 4.2 canonical units, the region where the effects are stronger. These figures show clearly how the J_2 has a great influence in the modification of the trajectories in that region.

Figure 10 shows the changes in the trajectories of the spacecraft using two different values for J_2: 0 and 0.147. It is shown a change from the type of trajectory K (green) to I (red). In this example, the velocity at the periapsis is 4.12 canonical units and the angle of approach Ψ is 330°.

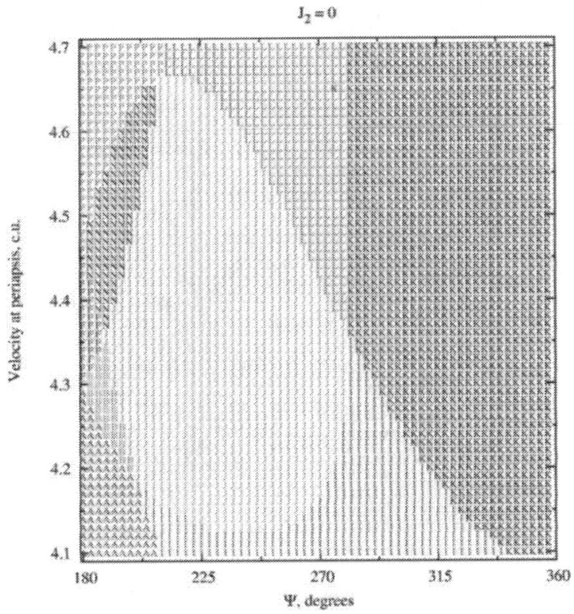

Figure 2. $J_2 = 0$.

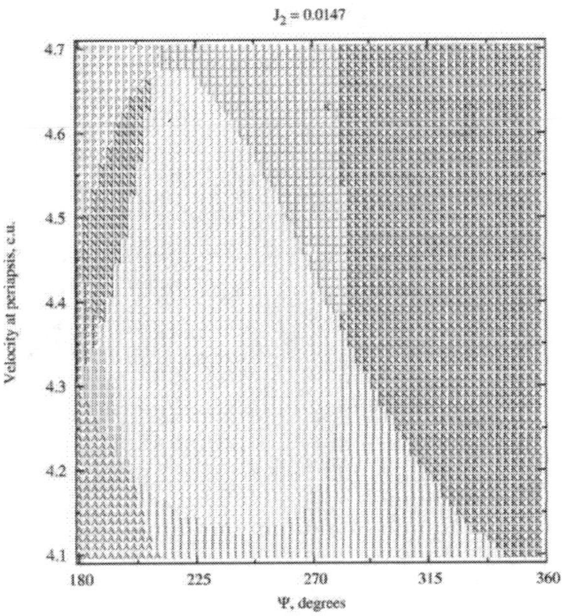

Figure 3. $J_2 = J_2$ of Jupiter.

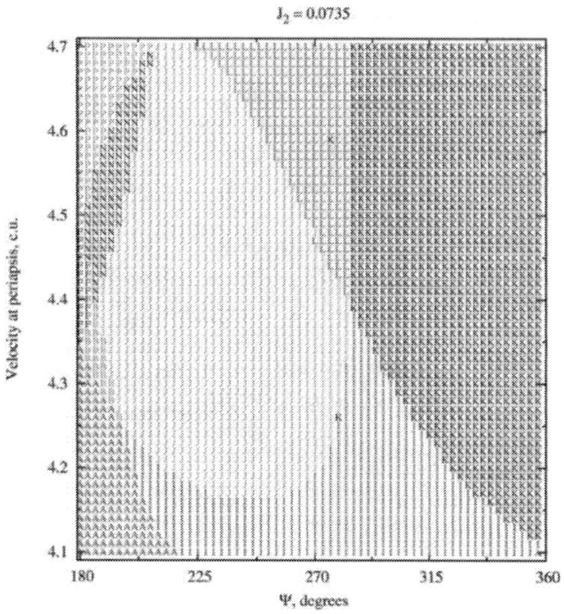

Figure 4. J_2 = 5 times J_2 of Jupiter.

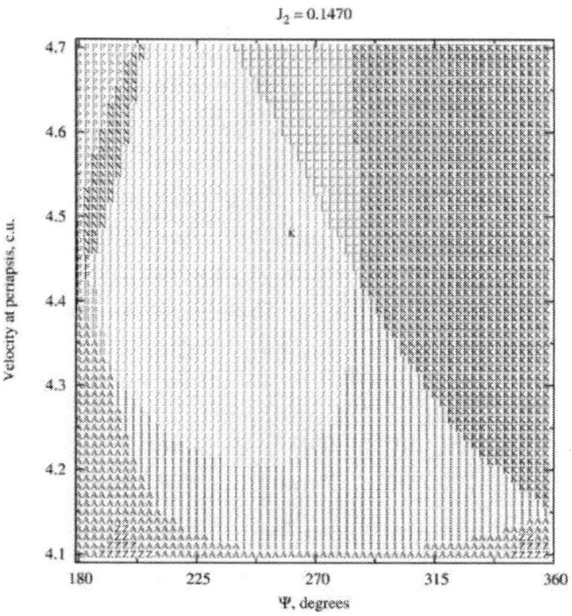

Figure 5. J_2 = 10 times J_2 of Jupiter.

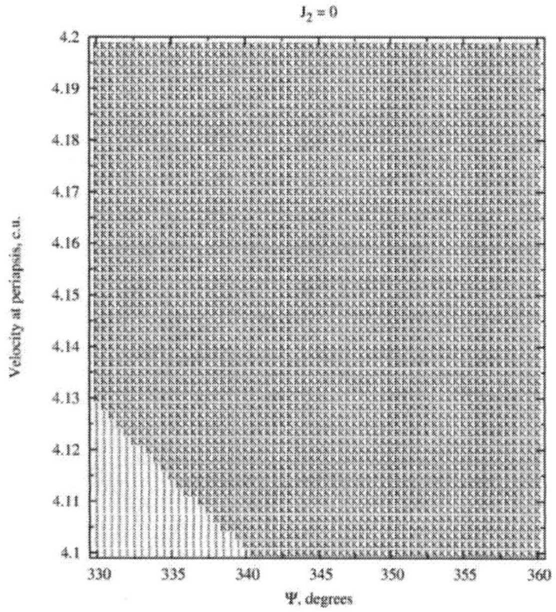

Figure 6. $J_2 = 0$

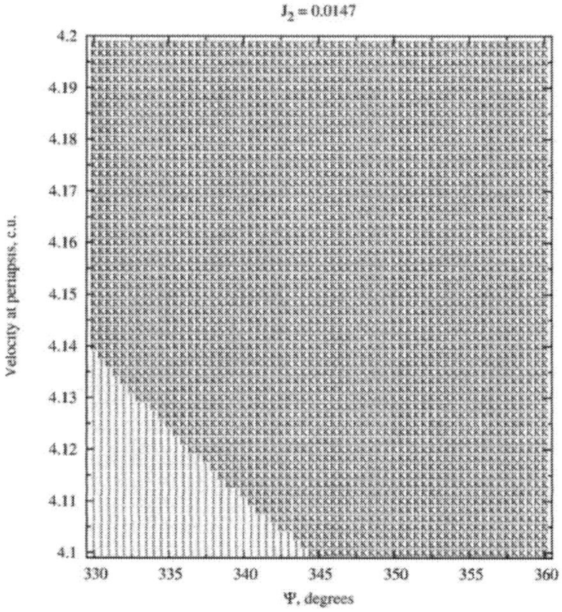

Figure 7. J_2 = J_2 of Jupiter

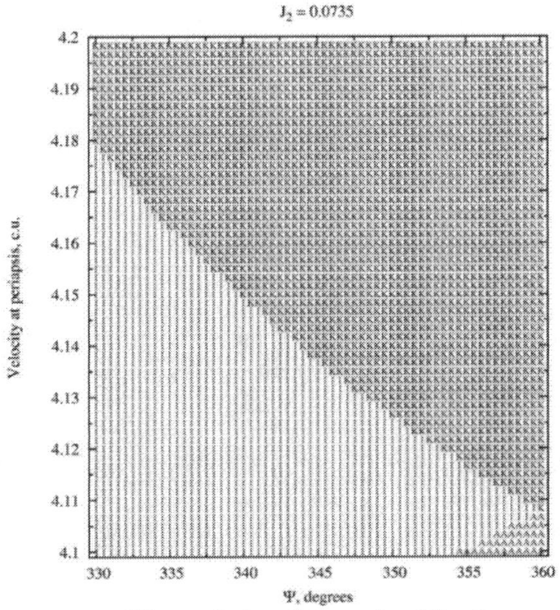

Figure 8. J_2 = 5 times J_2 of Jupiter

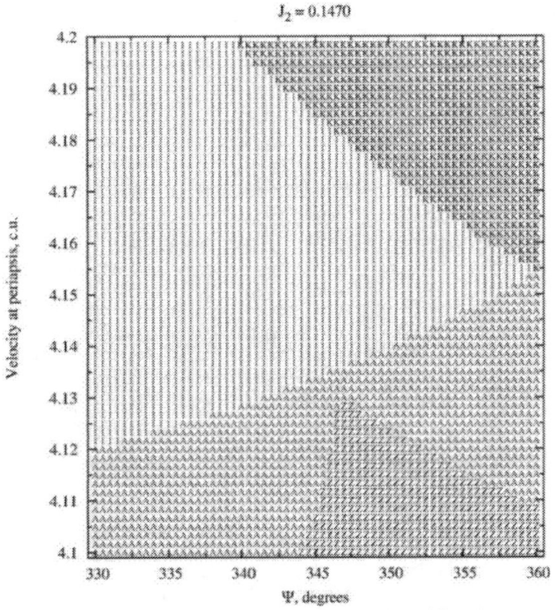

Figure 9. J_2 = 10 times J_2 of Jupiter

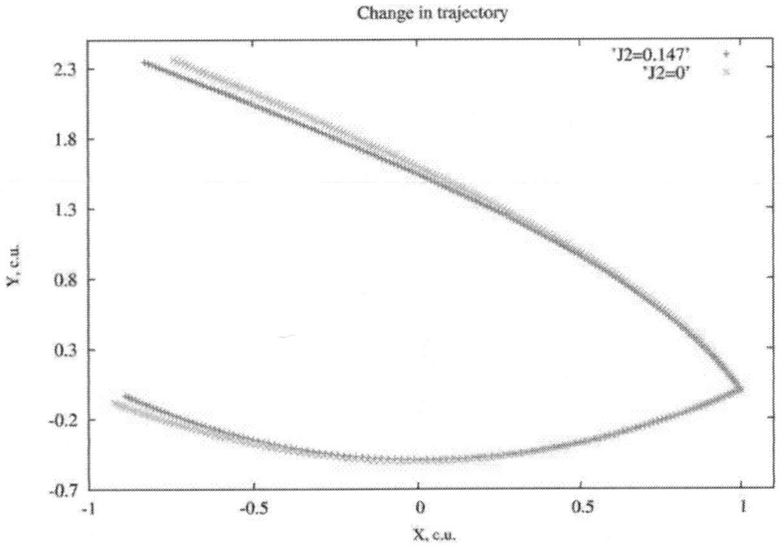

Figure 10. Changes in trajectories due to different values for J_2.

4. Conclusions

It was showed that the inclusion of the oblateness of the planet causes a variation in the trajectories of the spacecraft. Several values for J_2 were used and it was observed a modification of the trajectories, in particular in the border lines between different families of orbit, where there are strong changes in the types of orbits caused by the oblateness of the planet. The physical reason for this variation is that the oblateness changes the mass distribution of the body, reducing the average distances from each mass element to the spacecraft, which is equivalent to an increase in the mass of the planet, thereby increasing the effect of the passage of the vehicle.

5. References

[1] Broucke R A 1988 The celestial mechanics of gravity assist, *In: AIAA/AAS Astrodynamics Conf., Minneapolis,* 88

[2] D'Amario L A, Byrnes D V and Stanford R H 1982 *Jour. Guid. Cont. and Dyn.* 5 465-471

[3] Farquhar R W and Dunham D W 1981 *Jour. Guid. Cont. and Dyn.* 4(2) 192-196

[4] Formiga J K S and Santos D P S 2014 An analytical description of the three-dimensional swing-by *Comp. Ap. Mat.* 3-18

[5] Formiga J K S and Prado A F B A 2014 Studying sequences of resonant orbits to perform successive close approaches with the Moon *J. Braz. Soc. Mec. Sc. Eng.* v. 1 p. 1

[6] Flandro G 1966 Fast reconnaissance missions to the outer Solar System utilizing energy derived from the gravitational field of Jupiter *Ast. Acta* 12(4)

[7] Gomes V M and Prado A F B A 2010 A study of the impact of the initial energy in a close approach of a cloud of particles *WSEAS Trans. Mat.* v. 9, 811-820

[8] Gomes V M, Prado A F B A and Golebiewska J 2013 Dynamics of space particles and spacecrafts passing by the atmosphere of the earth *The Sc. World J.* v. 2013, 1-6

[9] Helton A F, Strange N J and Longuski J M 2002 Automated design of the europa orbiter tour *J. Spac. Roc.* 39(1).

[10] Longuski J M and Williams S N 1991 The last grand tour opportunity to pluto *J. Astr. Sc.* 39. 359-365

[11] Prado A F B A and Broucke R A 1995 Effects of atmospheric drag in swing-by trajectory *Acta Astr.* 36(6). 285-290

[12] Prado A F B A 1996 Powered swing-by *J. Guid. Cont. Dyn.* vol. 19, no 5, 1142-1147

[13] Prado A F B A 1997 Close approach trajectories in the elliptic restricted problem *J. Guid. Cont. Dyn.* 20(4), 797–802

[14] Prado A F B A 2007 A comparison of the patched-conics approach and the restricted problem for swing-bys *Adv. Space Res.* 40. 113-117

[15] Prado A F B A and Gomes V M 2014 Searching for capture and escape trajectories around jupiter using its galilean satellites *Comp. Ap. Mat.* v. 33, 1

[16] Sanchez D, Prado A F B A, Sukhanov A and Yokoyama T 2014 Optimal transfer trajectories to the haumea system *In: 13th International Conference on Space Operations 2014, 2014, Pasadena .13th International Conference on Space Operations 2014.* Reston: American Institute of Aeronautics and Astronautics. v. 1

[17] Santos D P S, Prado A F B A, Casalino L and Colasurdo G 2008 Optimal trajectories towards near-earth-objects using solar electric propulsion (sep) and gravity assisted maneuver *J. Aer. Eng., Sc. and App.* v. 1, 51-64

[18] Silva A F, Prado A F B A and Winter O C 2013 Powered swing-by maneuvers around the moon *J. Phy. Conf. Series* v. 465, p. 012001

[19] Strange N J and Longuski J M 2002 Graphical method for gravity-assist trajectory design *J. Spac. Roc.* v. 39, n. 1, 2002

[20] Striepe S A and Braun R D 1991 *J. Astr. Sc.* 39(3) 299-312

[21] Szebehely V G 1967 *Theory of Orbits* (Academic Press New York)

Three-dimensional Two-Body Tether System – Equilibrium Solutions

Denilson Paulo Souza dos Santos and Alessandra Ferreira

Centre for Mechanical and Aerospace Science and Technologies (C-MAST) – UBI – Portugal
Division of Space Mechanics and Control – INPE (National Institute for Space Research) – São José dos Campos – SP, Brazil

E-mail: denilson.paulo@gmail.com

Abstract. The tethered spacecraft system was studied. This is a three-body problem formed by the main body with larger mass, m_1 and m_2 are the smaller bodies and they are connected by a cable. The main body is orbited by the center of mass m_1 and m_2. The cable was assumed rigid and massless. The study goal is analyze the existing equilibrium to tether in two situations: using a control law for the cable size and constant cable size. Assuming a control law, the solutions were fixed values independent of eccentricity (e) and true anomaly (ν) of the system. For the constant case there are periodic solutions, continuous and discontinuous for different values of e and ν. The results obtained are angles functions which describe the m_1 and m_2 plane and out plane motion.

1. Introduction

The space tether systems problem has been studied in several published articles. Beletsky and Levin [1] begins by setting the scene for tethers in space. They summarized possible applications and also discussed fact and fiction, analyzing clearly the main parameters and applications for tethers system as the material density, the effective forces, orbital dynamics, mechanics models, attitude and possible disturbances for flexible tethers with end masses, massless and massive variations. Pearson (1979) [2] considered the applications lunar spacecraft formulations for data transportation and communication, using the space elevators concept, the three bodies dynamics, the Lagrangian points $L1$ and $L2$, exploring stability. The space elevator idea as a convenient and low cost way to access space was first formulated in 1960 by the Russian Yuri Artsutanov ([3] and [4]) in "Into the Cosmos without Rockets". This study analyzed the possibility of creating a space elevator connecting the Earth. The minimum requirement for a space elevator is obviously a cable strong enough to support its own weight when hanging from geostationary orbit.

Burov et al. ([5] – [9]) studied the tethers motion considering dumbbell oscillations, bodies in the central field of vibrating forces, showing stability solutions for angles on the chaotic dynamics in elliptical orbit. Analyzing the problem in other aspect of Moon-tethered pendulum there are works considering the two-body tethered system uniform rotations in planar motion and tether length control [10], and Moon-tethered system with variable tether length in restricted three-body problem [11]. In both studies [10] and [11] were suggested methods of controlling the geometric configuration. Other available studies in the literature about tether can be seen in [12] to [16].

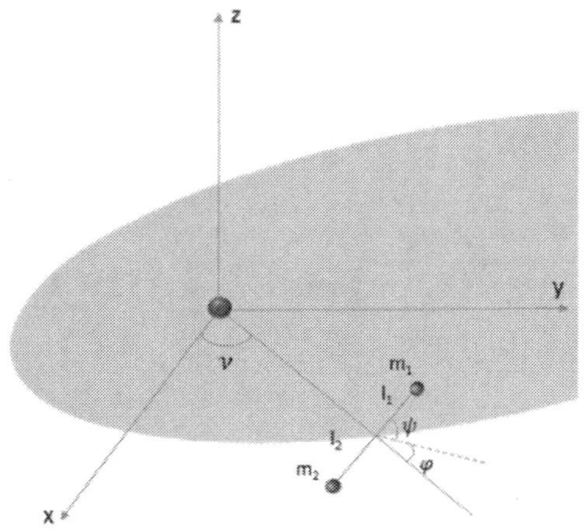

Figure 1. Dumbbell-like system model

This work is a study for a tether system with two bodies connected by a cable, which moves your center of mass in plane and around the main body. The equilibrium analysis is made for m_1 and m_2 plane and out plane movement. Based on the papers mentioned above and in particular the work [10] and [11].

2. Mathematical model

The model is a dumbbell-like system, with the two mass points (m_1 and m_2) connected with a rigid and massless tether. The system's center of mass is described by a Keplerian orbit fixed on an inert axis in primary body's center [13].

The angle ν is the tether true anomaly around primary body. The variable ρ is the distance from primary body to center of mass (CM), l_1 and l_2 are tether lengths from the mass points m_1 and m_2 to CM, respectively. φ is the angle between the plane tether projection and ρ, it describe the plane motion. The ψ is the angle between the tether and primary body plane. It describe the m_1 and m_2 out plane motion. The system geometry is presented in Figure 1.

2.1. System Positions

Since the model chosen is based on masses points and a massless tether. This allows the model to be described by Keplerian motion:

$$l_1 = \frac{m_2 l}{m_1 + m_2} \tag{1}$$

$$l_2 = \frac{m_1 l}{m_1 + m_2} \tag{2}$$

$$\rho = \frac{p}{(1 + e\cos(\nu))} \tag{3}$$

being p the focal parameter, e the eccentricity and ν the true anomaly. For the coordinates system the components of center of mass position vector (x_0, y_0, z_0) and components of point

Journal of Physics: Conference Series **641** (2015) 012009 doi:10.1088/1742-6596/641/1/012009

mass i position vector (x_i, y_i, z_i) are:

$$\begin{cases} x_0 = \rho\cos(\nu) \\ y_0 = \rho\sin(\nu) \\ z_0 = 0 \\ x_1 = x_0 + l_1\cos(\nu+\varphi)\cos(\psi) \\ y_1 = y_0 + l_1\sin(\nu+\varphi)\cos(\psi) \\ z_1 = l_1\sin(\psi) \\ x_2 = x_0 - l_2\cos(\nu+\varphi)\cos(\psi) \\ y_2 = y_0 - l_2\sin(\nu+\varphi)\cos(\psi) \\ z_2 = -l_2\sin(\psi) \end{cases} \tag{4}$$

2.2. Lagrange Motion Equations

The Lagrange Motion Equations, second-order ordinary differential equations, which describe the motions mechanical systems under the action of forces, can be obtained from the Lagrangian of the system given by $L = T - V$. Being T the kinetic energy and V potential energy.

For the following analysis, the generalized coordinates are φ and ψ. The system is only under the gravity-gradient. It is forces a conservative system. Two equations can be obtained based on those coordinates.

Defining φ or l allows to control the system, as can be seen in [2]. It has been chosen to define φ and consecutively obtain the l behavior.

$$\cos(\psi)\big(4p^3\dot{l}\cos(\psi)\,(\dot{\nu}+\dot{\varphi}) + l\big(3\mu_0\sin(2\varphi)\cos(\psi)(1+e\cos(\nu))^3 + \\ + 2p^3\big(-2\dot{\psi}\sin(\psi)\,(\dot{\nu}+\dot{\varphi}) + \cos(\psi)\,(\ddot{\nu}+\ddot{\varphi})\big)\big)\big) = 0 \tag{5}$$

$$\sin(2\psi)\left(3\mu_0\cos^2(\varphi)(1+e\cos(\nu))^3 + p^3\,(\dot{\nu}+\dot{\varphi})^2 + 2p^3\ddot{\psi}\right) + 4p^3\frac{\dot{l}}{l}\dot{\psi} = 0 \tag{6}$$

The motion in the spherical coordinates, ψ, φ, ν and l, where l represents the cable length, ν and φ in-plane and ψ out-plane rotation, see Figure 1. The equations 5 and 6 have been written in the time domain. Taking the true anomaly as an independent variable, assuming l behaves according to $l(\nu) = \frac{\eta(\nu)l_0}{1+e\cos(\nu)}$, being l_0 the tether initial length and $\eta(\nu)$ a new variable introduced to describe the cable behavior. The equations system is rewritten in ν domain as:

$$\frac{1}{2}l_0\cos(\psi)(1+e\cos(\nu)) \\ \left(\left(\cos(\psi)\left(\frac{3\sin(2\varphi)}{1+e\cos(\nu)} + 2\varphi''\right) - 4\,(\varphi'+1)\,\psi'\sin(\psi)\right) + \frac{4\eta'\,(\varphi'+1)\cos(\psi)}{\eta}\right) = 0 \tag{7}$$

$$2l_0(1+e\cos(\nu)) \\ \left(\frac{2\eta'\psi'}{\eta} + \psi''\right) + \frac{1}{2}l_0\sin(2\psi)\left(2\varphi'\,(\varphi'+2)\,(1+e\cos(\nu)) + 2e\cos(\nu) + 3\cos(2\varphi) + 5\right) = 0 \tag{8}$$

Assuming that the length l is also parameterized with the true anomaly ν, the equations 7 and 8 are the tethers oscillations equations.

3. Equilibrium points

To find equilibrium points analyzes the system with the conditions following: $\varphi'[\nu] = 0$, $\eta'[\nu] = 0$, $\psi'[\nu] = 0$ replaced in Equation 7 and 8.

$$\varphi'' = -\frac{3\sin(2\varphi)}{2(1+e\cos(\nu))} \tag{9}$$

$$\psi'' = -\frac{\sin(2\psi)(5+2e\cos(\nu)+3\cos(2\varphi))}{4(1+e\cos(\nu))} \tag{10}$$

The equilibrium solutions to the tether in this configuration, imposes $\varphi'' = 0$ and $\psi'' = 0$ (Eq.9 and 10), the solutions obtained are: i) in plane $\psi = k\pi$ and $\varphi = \{k\pi, \frac{\pi}{2} + k\pi\}$; ii) out plane $\psi = \frac{\pi}{2} + k\pi$ and $\varphi = \{k\pi, \frac{\pi}{2} + k\pi\}$, for $k \in Z$. See Figures 2 and 3.

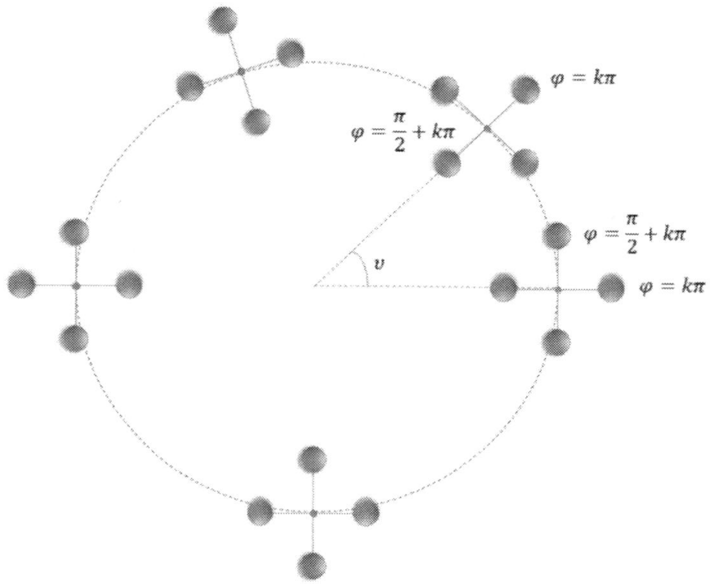

Figure 2. The geometry solutions in plane $\psi = k\pi$ and $\varphi = \{k\pi, \frac{\pi}{2} + k\pi\}$

The Eq. 11 and 12 for the system equilibrium points were obtained by substituting variables in Eq. 5 and 6, to change the time domain to the ν domain and assuming a constant l. The solutions in plan (φ angle) for eccentricities $e \leq 0.75$ are continuous and periodic (in true anomaly ν), see Figures 4 and 5, for eccentricity $e > 0.75$ there is no physically possible motion because the function $\varphi(\nu)$ is discontinuous, however periodic.

$$\varphi'' = \frac{4e\sin(\nu) - 3\sin(2\varphi)}{2(1+e\cos(\nu))} \tag{11}$$

$$\psi'' = -\frac{\sin(2\psi)(5+2e\cos(\nu)+3\cos(2\varphi))}{4(1+e\cos(\nu))} \tag{12}$$

Figures 4 and 5 presented the solution for $\sin(2\psi) = 0$ in the Eq. 11 and 12, which implies $\psi = \{k\pi, \frac{\pi}{2} + k\pi\}$ (Figure 2 and 3).

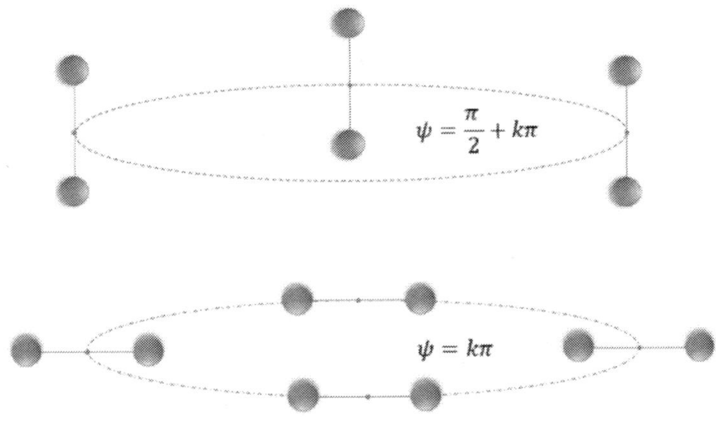

Figure 3. The geometry solutions out plane $\psi = \{k\pi, \frac{\pi}{2} + k\pi\}$ and $\varphi = \{k\pi, \frac{\pi}{2} + k\pi\}$, for $k \in Z$.

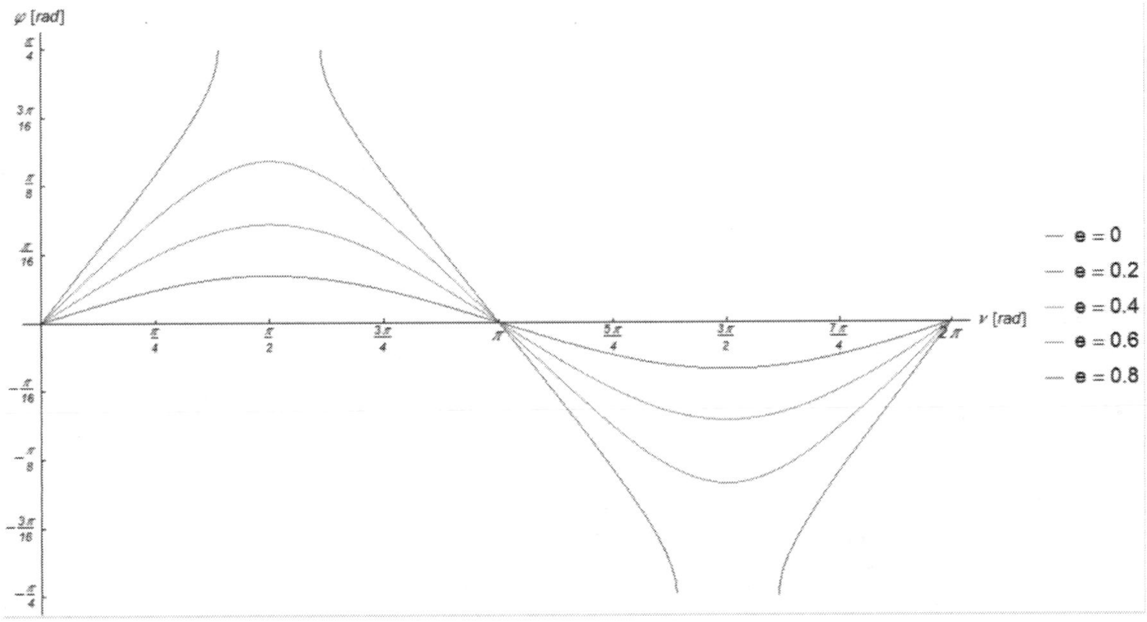

Figure 4. The periodical solutions in plane for eccentricities of the reference orbit.

4. Conclusion

The equilibrium of two mass points connected by a cable, orbiting a common center of mass, the center of mass fixed in line with the primary body was searched. The generalized coordinates of motion Lagrange Equation describing the motion in plane (φ) and out plane (ψ). This is a conservative system. Time domain change to the ν domain, due to the center of mass has a Keplerian orbit.

Mathematically, a point is in equilibrium when its speed and acceleration are equal to zero, then assume $\varphi' = \psi' = l' = \eta' = 0$ and $\varphi'' = \psi'' = l'' = \eta'' = 0$. The first case analysed was l assuming a control law dependent of e, ν, cable initial value (l_0) and a parameter η. The solutions are fixed values and independent e and ν. For constant l the equilibrium is dependent e and ν and they are continuous and periodic for $e \leq 0.75$, when e larger than this value the

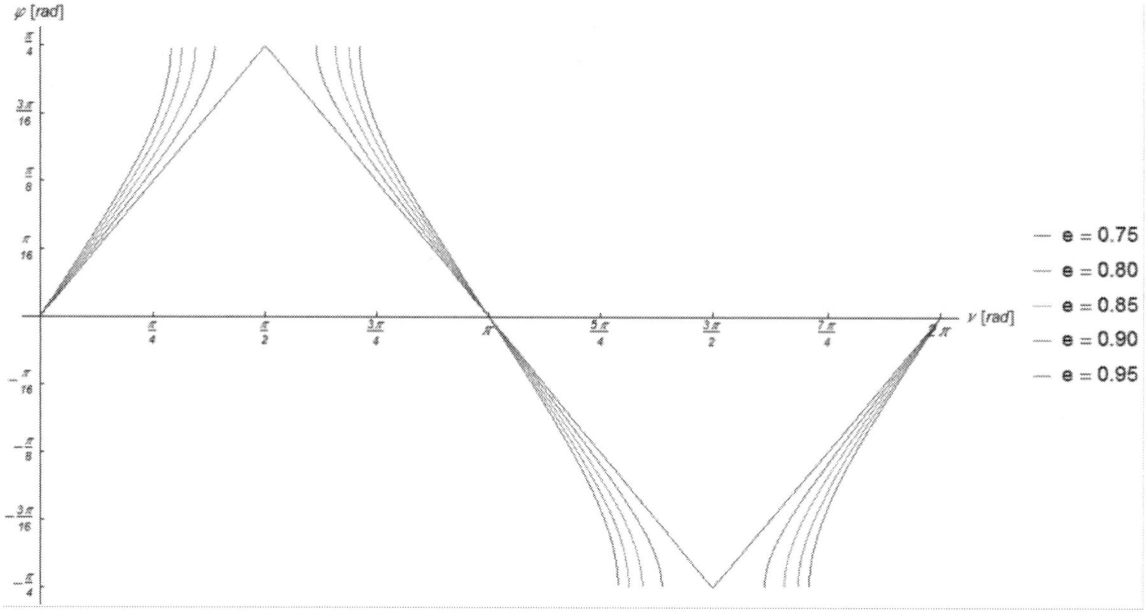

Figure 5. The periodical solutions in plane for eccentricities $e \geq 0.75$.

solutions become discontinuous in true anomaly, not allowing motion tether.

Acknowledgments

This work was accomplished with the support of CAPES – Brazil, under Contract BEX 10689-13-3, CNPQ - Project 401373/2012-6, SWE 202779/2014-9, INPE – National Institute for Space Research – Brazil, Centre for Mechanical and Aerospace Science and Technologies (C-MAST) – UBI – Portugal and especially Prof. $Dr^{\underline{a}}$ Anna D. Guerman – UBI– Portugal.

References

[1] Belestsky, Vladimir V.; Levin, Evgenii M., 1993. Dynamics of Space Tethers Systems. San Diego - California: Advances in the Astronautical Sciences, Vol. 83 - American Astronautical Society ISSN 0-87703-370-6.
[2] J. Pearson, Anchored Lunar Satellites for Cislunar Transportation and Communication, Journal of the Astronautical Sciences, Vol. XXVII, No. 1, 1979, pp. 39-62.
[3] Artsutanov, Y. N., "Into the Cosmos without Rockets," Znanije-Sila 7, 25, 1969.
[4] Artsutanov, Y. N, The Earth-to-Moon Highway, Technics to Youth, No. 4, 21, 35, 1979 (in Russian).
[5] A.A. Burov and I.I. Kosenko, On relative equilibria of an orbital station in regions near the triangular libration points, Doklady Physics, Vol. 52, Issue 9, 2007, pp.507-509.
[6] A.A. Burov, M. Pascal, and S.Ya. Stepanov, The gyroscopic stability of the triangular stationary solutions of the generalized planar three-body problem, Journal of Applied Mathematics and Mechanics, Vol. 64, Issue 5, 2000a, pp. 729-738.
[7] A.A. Burov and H. Troger, The relative equilibria of an orbital pendulum suspended on a tether, Journal of Applied Mathematics and Mechanics, Vol. 64, Issue 5, 2000b, pp. 723-728.
[8] A. Burov, O. I. Kononov, and A. D. Guerman, Relative equilibria of a Moon - tethered spacecraft,Advances in the Astronautical Sciences, v. 136, 2011a, pp. 2553-2562.
[9] A. Burov and I.I. Kosenko, Plane oscillations of a body with variable mass distribution in an elliptic orbit, Proc. of ENOC 2011, July 24-29, 2011b, Rome, Italy.
[10] A.A. Burov, I.I. Kosenko, and A. D. Guerman, Dynamics of a moon-anchored tether with variable length. Advances in the Astronautical Sciences, 2012, Vol. 142, pp.3495-3507.
[11] Burov, A. A., Guerman, A. D., Kosenko, I. I. (2013). Equilibrium configurations and control of a moon-anchored tethered system. Advances in the Astronautical Sciences, 146, 251-266.

[12] Bainum, Peter M., and V. K. Kumar. "Optimal control of the shuttle-tethered-subsatellite system." Acta Astronautica 7.12 (1980): 1333-1348.
[13] Misra, A. K., Z. Amier, and V. J. Modi. "Attitude dynamics of three-body tethered systems." Acta Astronautica 17.10 (1988): 1059-1068.
[14] Kalantzis, S., Modi, V. J., Pradhan, S., Misra, A. K. (1998). Dynamics and control of multibody tethered systems. Acta astronautica, 42(9), 503-517.
[15] Misra, A. K. "Dynamics and control of tethered satellite systems." Acta Astronautica 63.11 (2008): 1169-1177.
[16] V.A. Chobotov, D.L. Mains, Tether satellite system collision study, Acta Astronautica, 44 (712) (1999), pp. 543551

Journal of Physics: Conference Series **641** (2015) 012010 doi:10.1088/1742-6596/641/1/012010

A study of optimal low-thrust limited-power transfers between coplanar orbits with small eccentricities

Francisco das Chagas Carvalho[1] and Sandro da Silva Fernandes[2]

[1] Departamento de Física, Instituto Tecnológico de Aeronáutica
São José dos Campos – 12228-900 – SP – Brasil
[2] Departamento de Matemática, Instituto Tecnológico de Aeronáutica
São José dos Campos – 12228-900 – SP – Brasil

E-mail: fchagas.carvalho@gmail.com

Abstract. In this paper, an analytical solution, which includes the short periodic terms, for the problem of optimal time-fixed low-thrust limited power transfers (no rendezvous), in an inverse-square force field, between coplanar orbits with small eccentricities is described. This solution is derived through canonical transformation theory. Some numerical results based on the analytical solution are presented for a preliminary analysis of interplanetary missions.

1. Introduction

The study presented in the paper has been motivated by the renewed interest in the use of low-thrust propulsion systems in space missions in the last thirty years. Important space missions have made use of low-thrust propulsion systems: NASA-JPL Deep Space 1 and ESA-SMART1. Deep Space 1 was the first interplanetary spacecraft to utilize Solar Electric Propulsion. It was developed by NASA in the New Millennium program to testing new technologies for future Space and Earth science programs. It was launched on October 24, 1998. Deep Space 1 mission terminated on December 18, 2001, when its fuel supply exhausted. SMART-1 was the first of a series of ESA Small Missions for Advanced Research in Technology. It was used to test solar electric propulsion and other deep-space technologies. It was launched on September 27, 2003. SMART-1 mission ended on September 3, 2006, when the spacecraft, in a planned manoeuvre, impacted the lunar surface. Interesting details about these space missions can be found in [2], [11] and [12].

Low-thrust electric propulsion systems are characterized by high specific impulse and low-thrust capability (the ratio between the maximum thrust acceleration and the gravity acceleration on the ground is small, between 10^{-4} and 10^{-2}) and have their greatest benefits for high-energy planetary missions. Several researchers have obtained numerical and analytical solutions for several maneuvers involving specific initial and final orbits and specific thrust profiles ([1], [4], [5], [6], [7], [10]). In the analytical studies, averaging techniques are applied and solutions of the averaged equations are obtained such that only secular behavior of the optimal solutions is discussed. Few works discuss the inclusion of periodic terms which are, in general, considered only for transfers between close orbits.

In a previous work (Da Silva Fernandes and Carvalho, [4]), the authors presented a complete analytical solution, which includes the short periodic terms, for the problem of optimal low-thrust limited-power transfers between arbitrary elliptic coplanar orbits. But, this analytical solution becomes singular for circular orbits. So, the main purpose of this paper is to present a complete first order

Journal of Physics: Conference Series **641** (2015) 012010 doi:10.1088/1742-6596/641/1/012010

analytical solution which eliminates the singularity for circular orbits and can be applied for time-fixed transfers between coplanar orbits with small eccentricities. This study is particularly interesting because the orbits found in practice often have a small eccentricity and the problem of transferring a vehicle from a low Earth orbit to a high Earth orbit is frequently found.

2. Formulation of the optimization problem

For completeness, in this section the general planar low-thrust limited power transfer problem is formulated as a Mayer problem of optimal control theory with the radial distance and the radial and circumferential components of the velocity as state variables, and, the maximum Hamiltonian function which governs the optimal trajectories is derived by applying the Pontryagin Maximum Principle (Pontryagin et al, [9]). This formulation of the transfer problem provides a straightforward physical interpretation to the optimal thrust acceleration in terms of the adjoint variables.

For a low-thrust limited-power propulsion system – LP system, the fuel consumption is described by the variable J defined as (Marec, [8])

$$J = \frac{1}{2}\int_{t_0}^{t} \gamma^2 dt\,, \tag{1}$$

where γ is the magnitude of the thrust acceleration vector **γ**, used as a control variable. The consumption variable J is a monotonic decreasing function of the mass m of the space vehicle,

$$J = P_{\max}\left(\frac{1}{m} - \frac{1}{m_0}\right),$$

where $P_{\max}$ is the maximum power and m_0 is the initial mass. The minimization of the final value of the fuel consumption, J_f, is equivalent to the maximization of m_f.

Consider the motion of a space vehicle M, powered by a limited-power engine in an inverse-square force field. At time t, the state of a space vehicle M is defined by the radial distance r from the center of attraction, the radial and circumferential components of the velocity, v_r and v_s, and the fuel consumption J.

The optimization problem can be formulated as a Mayer problem of optimal control as follows (Marec, [8]): It is proposed to transfer the space vehicle M from the initial state $(r_0, v_{r0}, v_{s0}, 0)$ at time t_0 to the final state $(r_f, v_{rf}, v_{sf}, J_f)$ at time t_f, such that the final consumption variable J_f is a minimum. The duration of the transfer $t_f - t_0$ is specified. In the two-dimensional formulation, the state equations are given by

$$\frac{dr}{dt} = v_r \qquad \frac{dv_r}{dt} = \frac{v_s^2}{r} - \frac{\mu}{r^2} + R \qquad \frac{dv_s}{dt} = -\frac{v_r v_s}{r} + S \qquad \frac{dJ}{dt} = \frac{1}{2}\left(R^2 + S^2\right) \tag{2}$$

where μ is the gravitational parameter (for the Earth, $\mu = 3.986\times10^5 km^3/s^2$), R and S are the radial and circumferential components of the thrust acceleration vector, respectively. The performance index is then defined by

$$IP = J(t_f) \tag{3}$$

For LP system, it is assumed that there are no constraints on the thrust acceleration vector.

According to (Da Silva Fernandes et al, [3]), the optimal trajectories are governed by the maximum Hamiltonian H^*,

$$H^* = v_r p_r + \left(\frac{v_s^2}{r} - \frac{\mu}{r^2}\right) p_{v_r} - \frac{v_r v_s}{r} p_{v_s} + \frac{1}{2}\left(p_{v_r}^2 + p_{v_s}^2\right), \tag{4}$$

with the optimal thrust acceleration given by

$$R^* = p_{v_r} \qquad S^* = p_{v_s}. \tag{5}$$

Note that the components of the optimal thrust acceleration are given by the adjoint variables to the components of the velocity vector; in this case, the acceleration is modulated (Marec, [8]).

3. Analytical Solution

A first order analytical solution for the system of differential equations governed by the maximum Hamiltonian, which includes short periodic terms, can be derived though canonical transformation theory, as described in Da Silva Fernandes et al [3]. The optimal low-thrust power-limited trajectories for transfers between coplanar orbits with small eccentricities in an inverse-square force field are then described in a set of nonsingular elements by the following equations:

$$\begin{aligned} a''(t) = a'''(t) + \sqrt{\frac{a'''^5}{\mu^3}} \Big\{ & \left[8h'''\sin\ell''' - 8k'''\cos\ell'''\right] a'''^2 p_{a'''} + \left[4a'''\sin\ell''' - 2a'''k'''\cos 2\ell''' \right. \\ & \left. + 2a'''h'''\sin 2\ell'''\right] p_{h'''} + \left[-4a'''\cos\ell''' - 2a'''h'''\cos 2\ell''' - 2a'''k'''\sin 2\ell'''\right] p_{k'''} \Big\}\Big|_{t_0}^{t} \end{aligned} \tag{6}$$

$$\begin{aligned} h''(t) = h'''(t) + \sqrt{\frac{a'''^5}{\mu^3}} \Big\{ & \left[4a'''\sin\ell''' - 2a'''k'''\cos 2\ell''' + 2a'''h'''\sin 2\ell'''\right] p_{a'''} \\ & + \left[-2h'''\sin\ell''' - 2k'''\cos\ell''' + \frac{3}{4}\sin 2\ell''' + \frac{2}{3}h'''\sin 3\ell''' - \frac{2}{3}k'''\cos 3\ell'''\right] p_{h'''} \\ & + \left[2h'''\cos\ell''' - 2k'''\sin\ell''' - \frac{3}{4}\cos 2\ell''' - \frac{2}{3}h'''\cos 3\ell''' - \frac{2}{3}k'''\sin 3\ell'''\right] p_{k'''} \Big\}\Big|_{t_0}^{t} \end{aligned} \tag{7}$$

$$\begin{aligned} k''(t) = k'''(t) + \sqrt{\frac{a'''^5}{\mu^3}} \Big\{ & \left[-4a'''\cos\ell''' - 2a'''h'''\cos 2\ell''' - 2a'''k'''\sin 2\ell'''\right] p_{a'''} \\ & + \left[2h'''\cos\ell''' - 2k'''\sin\ell''' - \frac{3}{4}\cos 2\ell''' - \frac{2}{3}h'''\cos 3\ell''' - \frac{2}{3}k'''\sin 3\ell'''\right] p_{h'''} \\ & + \left[2h'''\sin\ell''' + 2k'''\cos\ell''' - \frac{3}{4}\sin 2\ell''' - \frac{2}{3}h'''\sin 3\ell''' + \frac{2}{3}k'''\cos 3\ell'''\right] p_{k'''} \Big\}\Big|_{t_0}^{t} \end{aligned} \tag{8}$$

with a''', ..., $p_{k'''}$ given by

$$a'''(t) = \frac{a_0'''}{1 + \frac{4a_0'''}{\mu}\left(\frac{1}{2}\mathrm{E}t^2 - a_0''' p_{a_0'''} t\right)} \tag{9}$$

Journal of Physics: Conference Series **641** (2015) 012010 doi:10.1088/1742-6596/641/1/012010

$$h'''(t) = h_0''' + \sqrt{\frac{5}{2}}\frac{p_{h'''}}{C}\left\{\tan^{-1}\left(\left(\frac{4\mu E}{5C^2 a_0'''} - 1\right)^{1/2}\right) - \tan^{-1}\left(\left(\frac{4\mu E}{5C^2 a'''} - 1\right)^{1/2}\right)\right\} \tag{10}$$

$$k'''(t) = k_0''' + \sqrt{\frac{5}{2}}\frac{p_{k'''}}{C}\left\{\tan^{-1}\left(\left(\frac{4\mu E}{5C^2 a_0'''} - 1\right)^{1/2}\right) - \tan^{-1}\left(\left(\frac{4\mu E}{5C^2 a'''} - 1\right)^{1/2}\right)\right\} \tag{11}$$

$$p_{a'''}^2 = \left(\frac{a_0'''}{a'''}\right)^3 p_{a_0'''}^2 + \frac{5}{8}C^2\left(\frac{a_0'''}{a'''^3} - \frac{1}{a'''^2}\right) \tag{12}$$

$$p_{h'''} = p_{h_0'''}, \tag{13}$$

$$p_{k'''} = p_{k_0'''}, \tag{14}$$

with C and E given in terms of the initial conditions by

$$C^2 = p_{h_0'''}^2 + p_{k_0'''}^2, \tag{15}$$

$$4\mu E = a_0'''\left(8\left(a_0''' p_{a_0'''}\right)^2 + 5\left(p_{h_0'''}^2 + p_{k_0'''}^2\right)\right). \tag{16}$$

The initial conditions for the state variables are $a'''(0) = a_0'''$, $h'''(0) = h_0'''$ and $k'''(0) = k_0'''$. Note that the mean latitude ℓ''' in equations above is given by

$$\ell'''(t) = \ell'''(t_0) + \int_{t_0}^{t}\left[\sqrt{\frac{\mu}{a'''^3}} + \frac{7}{4}\frac{a'''}{\mu}\left(k''' p_{h'''} - h''' p_{k'''}\right)\right]dt. \tag{17}$$

The set of nonsingular orbital elements are related to the classical orbital elements through the equations

$$h = e\cos\omega \qquad k = e\sin\omega \qquad \ell = M + \omega,$$

where e is the eccentricity, ω is the argument of periapsis and M is the mean anomaly, and, a denotes the semi-major axis. For simplicity, primes are omitted. Primes are used to denote new variables introduced through the canonical transformations built to obtain the analytical solution.

4. Numerical Results

In this section, the analytical solution is applied in a preliminary analysis of some interplanetary missions considering time-fixed transfers from Earth to Venus ($\rho = 0.7270$), and, from Earth to Mars ($\rho = 1.5236$). In this analysis, the following assumptions are considered:

1. The orbits of the planets are circular;
2. The orbits of the planets lie in the plane of the ecliptic;
3. The flight of the space vehicle takes place in the plane of the ecliptic;
4. Only the heliocentric phase is considered; that is, the attraction of planets on the spacecraft is neglected.

The results provided by the analytical theory are compared to numerical results obtained by solving the two-point boundary value through a neighboring extremals algorithm based on the state transition matrix. This algorithm has been chosen as the exact solution for each maneuver, in view of the accuracy obtained in fulfillment of the terminal constraints. A comparison with a linear theory is also presented. The values of the consumption variable J are presented in Table 1. The time evolution of eccentricity and semi-major axis are presented in Figures 1 to 2 for transfers with time of flight, $t_f - t_0 = 25$ time units. The results are expressed in canonical units.

Table 1. Consumption variable J.

$\rho = r_f / r_0$	$t_f - t_0$	$J_{Analytical}$	$J_{neighboring}$	J_{Linear}
0.7270	25.0	5.9706×10^{-4}	5.9852×10^{-4}	5.7883×10^{-4}
	50.0	2.9864×10^{-4}	2.9901×10^{-4}	2.8941×10^{-4}
	75.0	1.9910×10^{-4}	1.9926×10^{-4}	1.9294×10^{-4}
	100.0	1.4933×10^{-4}	1.4940×10^{-4}	1.4470×10^{-4}
	125.0	1.1947×10^{-4}	1.1949×10^{-4}	1.1576×10^{-4}
1.5236	25.0	7.2912×10^{-4}	7.2468×10^{-4}	6.8706×10^{-4}
	50.0	3.6143×10^{-4}	3.6093×10^{-4}	3.4272×10^{-4}
	75.0	2.4058×10^{-4}	2.4044×10^{-4}	2.2793×10^{-4}
	100.0	1.8034×10^{-4}	1.8028×10^{-4}	1.7067×10^{-4}
	125.0	1.4424×10^{-4}	1.4421×10^{-4}	1.3646×10^{-4}

From the results presented in Table 1 and Figures 1 – 2, major comments are as follows:

The analytical solution gives a very good approximation to the values of the consumption variable J for all transfers. The absolute relative differences vary between 0.00% and 0.65%. The larger differences occur for transfers with lesser time of flight. The time evolution of eccentricity and semi-major axis shows good agreement between analytical and numerical results.

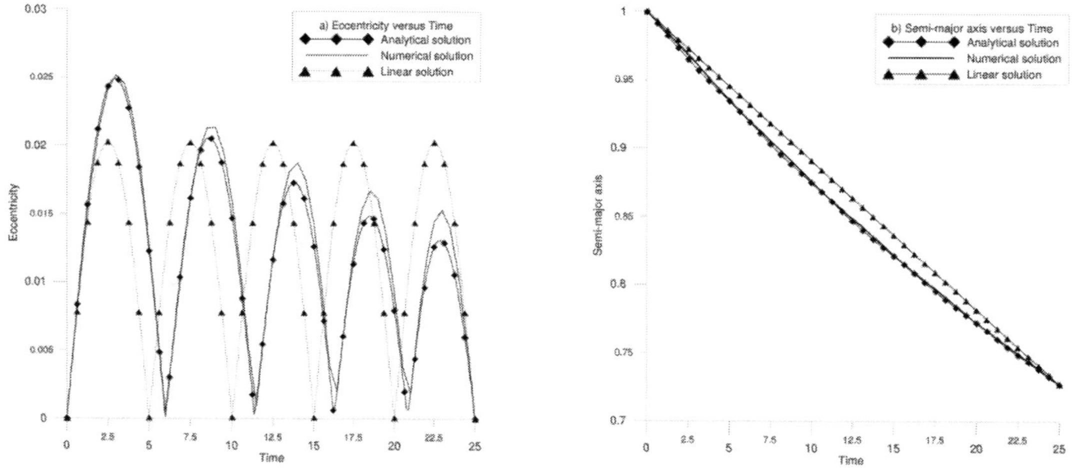

Figure 1. Eccentricity and semi-major axis for $\rho = 0.7270$ and $t_f - t_0 = 25.0$.

Journal of Physics: Conference Series **641** (2015) 012010 doi:10.1088/1742-6596/641/1/012010

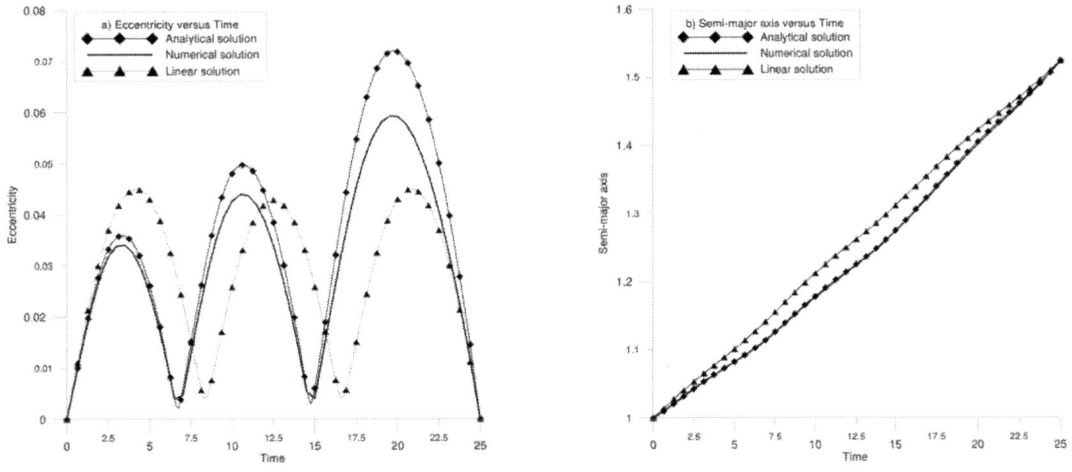

Figure 2. Eccentricity and semi-major axis for $\rho = 1.5236$ and $t_f - t_0 = 25.0$.

5. Conclusion

In this paper, a first order analytical theory for optimal time-fixed low-thrust limited-power transfers is described and applied to a preliminary analysis of some interplanetary missions from Earth to Venus, and, Earth to Mars. Numerical results show good agreement between analytical and numerical results.

6. Acknowledgements

This research has been supported by CNPq under contract 304913/2013-8 and FAPESP under contract 2012/21023-6.

7. References

[1] Bonnard, B, Caillau J B, Dujol, R 2006 Averaging and optimal control of elliptic Keplerian orbits with low propulsion. *Systems Control Lett.* 55, 755 – 760

[2] Camino, O, Alonso, M, Blake, R et al 2005 SMART-1: Europe's lunar mission paving the way for new cost effective ground operations (RCSGSO) in *Proceedings of the 6th International Symposium on Reducing the Costs of Spacecraft Ground Systems and Operations* (RCSGSO), 14-17 June **2005**, ESA/ESOC, Darmstadt, Germany

[3] Da Silva Fernandes, S, Carvalho, F C, Moraes, R V 2015 Optimal low-thrust transfers between coplanar orbits with small eccentricities. *Computational and Applied Mathematics*, DOI: 10.1007/s40314-015-0249-9

[4] Da Silva Fernandes, S, Carvalho, F C 2008 A first-order analytical theory for optimal low-thrust limited-power transfers between arbitrary elliptical coplanar orbits. *Mathematical Problems in Engineering*, vol. **2008**, 30 pp. (Article ID 525930)

[5] Geffroy S, Epenoy, R 1997 Optimal low-thrust transfers with constraints – generalization of averaging techniques. *Acta Astronaut.* 41 (3), 133 – 149

[6] Huang, W 2012 Solving coplanar power-limited orbit transfer problem by primer vector approximation method. International Journal of Aerospace Engineering, vol. **2012**, 9 pp., (Article ID 480320)

[7] Kiforenko, B N 2005 Optimal low-thrust orbital transfers in a central gravity field. *Int. Appl. Mech.* 41 (11), 1211 – 1238

[8] Marec, J. P. *Optimal Space Trajectories*. Elsevier, New York, USA, 1979

[9] Pontryagin, L S, Boltyanskii, V G, Gamkrelidze, R V et al 1962 *The Mathematical Theory of Optimal Control Process*. JohnWiley & Sons, New York, USA

[10] Quarta, A A, Mengali, G 2013 Trajectory approximation for low-performance electric sail with constant thrust angle. *J. Guid. Control Dyn.* 36 (3), 884 – 887

[11] Racca, G D, Marini, A, Stagnaro, L et al 2002 SMART-1 mission description and developments status. *Planet. Space Sci.* 50, 1323 – 1337

[12] Rayman, M D, Varghese, P, Lehman, D et al 2000 Results from the Deep Space 1 technology validation mission. *Acta Astronaut.* 47 (2–9), 475 – 487

Journal of Physics: Conference Series **641** (2015) 012011 doi:10.1088/1742-6596/641/1/012011

Searching for less perturbed elliptical orbits around Europa

J Cardoso dos Santos[1], J P S Carvalho[2], A F B A Prado[3], R Vilhena de Moraes[3,4]

[1]Universidade Estadual Paulista (UNESP), Department of Mathematics, Guaratinguetá, Brazil.
[2]Universidade Federal do Recôncavo Baiano (UFRB), CETENS, Feira de Santana, Brazil.
[3]Instituto Nacional de Pesquisas Espaciais (INPE), DEMEC, São José dos Campos, Brazil.
[4]Universidade Federal de São Paulo (UNIFESP), ICT, São José dos Campos, Brazil.

E-mail: [1]`josuesantosunesp@gmail.com`

Abstract. Space missions intending to visit Europa, one of the famous Galilean's moons of Jupiter, are among the most important topics in space activities today. There is an increasing interest in the scientific community to send spacecrafts to be inserted into Europa's orbit, with goals like mapping its surface and gravitational field. From the quality of the observations until the orbital maneuvers, the required aspects for the success of the mission will depend on the orbits used by the spacecraft. The present work searches for less perturbed elliptical orbits around Europa, because they are very important, since these orbits are expected to be more stable to place the spacecraft. The development of the study is based on the net effects of the perturbing forces over the time, evaluated by the integral of those forces with respect to the time. The value of this integral depends both on the dynamical model and the orbit of the spacecraft. Jupiter's third-body perturbation and the J_2 and J_3 terms of the gravitational potential of Europa are the perturbing forces considered. The results presented here are obtained by performing numerical integrations of the perturbing forces, and they show the locations of the less perturbed orbits.

1. Introduction

Currently, there exist plans to send spacecrafts to study the planetary moons of our Solar System. Europa is among the group of bodies with greater potential to receive scientific missions in the next decade[1]. The possibilities to explore the inner oceans and the search for organic molecules to better understand the complexity of the existence of life in the universe are among the most important objectives. Nowadays, the most advanced project under development for this purpose is the Jupiter Icy Moon Explorer (JUICE). This one and other similar types of missions require near-circular, near-polar orbits [1].

The present paper has the objective of mapping elliptical orbits around Europa, in order to search for the less perturbed ones, which can be used in some types of missions. Less perturbed orbits have smaller variations of Keplerian elements. In the majority of the cases, they demand less station keeping maneuvers for these types of space missions. The search is based on the integral of the perturbing forces over the time, because this quantity can express the net velocity variation that the spacecraft received from the perturbation forces. Several previous works in the literature used this technique to study problems related to orbital maneuvers, fuel consumption, maneuvering time, and so forth [2–10].

[1] NASA's plans to Europa: http://www.jpl.nasa.gov/missions/europa-mission/

Journal of Physics: Conference Series **641** (2015) 012011 doi:10.1088/1742-6596/641/1/012011

For the present study, the third-body perturbation due to Jupiter and the terms J_2 and J_3 of the gravitational potential of Europa are considered. Jupiter is assumed to be in an orbit coplanar with the Europa's equatorial plane. This will be the reference plane. We did not find information in the literature regarding the influence of the other Galilean moons and the Sun in the dynamics of a spacecraft around Europa. Therefore, we performed computations of the direct maximum perturbations from these bodies and the minimum direct perturbation due to Jupiter. The results obtained shows that the perturbation due to the Sun has maximum order of magnitude of $8.8\times10^{-6}\%$ of the perturbation due to Jupiter. Little influences, with the order of magnitude of 0.033%, 0.025%, 0.002% were inferred for Io, Ganymede and Callisto, respectively. The magnitude of these forces were obtained from direct evaluation using the Newton's law of gravitation.

Numerical simulations are performed to map the effects of these perturbations over the spacecraft orbits around Europa. Such mapping is shown as function of the orbital inclination and the semi-major axis. Several characteristics of this system are shown in the present study, in particular orbits with minimum perturbation effects. Low-altitude, near-polar, almost frozen (AF) orbits are very desirable for scientific missions, and they can be found using those minimum values as guidance. This study can be applied to search for orbits of spacecrafts around other celestial bodies. Thus, this paper is structured as follows. In Section 2 we present the theoretical principles used for the numerical simulations. Section 3 is devoted to the analysis of the numerical results, which present near-circular orbits. The theory developed here is used to study the behavior of a spacecraft orbiting the planetary moon Europa, where Jupiter is the disturbing body. We present plots showing the Keplerian elements of the simulated orbits. Final comments are presented in Section 4.

2. Perturbation Integral, force functions and disturbing potentials

The main effect of the forces involved in the system is to change the velocity V of the spacecraft, according to the physical law of impulse:

$$\int_0^T F dt = PI \tag{1}$$

where $PI = \Delta T$ is the variation of kinetic energy per unit of mass.

This integral will be called *Perturbation Integral PI*, due to its physical meaning, and will be measured in km/s. This concept was presented in [7] and this expression will be used to evaluate the changing on the orbital velocity of the spacecraft. The larger is this value, the more are the variations of the Keplerian elements and, consequently, the larger is the fuel consumption required for the station keeping in the majority of the cases. It makes these orbits less interesting for practical applications. The value of this integral is a characteristic of the dynamical model considered and the orbit of the spacecraft.

2.1. Force Function due to the disturbing body

It is possible to expand the potential due to a third-body in Legendre polynomials. The disturbing potential R_2 expanded up to the second order can be written in the form [3, 11]:

$$R_2 = \frac{1}{2} N^2 r^2 (3\cos^2(S) - 1) \tag{2}$$

where r is the norm of the position vector of the spacecraft, N is the mean motion of Jupiter and S is the angle between the position vectors of the spacecraft and Jupiter, both orbiting Europa.

According to [12], the relation between the angle S and the true anomaly (f) of the spacecraft is given by $\cos(S) = \alpha\cos(f) + \beta\cos(f)$, where the coefficients α and β are presented in [13].

Journal of Physics: Conference Series **641** (2015) 012011 doi:10.1088/1742-6596/641/1/012011

2.2. Force function due to the non-sphericity of Europa
Analogously to the case of R_2, it is developed now the force functions for the J_2 and J_3 (R_{J2} and R_{J3}, respectively) terms of the potential due to the non-sphericity of Europa. However, these expressions are too large to be presented here in details.

The potential of a massive body at a distance r and the force per unit of mass are related by [14]:

$$F = \frac{\partial R}{\partial r} \tag{3}$$

where $R = R_2 + R_{J2} + R_{J3}$ is the disturbing potential. All the force functions due to the disturbing potentials (R_2, R_{J2} and R_{J3}) considered in the present paper will be obtained using this last expression. So, Equation (1) is numerically integrated using a computer code developed to analyze the influence of each perturbation (R_2, R_{J2} and R_{J3}) on the orbit of the spacecraft around Europa. All the computer codes developed for this work are developed for the software Maple®.

3. Numerical Simulations and Results

The data used in the simulations are presented in Table 1 that shows the mean Keplerian elements of Jupiter, considered in a planar elliptic orbit around Europa, being these Keplerian elements denoted by the subscript *J*. The Table 1 also presents the numerical data used for evaluating the perturbations due to the non-sphericity of Europa and its mean radius, which is a value of reference for the semi-major axes used in the numerical simulations. These values were hundreds of *km* added to such mean radius.

It was developed a computer code, as mentioned previously, in order to evaluate the effect of these force functions in changing the velocity of the orbital trajectory of a spacecraft around Europa (and consequently its orbit). The effects of each *PI* due to the different disturbing potentials were calculated by this program. The results presented in Figures 1 and 2 show several characteristics that depends on the orbit of the spacecraft, like:

1) The role of each individual term of the perturbing forces (Figure 1);

2) As expected, the effects of the Jupiter perturbation increase as the orbital semi-major axis increase. On the other hand, the resulting effects from the perturbations due to J_2 and J_3 increase as the orbital semi-major axis decreases. (Figure 1);

3) The forces have different signs with respect to each other. Depending on the position of the spacecraft, some forces tend to destroy the Keplerian orbit, while others give assistance to the propulsion system to keep the nominal orbit. In such cases, the fuel consumption is reduced (Figure 1);

4) The existence of orbits with minimum value for this index with respect to the semi-major axis and inclination, for fixed values of inclination and semi-major axis, respectively (Figure 2);

5) The orbits having the semi-major axes ranging from 1760.8 up to 1960.8 km present a flat minimum, i.e., regions with inclinations from 75 to 105 deg present approximately constant values for *PI*. In special, the median value of such range (a = 1860.8 km) presents the lower

Table 1. Mean Keplerian elements of Jupiter around Europa (see: http://ssd.jpl.nasa.gov/) and harmonics J_2 and J_3 of Europa (see: [15] and http://ssd.jpl.nasa.gov/).

e_J	$a_J(km)$	$\omega_J(deg)$	$\Omega_J(deg)$	$M_J(deg)$
0.0094	671100	88.970	219.106	171.016
	J_2	J_3	Europa's Mean radius	
	4.355x10^{-4}	1.3784x10^{-4}	1560.8 *km*	

values for PI, which means that it is the semi-major axis presenting less net perturbations. This result quantifies the qualitative result highlighted in the item (2) (Figure 2);

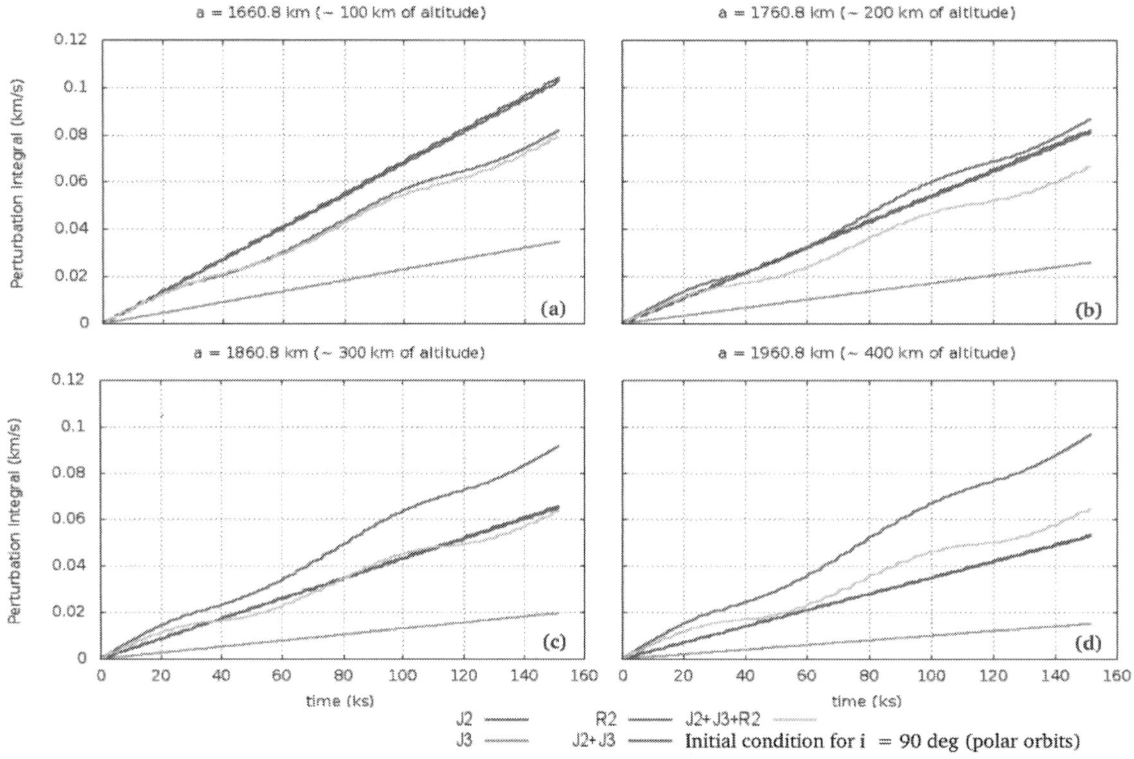

Figure 1. PI as a function of time for $i = 90°$ and different semi-major axes, as denoted in each panel, for a spacecraft orbiting Europa under influence of its J_2 and J_3 terms of the gravity field and the Jupiter's disturbing potential.

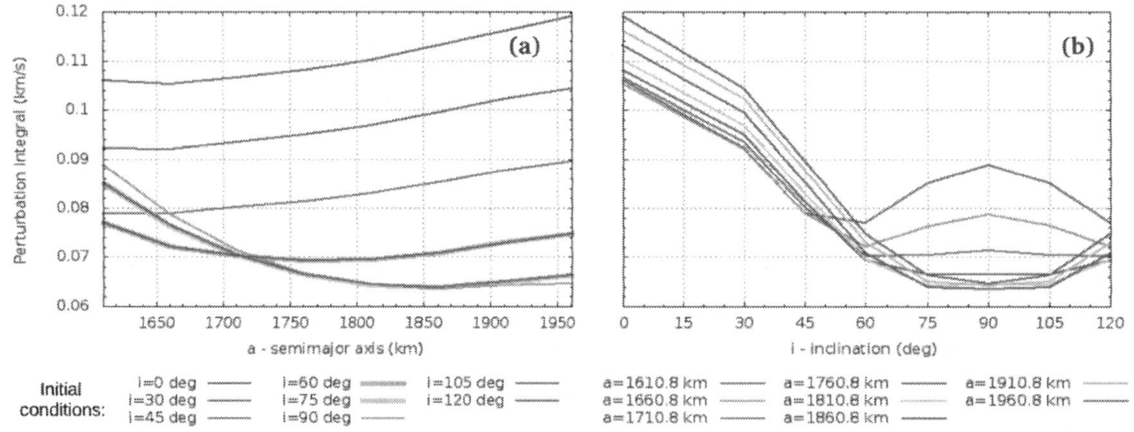

Figure 2. PI as a function of the spacecraft's semi-major axis (a) and orbital inclination (i). The mean radius of Europa in Table 1 is a value of reference for the altitudes considered.

6) In all the results, the equatorial orbits for the spacecraft have *PI* with higher values compared to the other ones, due to the fact that the closer the spacecraft is to the bulge, the higher is the perturbations due to the oblateness. The desired orbits for the present study have high inclinations, which allow a better coverage of the Europa's surface (Figure 2);

7) The existence of symmetry with respect to the orbital inclination (Figure 2.b).

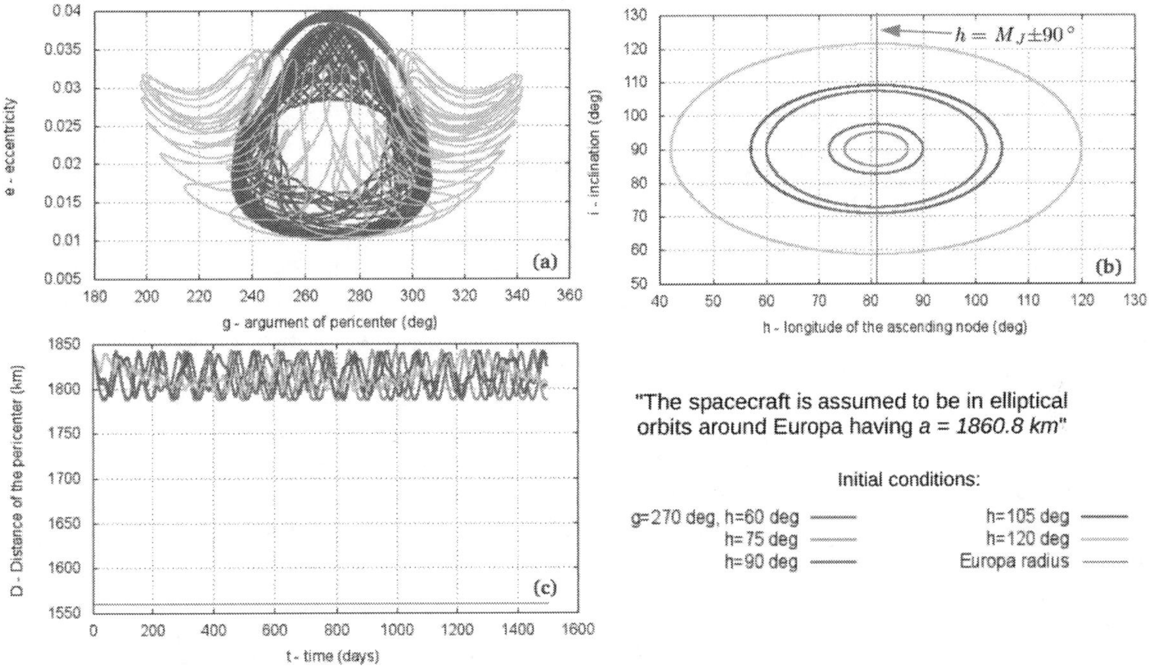

Figure 3. Time evolution of a spacecraft orbiting Europa. Initial conditions: $i = 90°$, $e = 0.01$, $g = 270°$, $a = 1860.8$ km. Integrations are performed for a period of 1500 days.

The *PI* can also be used as a good tool to search initial conditions for the Keplerian elements that will produce less perturbed orbits. Figures 1 and 2 show that lower values of *PI* are obtained with $a = 1860.8$ km and $i = 90°$ (polar orbits). Thus, orbits with different eccentricities are obtained using these values in a dynamical model presented in [16, 17]. This model presents the spacecraft's evolution around Europa subject to perturbation from Jupiter, assumed to be in a circular orbit. It is good approximation once $e_J = 0.0094$. The results are shown in Figure 3 and present near-circular ($e \approx 0$), near-polar (i ranging close to $90°$) AF orbits with altitudes near 300 km ($a \approx 1860.8$ km). These results were obtained by numerical integration of the Lagrange planetary equations where the disturbing potential $R = R_2 + R_{J2} + R_{J3}$ was taken into account. The analysis performed with the force function was used to obtain initial conditions that generated the AF orbits. Figure 3.a shows the eccentricity and the argument of pericenter librating around the equilibrium point for the initial value of $g = 270°$. Note that the orbits have a variation of the eccentricity around $\Delta e = 0.03$, but these orbits do not collide with the Europa's surface, where the distance of the pericenter versus time is visible in Figure 3.c. Frozen orbits are orbits having no secular perturbations in the eccentricity, inclination and argument of pericenter [18]. These orbits are periodic, except for the orbital plane of precession, and are therefore called frozen orbits [19]. Once the present orbits have a significant range in the argument of pericenter g, they cannot be considered frozen orbits, but we consider them as AF orbits ($\Delta e \approx 10^{-2}$ and $\Delta i \leq 10°$ can be smaller for values of $h \approx M_J \pm 90°$).

Average methods were not considered in the present paper ([16] did not used it for the Jupiter's motion), so the results are dependent of the initial conditions of the bodies involved. Simulations not showed here due to the limitation of space confirm that the less perturbed orbits depend on the initial position of Jupiter, so the results shown in here are valid for $M_J = 171.016\,^\circ$ (see Figure 3.b). New calculations needs to be done if different values are used. New numerical simulations confirmed that the less perturbed orbits follows the rule $h = M_J \pm 90\,^\circ$. It comes from the geometry of the system. This rule gives the orbit that has a maximum averaged distance Jupiter-spacecraft, so the gravitational perturbation from Jupiter is minimized.

4. Conclusions

In the present paper, the method of *Perturbation Integral* is used to evaluate the effects of some perturbations over the spacecraft's orbit and to obtain orbits that are less perturbed. A study is performed to know the effects of the main terms of the potential of Europa (J_2 and J_3) and the third-body perturbation due to Jupiter. The disturbing body is considered to be in an elliptical-planar orbit. Low-altitude, near-polar, AF orbits which are very desirable for scientific missions are found using those minimum values as guidance. This study can be applied to study orbits for spacecrafts around other celestial bodies where the existence of higher eccentricities for the disturbing body will require a more accurate model for the spacecraft's orbital evolution [20]. Theoretical interests of this type of research are: studying the effect of the non-sphericity of Europa and other celestial bodies, third-body perturbations, force functions, frozen orbits, lifetime. Practical interests are: applications to science missions to planetary moons and other small bodies of the Solar System, like dwarf planets and asteroids. The present work is an extension of the work developed in [3], which is based in the theory presented by [7].

5. Acknowledgments

The authors are wishing to acknowledge support from UNESP, UFRB, INPE, UNIFESP, CNPq (contract 306953/2014-5) and FAPESP (processes 2013/26652-4, 2011/05671-5 and 2012/21023-6). We also thank the relevant contributions in the work of revision of the present paper. Special thanks to God for all the opportunities given during the development of this study.

References

[1] ESA **2014** *JUICE definition study report* **1** *sci.esa.int/juice/54994-juice-definition-study-report/*.
[2] Carvalho J P S *et al.* **2013** *PSBSACM* **1** 1 DOI 10.5540/03.2013.001.01.0009.
[3] Carvalho J P S, Vilhena de Moraes R and Prado A F B A **2014** *MPE* DOI: 10.1155/2014/529716.
[4] Gonçalves L D *et al.* **2013** *PSBSACM* **1** 1 DOI: 10.5540/03.2013.001.01.0095.
[5] Oliveira T C, Prado A F B A and Rocco E M **2014** *Math. Mod. in Engin. and Comp. Sci.* **1** pp. 90-95.
[6] Oliveira T C and Prado A F B A **2014** *Acta Astronautica* **104** 1 DOI:10.1016/j.actaastro.2014.06.035.
[7] Prado A F B A **2013** *MPE* DOI: 10.1155/2013/415015.
[8] Prado A F B A **2013** *PSBSACM* **1** 1 DOI: 10.5540/03.2013.001.01.0006.
[9] Prado A F B A **2014** *ASR* **53** pp. 877-89. DOI: 10.1016/j.asr.2013.12.034.
[10] Sanchez D M, Prado A F B A and Yokoyama T **2014** *ASR* **54** pp. 1008-18. DOI: 10.1016/j.asr.2014.06.003.
[11] Carvalho J P S *et al.* **2012** *ASR* **49** 5 pp. 994-1006. DOI: 10.1016/j.asr.2011.11.036.
[12] Broucke R A **2003** *Journal of Guidance, Control and Dynamics* **26** 1 pp. 27-32. DOI: 10.2514/2.5041.
[13] Domingos R C, Vilhena de Moraes R and Prado A F B A **2008** *MPE* DOI: 10.1155/2008/763654.
[14] McCuskey S W **1963** *Introd. to Celestial Mechanics*(Addison-Wesley Publ. Comp., Reading-MA, USA).
[15] Lara M and Russell R **2006** *2006 AAS/AIAA SMM* pp. 22-26. http://hdl.handle.net/2014/38715.
[16] Carvalho J P S, Vilhena de Moraes R and Prado A F B A **2013** *MPE* DOI: 10.1155/2013/182079.
[17] Cardoso dos Santos J, Carvalho J P S and Vilhena de Moraes **2013** *PSBSACM* **1** 1 DOI: 10.5540/03.2013.001.01.0189.
[18] Brouwer D **1959** *Astronomical Journal* **64** 1274 pp. 378-397. DOI: 10.1086/107958.
[19] Ceccaroni M, Biscani F and Biggs J **2014** *Solar System Research* **48** DOI: 10.1134/S0038094614010031.
[20] Cardoso dos Santos J *et al.* **2015** *Advances in Space Research* **submitted**.

Study of the influence of computational parameters on the formation of a giant gaseous planet

R A Moraes[1] **and E Vieira Neto**[1]

[1]Universidade Estadual Paulista - UNESP - Campus de Guaratinguetá, SP, Brasil

E-mail: ricardo.moraes07@gmail.com

Abstract. Today, computational simulatios are the best method to describe the formation of our Solar System, however this kind of simulations contain numerical errors caused by boundary conditions and choises of the grid, for instance. In this paper we will present results from several hidrodynamical simulations of the formation of a planet like Jupiter able to migrate using different computational parameters. We will compare the density profile, the semimajor axis and the eccentricity of the planet for those different parameters and measure the effect of them. We will not extend our results for planets with initial eccentric orbits in this paper.

1. Introduction

One of the first hydrodynamical aproach for the formation of planets was given by [1], where the authors proposed a exponential model for the modeling of a gas nebula around the planet. Since there, with the advent of powerful computers it was developed a great number of computational facilities, which allow us to simulate with more accurate the formation of planets. Some examples are FARGO [5], ZEUS [8] and NIRVANA [10]. In [2] the authors compare the performance of several numerical integrators, including the ones cited here, in a simulation with two planets with mass comparable with Jupter and Neptune on circular orbits, the aim of the authors were invetigate the differences among the hydrodynamical codes and the reliability of the astrophysical hydrodynamic codes available at that time in the disk-planet problem, the authors was not interested in see the changes which appears when some numerical parameters are changed.

We will not perform simulations with many codes, but only with FARGO code, which the accurate is described in [5]. The pourpose of this paper is evaluate the effects of different parameters in a planet formation scenario, we will change the grid, the initial conditions, the accretion rate and the initial surface density (Σ_0).

This paper is organized as follows, in section 2 we describe the model set-up with all the parameters that was used, in the section 3 we presents the results of the simulations for the semimajor axis, eccentricity and density using different parameters and in the section 4 we draw our final conclusions about this work.

2. Numerical Model

In order to perform our hydrodymical simulations, we used the publically available FARGO (Fast Advection in Rotating Gaseous Objects). FARGO is by default a bidimentional hydrodynamical polar grid code centered on the star, based on the van Leer upwind algorithm on a staggered

Journal of Physics: Conference Series **641** (2015) 012012 doi:10.1088/1742-6596/641/1/012012

mesh. It is capable to solves the Navier-Stokes and continuity equations for a Keplerian disk subject to the gravity of the central object and that of embedded protoplanets as well [9].

To simulate the gaseous nebula around planet we work with cilindrical coordinates (r, θ, z), with simetry on the z component with the vertically integrated equations. We adopted the viscous forces per unit of area of Navier-Stokes to model the viscous force. The origin of the coordenate system was placed on the central star, while the planet is at 1 dimensioless unity far from the star. The equations of motion and the gravitational potential are given by [4] or [7] and will not be repeated here.

We simulated a bidimensional protoplanetary disk. The tidal and viscous torque was considered, but the self-gravity was neglected. We work with an acretive disk, which has an inner radius of 0.4 and an outer radius of 2.5 with thickness radius given by the ratio $H/r = 0.05$. In addiction, all simulations was performed considering a planet with initial circular orbit with the aim to keep the problem as simple as possible.

2.1. Initial and Inner Boundary Conditions

In all simulations performed here we considered a corotating frame and the planet has the characteristics of Jupiter, its mass is $M_p = 10^{-3}\ M_*$, where M_* is the mass of the star and it is in a circular orbit around the star.

In this paper we use three different inner boundary conditions such as non-refleting, open and rigid. The outer boundary condition is open for all simulations, thus hereafter we will refer to the inner boundary connditions as boundary conditions. According to [5] the rigid boundary conditions, also know as wall boundary condition, is a low realistic condition and is often used to test codes, the use of this condition implies to set to zero the radial velocity at the boundary, besides the flow can not leave the mesh, which means that the gaseous mass will be conserved. On the other hand, we have the open boundary conditions, which allows the gas to move outside the mesh, in this case the total gaseous mass decrease with the time. For last, the non-reflecting boundary conditions are applied in order to eliminate the reflected waves on the planet, transporting these waves to a "ghost cell", improving the results, these boundary conditions does not restrict the disk to conserve its mass or angular momentum.

We will also analyze the density of the disk, for this we use different values for the initial density and the accretion rate. In hydrodynamic simulations of a planet we a mass of Jupiter it is usual to consider the initial density as 6.4×10^{-4} because this initial value means that theres the equivalent of two Jupiter masses in the disk interior to the orbital radius of the protoplanet [6], we will change this parameter ten times in order to see what kind of changes this increase in the initial density can produce. The same argumentation can be invoked to the accretion rate, for standart we will use 10^{-6} and in one simulation this value will be increased to 10^{-3}, which means that the planet will grow fastly, in theory the density of the gas might be lower in this simulation.

2.2. Grid

We work with a polar grid composed of radial and azimuthal (sectors) zones. The radial and azimuthal zones are both equally spaced, in particular, to use the FARGO algorithm we need to have equally spaced zones in the azimuth. The total number of cells in the mesh can be found using $N_t = N_{rad} \times N_{sec}$, where N_{rad} and N_{sec} are the number of radial and azimuthal zones, respectively. The number of cells are directly proportional to the accuracy of the results, which means that more cells in the mesh will given us better results.

In our simulations we used three meshes with different number of cells, 150×450, 300×300 and 300×900.

3. Simulations and Results

In this section we will describe the parameters of the simulations that was performed in this paper and our results.

In table 1 we show all the simulations and the most important parameters to our analysis. In addition, we are working on a corotating frame, with a planet in a circular orbit around the star, the planet was placed in one dimensionless unit and feel all disk effects, being able to migrate. The mass of the planet is the same for all simulations, even the ones which the accretion rate is different and all the simulations were performed for 300 Jupiter's orbits.

Table 1. List of simulations. Columns indicate: run abreviation (Name); mass of the planet; initial surface density; mass acretion rate; grid; inner boundary condition.

Name	$M_p\,(10^{-3}M_*)$	$\Sigma_0\,(10^{-4})$	$\dot{M}_p\,(10^{-6}M_*)$	Grid($N_{rad} \times N_{sec}$)	Boundary Condition
out.i.non	1	6.34	1	150×450	Non-reflecting
out.vii.non	1	6.34	1	300×300	Non-reflecting
out.iv.non	1	63.40	1	300×900	Non-reflecting
out.v.non	1	6.34	1000	300×900	Non-reflecting
out.vi.non	1	6.34	1	300×900	Non-reflecting
out.i.open	1	6.34	1	150×450	Open
out.i.rigid	1	6.34	1	150×450	Rigid

In the next topics we will analyze the influence of the boundary conditions, the grid, the accretion and the initial surface density analyzing the density profile, the semimajor axis and the eccentricty of the planet.

3.1. Boundary Conditions

In Figure 1 we show the density of the disk for the simulation with non-reflecting (out.i.non), open (out.i.open) and rigid (out.i.rigid) boundary conditions, respectively. As the profiles are similar, for a better interpretation of the results we can see in panel A of Figure 2 the comparison of the density of the disk in the planet for the same boundary conditions.

The Figure 2 shows the comparison of the disk density profile (panel A), semimajor axis (panel B) and eccentricty (panel C) for different inner boundary conditions.

In the panel A of the Figure 2 the density profile are shown, we can see that the distribution of gas is similiar for the non-reflecting and open conditions. However, when it was applied the rigid conditions we found a denser region inside and close to the planet, outside the planet the difference among the density profiles is small, which means that the densities of the gas converges to the same value far from the planet, independent of the boundary condition. Analyzing the panel B we can observe the behavior of the semimajor axis of the planet, as expected by the theory is possible identify an inner migration of the planet, once again for the non-reflecting and open boundary conditions the difference for the results are hardly noticed and for the rigid condition the planet migrates faster after 100 Jupiter's orbits, at this time a gap is opened by the planet. The panel C shows the evolution of the eccentricity of the planet, which starts circular. It is possible to see a small difference among the results, but once the behavior is the same for the three cases, the diferences are irrelevant.

Journal of Physics: Conference Series **641** (2015) 012012 doi:10.1088/1742-6596/641/1/012012

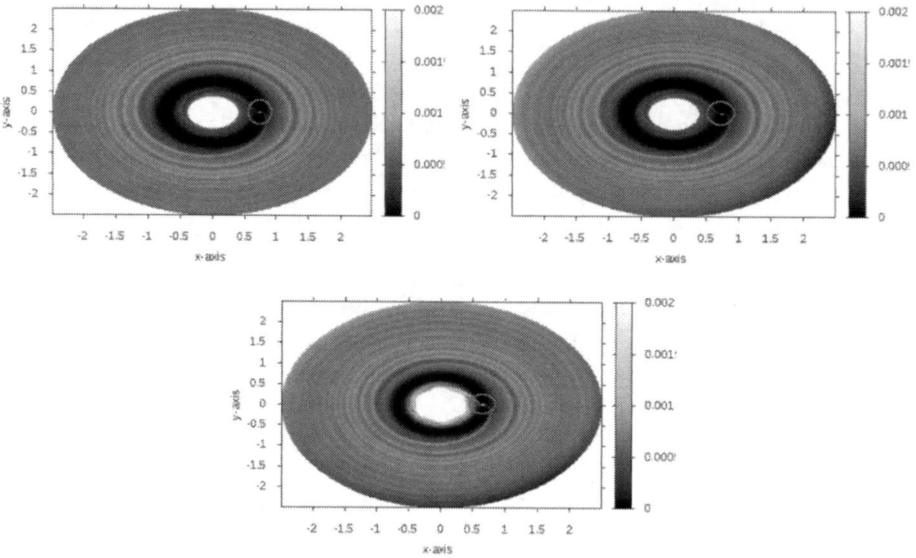

Figure 1. From the left to the right, density of disk for a simulation with non-reflecting, open and rigid boundary conditions, after 300 Jupiter's orbits. The red circle marks the location of the planet.

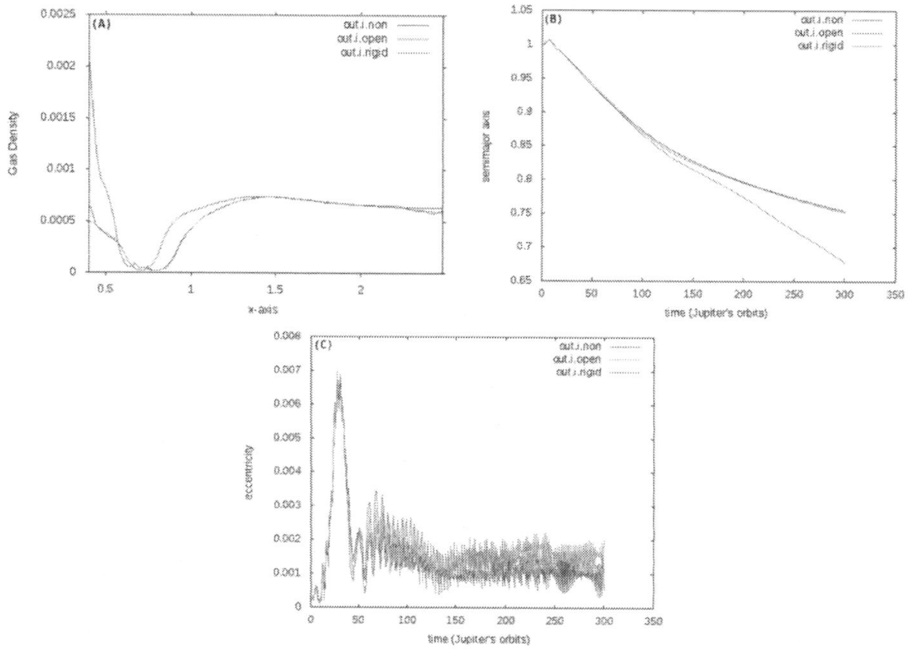

Figure 2. Comparison of the disk density profile (panel A), semimajor axis (panel B) and eccentricty (panel C) for non-reflecting (red line), open (green line) and rigid (blue line) boundary conditions after 300 Jupiter's orbits.

3.2. Grid

Figure 3 shows the comparison of the disk density profile (panel A), semimajor axis (panel B) and eccentricty (panel C) for different meshes. We used a low resolution grid, 150 × 450, with 67500 cells, an intermediate grid, 300 × 300, where the number of radial and azimuthal zones are equal, in this case the mesh have 90000 cells and a grid with higher resolution, 300 × 900, where the mesh have 270000 cells.

In practical terms we can say that a grid with more cells will give better results, because the number of cells is connected with the resolution of the results. On the other hand, more cells will produce more data to be processed, so the computational time will increase. All simulations discussed in the present topic was performed with non-reflecting boundary conditions due to its effcience to reproduce the formation of a planet [3].

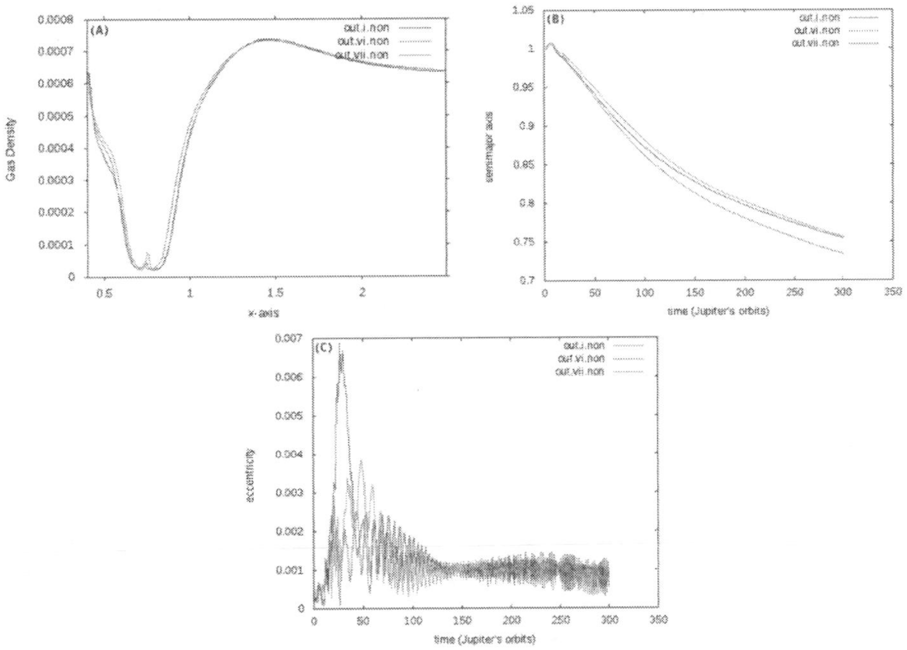

Figure 3. Comparison of the disk density profile (panel A), semimajor axis (panel B) and eccentricty (panel C) for three different grids: 150 × 450 (red line), 300 × 900 (green line) and 300 × 300 (blue line) after 300 Jupiter's orbits.

The simulation of the density profile (see panel A) shows that the profiles are slightly different, however, the grid with higher resolution (green line) the gas density around the planet has a peak compared with the other simulations, which means that the change grid has a strong influence over the results near the planet. The results for the semimajor axis are showed in panel B, the greatest variation between the simulations with the higher resolution (green line) and lower resolution (red line) appears around 100 Jupiter's orbits when the planet cleans its orbit opening a gap. This event starts on the vicinity of the planet, this may explain the difference of the results. The results of the simulation with a grid of 300 × 300 cells (blue line) may have been affected by the fact of the number of cells are equal, so the use of a grid with the same number of radial and azimuthal zones may not be a good choice. The panel C shows the comparison of the eccentricity among the three grids. The range of values reached by the simulations are not realy different, except for the simulation with the lower resolution which reaches a peak near 50

Journal of Physics: Conference Series **641** (2015) 012012 doi:10.1088/1742-6596/641/1/012012

Jupiter's orbit with $e \approx 0.007$ while the simulation with higher resolution hits its top value at $e \approx 0.004$.

3.3. Accretion

The comparisons of the disk density profile (panel A), semimajor axis (panel B) and eccentricty (panel C) for two different accretion rates are shown in Figure 4. The goal is to see if the rate of amount of gas that the planet aggregates during its formation will interfere in the mentioned parameters.

Theoretically, the accretion rate will affect the time of formation of the planet, which will accumulates mass faster if the accretion rate is higher. However the discussion about how a different rate of gas being accreted will change the semimajor axis or the eccentricity of the planet was not discussed yet.

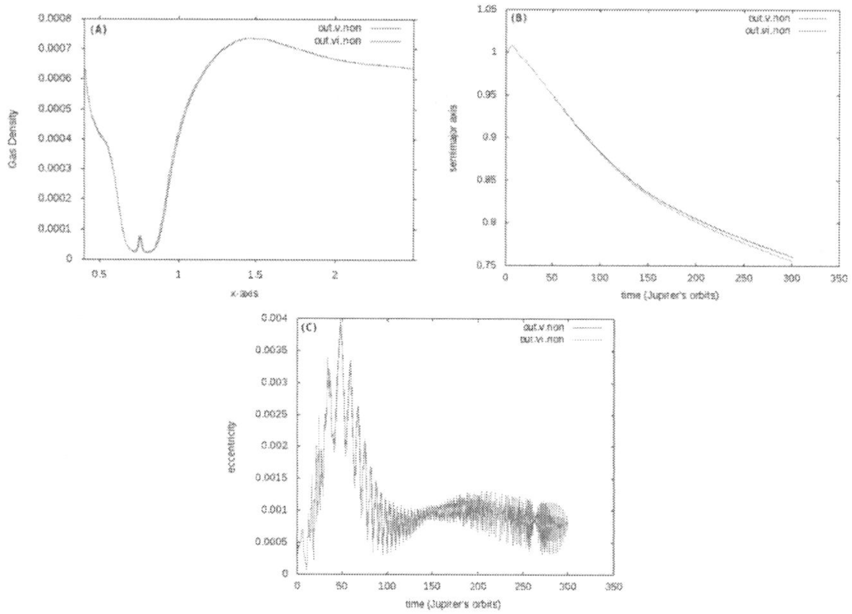

Figure 4. Comparison of the disk density profile (panel A), semimajor axis (panel B) and eccentricty (panel C) for two different accretion rates: $\dot{M}_p = 10^{-3}$ (red line) and $\dot{M}_p = 10^{-6}$ (green line) after 300 Jupiter's orbits.

Figure 4 shows that for all parameters studied here different accretion rates barely interfere in its behavior. In fact, the only variation that can be noticed appear in panel B, where it is possible to see that for a smaller acretion rate the planet migrates a slightly faster than the case with higher acretion rate.

3.4. Initial Surface Density

In this topic we analyze the disk density profile (panel A), semimajor axis (panel B) and eccentricty (panel C) for two initial surface density, the results are presented by Figure 5. The differents densities should alter all the disk structure and consequently the parameters of the planet.

Panel A of Figure 5 shows a result that was already expected, the density profile of the disk with greater initial density is denser than the profile of the disk with lower initial density. The

Journal of Physics: Conference Series **641** (2015) 012012 doi:10.1088/1742-6596/641/1/012012

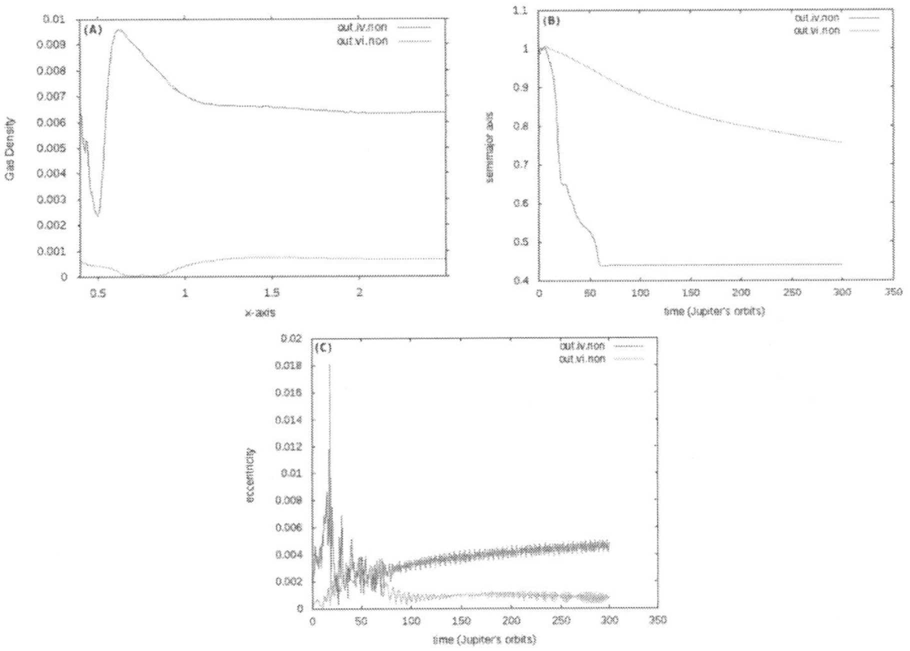

Figure 5. Comparison of the disk density profile (panel A), semimajor axis (panel B) and eccentricty (panel C) for two different initial surface density: $\Sigma_0 = 6.37 \times 10^{-3}$ (red line) and $\Sigma_0 = 3.67 \times 10^{-4}$ (green line) after 300 Jupiter's orbits.

main results are found for the semimajor axis and the eccentricity, panels B and C, respectively. For the denser disk, the semimajor axis decreases dramatically reaching the internal border of the disk immediately after 50 Jupiter's orbits. On the other hand, the eccentricity of the planet immersed in the denser disk increases moving away from values reached by the planet in the less dense disk, which means that the extra density brings disturbed effects for the disk.

4. Conclusion

For the three parameters evaluated the non-reflecting and the open boundary conditions does not change significantly the results, while the rigid boundary conditions have a denser density profile and the semimajor axis decreases faster, this happens mostly because on the rigid conditions scenario the gaseous mass is conserved.

The results for the grid show, as expected, that some effects close to the planet can be identified only with higher resolution, which leads us to say that the mesh with more cells is advised always that the aim of the simulation is study a small detailed region, also the option for the grid with the same number of radial and azimuthal zones presented poor results for the semimajor axis evolution.

For the parameters evaluated the change of the mass accretion rate does not shows significantly effect, so is possible to conclude that the accretion rate is not important for the density of the disk or for the evolution of the semimajor axis and eccentricty of the planet.

The simulations with two different initial densities was the one with more significantly contrast among the simulations performed in this paper, the result for the density of the disk was already expected, however, from the analysis of the semimajor axis and eccentricity we can conclude that the extra amount of gas in the disk drag the planet inside faster and bring disturbed effects to the disk making the eccentricity of the planet grows.

Acknowledgments
The authors are thankful to the FAPESP, process number 2013/24281-9, for the financial support.

References

[1] Adachi I, Hayashi C and Nakazawa K 1976 *Progress of Theoretical Physics* **56** 1756
[2] de Val-Borro M *et al* 2006 *MNRAS* **370** 529
[3] Kley W 1999 *MNRAS* **303** 696
[4] Lubow S H, Seibert M and Artymowicz P 1999 *ApJ* **526** 1001
[5] Masset F 2000 *Astronomy and Astrophysics* **141** 165
[6] Nelson R P, Papaloizou J C B, Masset F, Kley W, 2000, *MNRAS* **318** 18
[7] Moraes R A and Vieira Neto 2013 *Journal of Physics: Conference Series* **465**
[8] Stone J, Norman M L 1992 *ApjS* **80** 753
[9] Zhang X, Liu B, Lin D N C and Li H 2014 *ApJ* **797** 20
[10] Ziegler U, Yorke H W 1997 *Comput. Phys. Commun.* **101** 54

Journal of Physics: Conference Series **641** (2015) 012013 doi:10.1088/1742-6596/641/1/012013

Experiment on signal filter combinations for the analysis of information from inertial measurement units in AOCS

Maurício N Pontuschka[1], Ijar M da Fonseca[2] and Marco A A Melo[3]

[1]Computer Science Department, PUC-SP, Brazil
[2]Aeronautical Mechanics Division, ITA, Brazil Prof. Emeritus at EMC/INPE, Brazil
[3]Engineering Department, PUC-SP, Brazil Researcher LPS, USP-SP, Brazil

E-mail: tuska@pucsp.br

Abstract. The FDIR software subsystem may be part of the attitude and orbit control subsystem, AOCS. The AOCS quite often includes inertial navigation sensors being physically implemented by accelerometers and gyroscopes which provide electrical signals to the AOCS on-board software (OBSW) which, in its turn, generates the commands to the control actuators. In general, hardware like sensors and actuators present nonlinearities which sometimes make it difficult to properly interpret the output signals. In the scenario of space applications, filters are used to eliminate noise and to increase the reliability for the correct interpretation of those signals. In this paper we present a collection of filters used in inertial navigation subsystems enabling the fusion of data from sensors. Fundamentally, the filters are composed of the Kalman filter in its derivations. The filters can be used for state estimation of a system as well as for noise filtering. In this work the filters are configured with respect to their different orders of execution, their sampling rate, and their cutting-off frequency. The filter configurations can be changed by software so as to allow a flexible structure that can be adjusted for the best quality of output signal and consequently the best analysis of the satellite behaviour. The main purpose of this paper is to test the algorithm that combines several signal filters considered in this study. To accomplish this goal we developed an experiment encompassing an accelerometer and a wireless communication system so as to provide input signals to be filtered by the filtering algorithm.

1. Introduction

The scope of this paper is to develop an experiment to test and validate filter algorithms from IMU signals. The main objectives of the experiment are: (1) To test a collection of filters required for inertial navigation systems in an ordered configuration according to their executions; (2) To determine the individual parameter of each filter aiming at real time results; (3) To implement the Kalman filter, low pass filter, high-pass filter, moving average filter, and median filter for signal filtering; and (4) To generate real time graphical results associated with the output obtained from the experiment.

The AOCS and its associated FDIR subsystem are used in navigation control of satellites, rockets and others space vehicles [1] [14]. Those subsystems include inertial navigation sensors such as accelerometers and gyroscopes which provide electrical signals to the AOCS OBSW [9] [10]. Such sensors can be strategically combined to work together to obtain a more precise navigation [7]. In general, sensors present nonlinearities which sometimes make it difficult to properly interpret the signals. Thus, filters are used to eliminate noise and to increase the reliability of the interpretation of satellite sensor signals.

Three families of sensors are highlighted in the FDIR implementation: accelerometers, gyroscopes, and magnetometers. Accelerometers detect proper acceleration acting on the system. Gyroscopes detect the system angular velocity. By using gyroscope sensors it is possible to continuously calculate the orientation of the sensor while it is under acceleration. Magnetometers are not inertial sensors themselves, but they are usually part of a set of devices that are part of the AOCS, so they are under the surveillance of the FDIR software subsystem. Magnetometers give the orientation of the system with respect to the magnetic north. Such sensors provide the direction and the magnitude of the magnetic field vector for specific positions in orbit [8]. The signal analysis initiates with the signal reading, and then the readings are digitalized to be processed by the filters. Then the filtered signal can be processed and subjected to analysis.

The experiment, as shown on Figures 1 and 3, is comprised of equipment containing an accelerometer mounted on a rotating base and a wireless communication system. The experiment is assembled aiming at sensor signal filtering. The accelerometer generates the signal and provides it to the microprocessor. The processed signal is then sent to a computer by Bluetooth technology.

2. Experiment hardware

The processing unit is part of the experiment's communication system and it is comprised of one Arduino Uno microcontroller. The Arduino Uno is a processing device with input and output data that can be used in laboratory environments for embedded systems. It does not contain large processing capacity, but it is widely used in the academic environment.

The ADXL345 accelerometer provides the information from the three components of the force vector [4]. The sensor provides the digital information that represents the total force acting upon the experiment in the laboratory. The ADXL345 accelerometer is built in such a way that 255 represents 1 G.

The device used for communication between the computer and the experiment is a HC06 Bluetooth port module for Arduino[3]. It allows establishing a serial connection between the Arduino microcontroller and the computer over the Bluetooth protocol[1].

Figure 1. Equipment hardware.

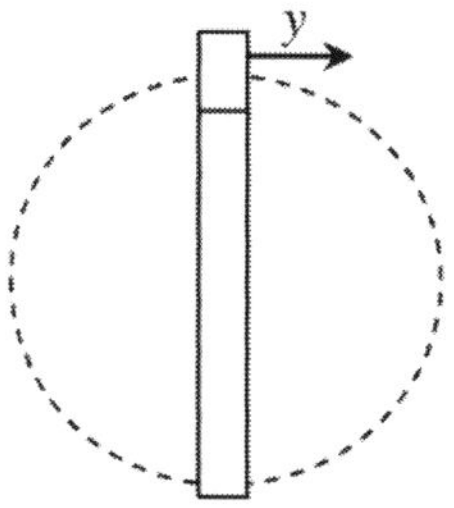

Figure 2. Accelerometer y-axis direction

Figure 3. Experiment assembly.

The equipment was designed with an Arduino processor equipped with an ADXL345 accelerometer [4] and a Bluetooth HC06 [13] assembled on a rotating base. The equipment is restricted to the planar motion as shown in Figure 2. The reading is done only in the y-axis.

3. Filter characteristics

This study addresses the following filters: (1) Kalman filter; (2) high pass filter; (3) low pass filter; (4) moving average filter; and (5) median filter.

[1] This module can operate in several rates, from 4.800bps to 1.382.400bps and the default value is 9.600bps. In this experiment the default rate of 9.600bps was used.

The Kalman filter was developed by Rudolf E. Kálmán [5]. The Kalman filter (KF) is the optimal Bayesian estimator that provides the global minimum mean square error estimate in a Gaussian and linear system. On the other hand, if the KF is used in a non-linear, non-Gaussian system, it also provides the Linear Minimum Mean Square Estimate (LMMSE), which is the best estimate that can be achieved from a linear filter.

According to Haykin [6], Kalman filter application for state-space formulation of linear dynamical systems provides a recursive solution to the linear optimal filtering problem. It applies to stationary as well as nonstationary environments. The recursive solution at each updated estimate of the state is computed from the previous estimate and the new input data. So only the previous estimate requires storage. Furthermore, in removing the need for storing the entire past observed data, the Kalman filter is computationally more efficient than computing the estimate directly from the entire past observed data at each step of the filtering process.

The low pass filter cuts the high frequency allowing only lower frequencies to pass. As shown in Figure 4, the main parameter of this filter is the v_I parameter. It consists of a number between 0 and 1 stating how much signal is allowed to pass as output signal [2].

$$H_d(v) = \sum_{n=-\infty}^{\infty} C_n e^{jn\pi v}$$

low pass : $C_0 = v_I$

$$C_n = \int_0^1 H_d(v).\cos(n\pi v)\,dv = \frac{\sin(n\pi v_1)}{n\pi}$$

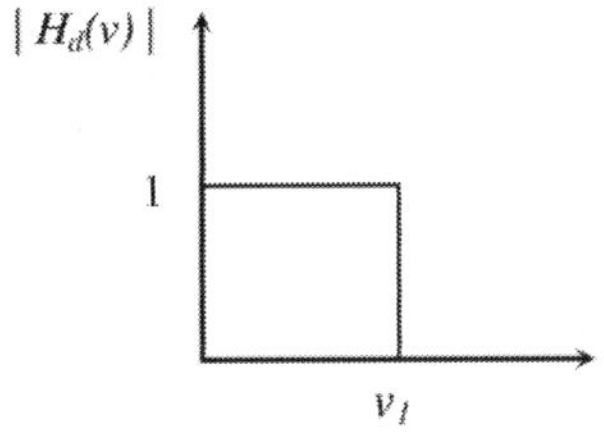

$H_d(v)$: is the response frequency of the magnitude of the digital filter or gain of the digital filter.
v : is the variable digital frequency, defined as the ratio between the analog frequency and the Nyquist rate.
v_I : is the digital cutting frequency.
C_n : is the Fourier coefficients of digital filter.

Figure 4. Low pass filter equation [2].

The high pass filters the low frequencies allowing only higher frequencies to pass. The main parameter of this filter is also the v_I parameter, which consists of a number between 0 and 1 as a signal informing whether to let pass as an output signal [2].

$$H_d(v) = \sum_{n=-\infty}^{\infty} C_n e^{jn\pi v}$$

high pass : $C_0 = 1 - v_I$

$$C_n = \int_0^1 H_d(v).\cos(n\pi v)\,dv = \frac{-\sin(n\pi v_1)}{n\pi}$$

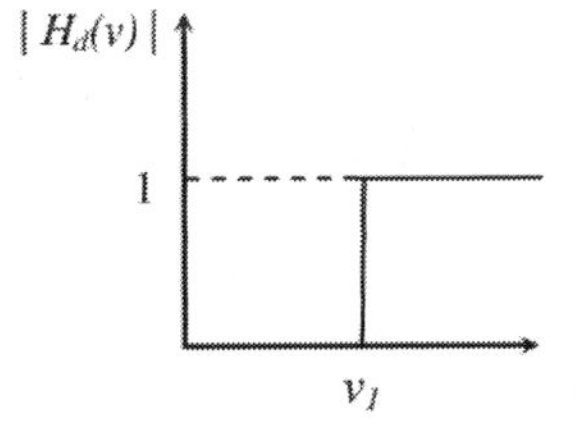

$H_d(v)$: is the response frequency of magnitude of the digital filter or gain of the digital filter.
v : is the variable digital frequency, defined as the ratio between the analog frequency and the Nyquist rate.
v_I : is the digital cutting frequency.
C_n : is the Fourier coefficients of digital filter.

Figure 5. High pass filter equation [2].

The geometric moving average algorithm can be categorized as low-pass filter. It produces a smoothing of curves representing the signal being processed. It is based on the use of an average signal range [11].

$$Yav(n) = Average[X(n-K), \dots, X(n), \dots, X(n+K)]$$

$Yav(n)$: is the temporal response of the moving average digital filter.
$X(n)$: is the temporal input of the moving average digital filter.
$X(n \pm K)$: is the temporal input of the moving average digital filter in the $\pm K$ instant.

Figure 6. Average filter equation.

The median filter is classified as low-pass filter, but unlike the moving average filter, it discards major changes in the signal. The algorithm sorts a signal range and then uses the median value. For this reason the range size must always be odd [11].

$$Ymed(n) = Median[X(n-K), \ldots, X(n), \ldots, X(n+K)]$$

$Ymed(n)$: is the temporal response of the median average digital filter.
$X(n)$: is the temporal input of the moving average digital filter.
$X(n \pm K)$: is the temporal input of the moving average digital filter in the $\pm K$ instant.

Figure 7. Median filter equation.

The scheduling process is based on a pipeline approach in which the loop of each algorithm is shared by all filter implementations. The scheduler controller is responsible for applying the filters using the correct order and parameters. This order can be configured without the need to recompile the software code.

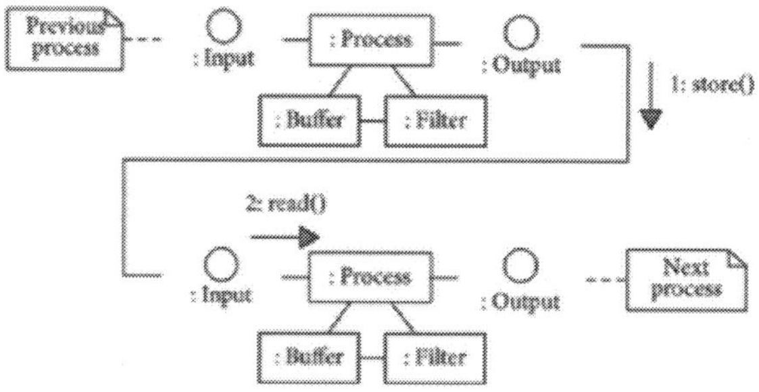

Figure 8. Basic scheme of the schedule approach.

The scheme illustrated by Figure 8 shows that each process has its own buffer and filter. It also shows that the output of a filter is connected to the next filter input. Each buffer is independent for each process; the data stored in a buffer has to be used only by the associated process [12].

In this experiment two software were developed: (1) Embedded software for the operation of the accelerometer and to transmit the signal to the computer; (2) Analysis software for the capture of transmission made by the experiment. It is also responsible for filtering the captured signal and providing a graphical visualization in real time. The embedded software was developed in C language with libraries for the Arduino environment. In general, the embedded software performs the initialization of the serial baud rate and establishes the connection. Then, the software loops which perform the reading of the "y" accelerometer axis and send the value of the reading to the serial port connected to the Bluetooth device. The analysis software was developed in Java language by using the Eclipse platform. The captured signal is analyzed on the computer where it is possible to perform the filter application. The system allows choosing which filter will be used and also the filter application order. A filter can be applied more than once if necessary. The pipeline approach was used while applying filters, allowing multiple filters to be applied in the same real-time analysis.

4. Results

In this work the software that was implemented, in addition to carrying out the filtering process, also shows the original signal (optional) and the processed signal graphically. The analysis can be performed in real time by viewing a representation of the forces that are sensed by the accelerometer.

The analysis software can show the signal that is processed by the combination of filters and, at the same time, compare that signal with the raw signal. Figures 9 to 14 present several tests of real signal readings from the device that was assembled.

Journal of Physics: Conference Series **641** (2015) 012013 doi:10.1088/1742-6596/641/1/012013

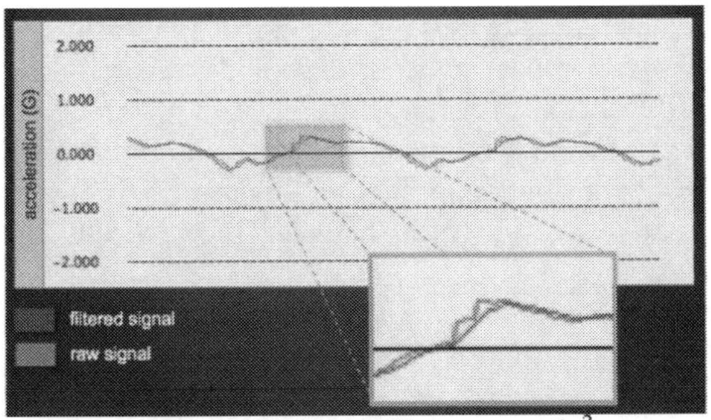

Figure 9. Kalman filter example[2].

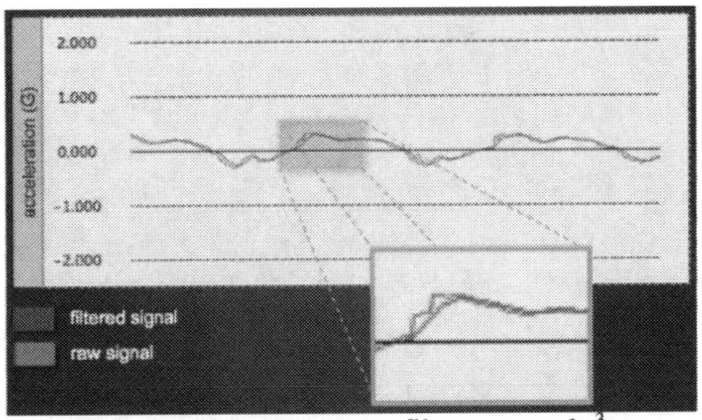

Figure 10. Low pass filter example[3].

As can be seen on the highlighted area, the signal wave (blue line) has been smoothed with respect to the noise that appears on the raw signal. The low pass is a category of filter that comprises an algorithm to filter the noise in a signal wave. As can be seen in Figure 10, the result is nearly the same as that for the Kalman filter.

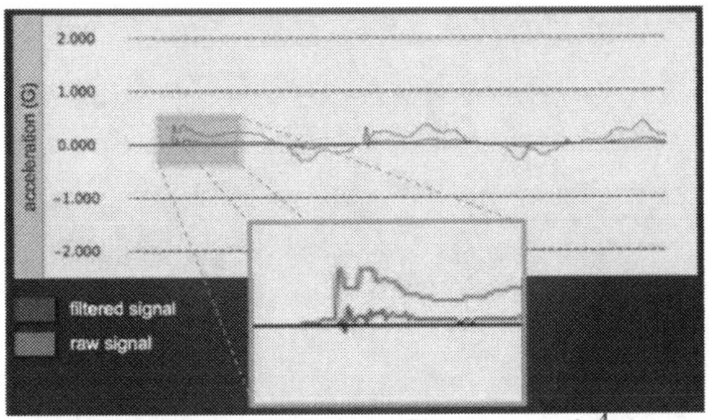

Figure 11. High pass filter example[4].

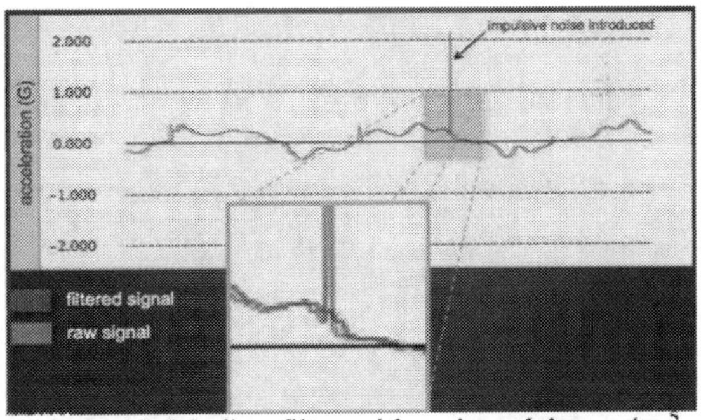

Figure 12. Median filter with an impulsive noise[5].

The high pass filter is a different kind of filter. In Figure 11 we see the high pass filter which filters the signal low frequency. Only the high frequencies can pass through that filter. The median filter illustrated in Figure 12 ignores abrupt changes caused by static electricity or other kinds of noise. In Figure 12 it is possible to see an abrupt change that is completely ignored by the median filter.

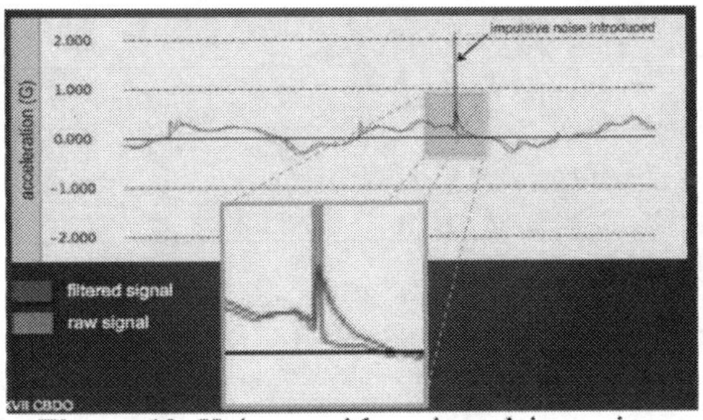

Figure 13. Kalman with an impulsive noise.

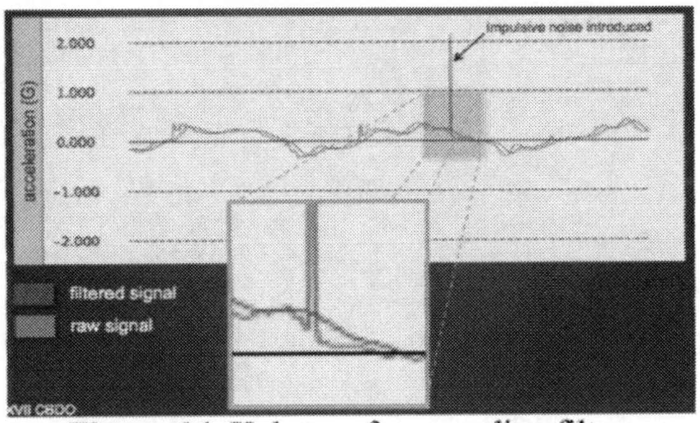

Figure 14. Kalman after a median filter.

[2] The Kalman filter parameters used in the experiment were: *process noise covariance* $Q=0.022$ and *measurement noise covariance* $R=0.617$.

[3] The low pass filter parameters used in the experiment were: $v1=0.02$ and *window size=7 bytes*.

[4] The high pass filter parameters used in the experiment were: $v1=0,98$ and *window size=7 bytes*.

[5] The median filter parameters used in the experiment were: *window size=7 bytes*.

In Figure 13 we used the Kalman filter in a signal associated with an impulsive noise application. It is possible to see that the filter did not ignore the impulsive noise. Instead, it provides a smoothed wave after the noise occurrence until the wave achieves the normal signal again.

In the example shown in Figure 14, two filters were combined. The median filter is applied in first place to remove the impulsive noise, and after that the Kalman filter is applied. In this case, we can see that the impulse was removed. For this case a small latency is observed between the raw and the filtered signal. The latency is associated with the use of two filters.

5. Conclusion

In this work several filter mathematical models were presented to be used in signal processing. An experiment containing an accelerometer was built to generate and process signals so as to test those filters. The experimental approach considered the combination of different filters execution order according to their applications. The results show that the combined approach can be an option for signal analysis.

References

[1] Bartys M, Patton R, Syfert M, Heras S de las, and Quevedo J 2006 *Introduction to the DAMADICS actuator FDI benchmark study*, (*Control Engineering Practice* vol 14) no. 6 (Elsevier) pp 577-596

[2] Chassaing R 1998 *Digital Signal Processing: Laboratory Experiments Using C and the TMS320C31 DSK* (Wiley)

[3] CuteDigi BMX Bluetooth to UART/I2C/USB Module (GEN II) Datasheet

[4] 2009-2013 *Digital Accelerometer ADXL345 Data Sheet* (*Analog Devices Inc.*)

[5] Kalman R E 1960 *A New Approach to Linear Filtering and prediction problems* (Journal of Basic Engineering 82) (ASME) pp 35-45

[6] Haykin S 2001 *Kalman Filtering and Neural Networks* (Willey)

[7] Isermman R 2005 *Model-based fault-detection and diagnosis – status and applications* (*Annual Reviews in Control* vol. 29) no 1 pp 71-85

[8] King A D 1998 *Inertial Navigation – forty years of evolution* (Gec Review vol 13) no 3

[9] Olive X 2012 *FDI(R) for satellites: how to deal with high availability and robustness in the space domain?* (*International Journal of Applied Mathematics and Computer Science* vol 22) no 1 pp 99-107

[10] Patton R J, Uppal F J, Simani S, and Polle B 2010 *Robust FDI applied to thruster faults of a satellite system* (*Control Engineering Practice* vol 18) no 9 pp 1093-1109

[11] Pitas I, Venetsanopoulos A N 1992 *Order Statistics in Digital Image Processing* (*Proceedings of the IEEE* vol 80) no 12

[12] Pontuschka M, Fonseca I et al *FDIR for the IMU Component of AOCS Systems* (*Mathematical Problems in Engineering* vol 2014)

[13] Scarfone K, Padgette 2008 J, *Guide to Bluetooth Security – Recommendations of the National Institute of the Standards and Technology* (*Computer Security Division Informatiion Technology Laboratory*, NIST) (Gaithersburg) pp 20899-8930

[14] Zhang Y and Jiang J 2008 *Bibliographical review on reconfigurable fault-torelant control systems* (*Annual Reviews in Control* vol 32) no 2 pp 229-252

Journal of Physics: Conference Series **641** (2015) 012014 doi:10.1088/1742-6596/641/1/012014

Out-of-plane orbital maneuvers using swing-bys with the Moon

J B S Neto[1], **A F B A Prado**[1] **and J K S Formiga**[1]

[1] National Institute for Space Research –INPE, Av dos Astronautas 1758, SJC-SP, Brazil.

E-mail: neto.jbs91@gmail.com.br

Abstract. This paper has the goal of showing some cases of plane change maneuvers using a swing-by with the Moon to decrease the magnitude of the impulses used, when compared to a classical Hohmann maneuver. The analytical model is based on the "patched-conics" approach, where a series of "two-body" problems is considered to build the whole maneuver. A study of the effects of the semi-major axes of the transfer orbits were made, to complete some previous studies made in the literature. The results show that, for some final inclinations, the use of the swing-by in the Moon is really advantageous.

1. Introduction

During the design phase of artificial satellite missions, one the most important factors is the cost. Optimization methods are usually applied to search for lower cost solutions from the conception to the end of the useful life of the equipment. More recently, studies are also made considering the later disposal of the spacecraft. Looking for solutions to reduce the magnitude of the impulses required in orbital maneuvers, several methods have been developed and used, such as the classical transfers of Hohmann and Bi-elliptical [1], both of them considering coplanar orbits. Now, if the objective is to make a plane change, it is necessary to apply some kind of propulsion with a component perpendicular to the orbital plane. This type of maneuver are usually very expensive in terms of fuel consumption and different options have been considered to reduce the fuel consumption. An alternative solution, already considered in the literature, is to use a gravity assisted maneuver (swing-by) with the Moon to make the out-of-plane part of the maneuver, as shown in [2].

The swing-by is a type of maneuver that does not use propulsion based on engines, but takes advantage of the gravitational influence of a celestial body when the spacecraft passes through its zone of influence [3-6]. The smaller body that is passing by the larger one is accelerated or braked by the gravity field of the celestial body and so its trajectory is modified. Successful missions, such as Voyager 1 and 2 [7], showed the efficiency of the swing-by in saving fuel. An example of the use of this maneuver to change the orbital inclination of a spacecraft is the Ulysses mission, presented in [8], where the spacecraft was placed in a near polar orbit around the Sun, in order to observe the poles of the Sun.

This paper aims to show that, for some desired final inclinations, a swing-by with the Moon (in the Earth-Moon system) can be more economical then using plane change classical maneuvers. The

Journal of Physics: Conference Series **641** (2015) 012014 doi:10.1088/1742-6596/641/1/012014

contribution of the present paper is to analyze the influence of the semi-major axis of the transfer orbit of the spacecraft, which was a point not considered in previous studies [2].

2. Methodology

The complete maneuver is divided in three parts. In the first phase an impulse is used to put the satellite in a close approach trajectory with the moon. In the second stage the three-dimensional swing-by is made and, in the last phase, two impulses are applied to take the satellite back to an orbit with the same semi-major axis and eccentricity of the initial orbit, but now with the final desired inclination.

The analytical model used is presented in [9], and it considers a system consisting of three bodies: a primary (M_1, the Earth), a secondary (M_2, the Moon) and a third body of negligible mass with its motion governed by the other two previous cited bodies (M_3, the artificial satellite). The three phases of the problem is dominated by the celestial mechanics of the two bodies, so it is possible to use the patched-conics approach. Initially, it is considered that the satellite is in an orbit around the Earth that is coplanar with the orbit of the Moon. For the calculations, it was assumed a canonical system of units, where: (i) the gravitational constant is equal to one; (ii) the unit of distances is the Earth-Moon distance (384400.05 km); (iii) the angular velocity of the primaries is equal to one; (iv) the mass of the moon is given by $\mu_L = m_2/(m_1+m_2)$ (where m_1 and m_2 are the masses of the Earth and Moon, respectively), and the mass of the Earth (μ_T) is given by $\mu_T = 1 - \mu_L$, so the total mass of the system is equal to one; (v) the time unit is set such that the period of the primaries is 2π.

2.1. Initial impulse

The impulse used in the first step to generate the transfer orbit (ΔV_1) and the eccentricity of this orbit (e_1) are given by:

$$\Delta V_1 = \left(\frac{2\mu_T}{a_0(1-e_0)} - \frac{\mu_T}{a_1}\right)^{1/2} - \left(\frac{2\mu_T}{a_0(1-e_0)} - \frac{\mu_T}{a_0}\right)^{1/2} \tag{1}$$

$$e_1 = 1 - \frac{a_0(1-e_0)}{a_1} \tag{2}$$

where a_1 is the semi-major axis of the transfer orbit and its minimum value is given by $a_1 = (1 + a_0)/2$, where a_0 is the semi-major axis of the initial orbit of the satellite around the Earth and e_0 is the eccentricity of this orbit.

2.2. The Swing-by maneuver

In the second step the calculation of the swing-by starts when the satellite approaches the Moon and its motion is governed by this body.

Figure 1 shows the geometry of the swing-by, where: r_{ap} is the distance of the closest approach of the satellite with the moon (perilune) and V_p is the velocity at this point; the angles α and β specify the periapsis position in the three-dimensional space and λ is the angle formed between the vector $\vec{V}_p$ and the horizontal plane passing by the perilune. Figure 2 shows the sum of the velocity vectors during the swing-by, where: $\vec{V}_\infty^+$ and $\vec{V}_\infty^-$ are the velocity vectors with respect to the moon before and after the swing-by, respectively, and the angle between them is 2δ, where δ is the so called deflection angle; $\vec{V}_i$ and $\vec{V}_o$ are the satellite velocity vectors before and after the swing-by, respectively, with respect to the Earth; $\vec{V}_2$ is the velocity vector of the Moon with respect to the Earth; γ is called flight-path-angle, which is the angle between $\vec{V}_2$ and $\vec{V}_i$; ϕ is the angle between $\vec{V}_2$ and $\vec{V}_\infty^-$.

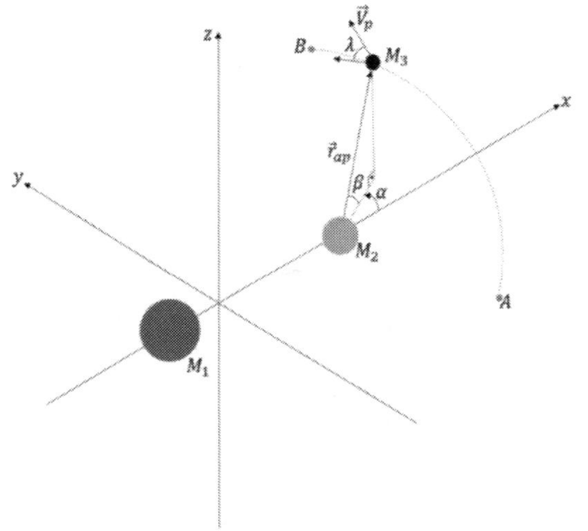

Figure 1. Geometry of the three-dimensional Swing-by.

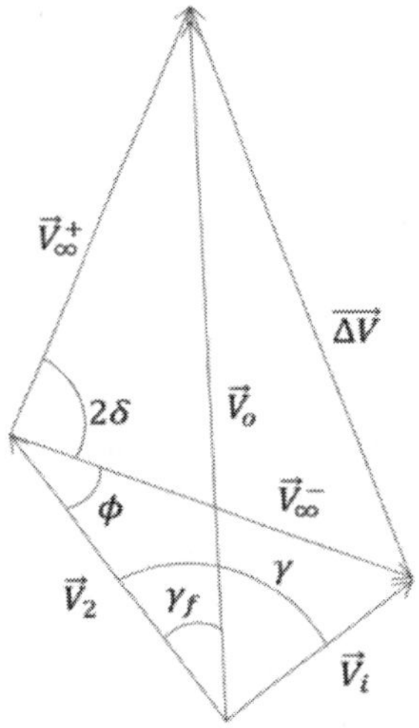

Figure 2. The sum of vectors during the swing-by.

Initially, it is calculated $\vec{V}_i$ and the true anomaly (θ):

$$V_i = \left(\mu_T\left(2 - \frac{1}{a_1}\right)\right)^{1/2} \tag{3}$$

$$\theta = \cos^{-1}\left[\frac{1}{e_1}\left(a_1\left(1 - e_1^2\right) - 1\right)\right] \tag{4}$$

It is now possible to calculate γ and the magnitude of the velocity of the satellite with respect to the moon (V_∞):

$$\gamma = \tan^{-1}\left[\frac{e_1 \sin\theta}{1 + e_1 \cos\theta}\right] \tag{5}$$

$$V_\infty = \left(V_i^2 + V_2^2 - 2V_iV_2\cos\gamma\right)^{1/2} \tag{6}$$

Here ϕ and δ are calculated by:

$$\phi = \cos^{-1}\left[-\frac{V_i^2 - V_2^2 - V_\infty^2}{2V_2V_\infty}\right] \tag{7}$$

$$\delta = \sin^{-1}\left[\left(1 + \frac{r_{ap}V_\infty^2}{\mu_L}\right)^{-1}\right] \tag{8}$$

The approach angle α and the angle λ are given below, respectively, by:

Journal of Physics: Conference Series **641** (2015) 012014 doi:10.1088/1742-6596/641/1/012014

$$\alpha = \pi + \phi + \delta \tag{9}$$

$$\lambda = \sin^{-1}[-\tan(\delta)\tan(\beta)] \tag{10}$$

The vectors $\vec{V}_i$ and $\vec{V}_o$ are given by [9]:

$$\begin{aligned} \vec{V}_i &= \vec{V}_\infty^- + \vec{V}_2 = V_\infty \sin\delta(\cos\beta\cos\alpha, \cos\beta\sin\alpha, \sin\beta) + \\ &+ V_\infty \cos\delta(-\sin\gamma\sin\beta\cos\alpha - \cos\gamma\sin\alpha, -\sin\gamma\sin\beta\sin\alpha + \\ &+ \cos\gamma\cos\alpha, \cos\beta\sin\gamma) + (0, V_2, 0) \end{aligned} \tag{11}$$

$$\begin{aligned} \vec{V}_o &= \vec{V}_\infty^+ + \vec{V}_2 = -V_\infty \sin\delta(\cos\beta\cos\alpha, \cos\beta\sin\alpha, \sin\beta) + \\ &+ V_\infty \cos\delta(-\sin\gamma\sin\beta\cos\alpha - \cos\gamma\sin\alpha, -\sin\gamma\sin\beta\sin\alpha + \\ &+ \cos\gamma\cos\alpha, \cos\beta\sin\gamma + (0, V_2, 0) \end{aligned} \tag{12}$$

$\vec{V}_i$ and $\vec{V}_o$ allows the calculation of the impulse generated by the swing-by ($\Delta\vec{V}$), given by [9]:

$$\Delta\vec{V} = \vec{V}_o - \vec{V}_i = -2V_\infty \sin\delta(\cos\alpha\cos\beta, \cos\beta\sin\alpha, \sin\beta) \tag{13}$$

$$\Delta V = \left|\Delta\vec{V}\right| = 2V_\infty \sin\delta \tag{14}$$

The calculation of the energy change due to the swing-by is given by:

$$\Delta E = \frac{1}{2}\left(V_o^2 - V_i^2\right) = -2V_2 V_\infty \cos\beta\sin\alpha\sin\delta \tag{15}$$

The angular momentum vectors before ($\vec{C}_i$) and after the swing-by ($\vec{C}_o$) are given by [9]:

$$\begin{aligned} \vec{C}_i &= \vec{R}\times\vec{V}_i = d_{TL}V_\infty(0, -\sin\beta\sin\delta + \cos\beta\cos\delta\sin\lambda, \frac{V_2}{V_\infty} + \cos\alpha\cos\delta\cos\lambda + \\ &+ \cos\beta\sin\alpha\sin\delta - \cos\delta\sin\alpha\sin\beta\sin\lambda) \end{aligned} \tag{16}$$

$$\begin{aligned} \vec{C}_o &= \vec{R}\times\vec{V}_o = d_{TL}V_\infty(0, \sin\beta\sin\delta - \cos\beta\cos\delta\sin\lambda, \frac{V_2}{V_\infty} + \cos\alpha\cos\delta\cos\lambda - \\ &- \cos\beta\sin\alpha\sin\delta - \cos\delta\sin\alpha\sin\beta\sin\lambda) \end{aligned} \tag{17}$$

where $\vec{R} = (d_{TL}, 0, 0)$ is the position vector of the Moon with respect to the Earth, and d_{TL} is the Earth-Moon distance. Then, the magnitude of the angular momentum change (ΔC) is given by [9]:

$$\begin{aligned} \Delta C &= \left|\vec{C}_o - \vec{C}_i\right| = \left|2d_{TL}V_\infty \sin\delta(0, \sin\beta, -\cos\beta\sin\alpha)\right| \\ &= 2dV_\infty \sin\delta\left(\cos^2\beta\sin^2\alpha + \sin^2\beta\right)^{1/2} \end{aligned} \tag{18}$$

Journal of Physics: Conference Series **641** (2015) 012014 doi:10.1088/1742-6596/641/1/012014

After the swing-by maneuver is completed it is possible to calculate the final inclination (i_o) by [9]:

$$Cos(i_o) = \frac{C_{oZ}}{|\vec{C}_o|} = \left(1 + \left(\frac{\sin\beta\sin\delta - \cos\beta\cos\delta\sin\lambda}{\frac{V_2}{V_\infty} + \cos\alpha\cos\delta\cos\lambda - \cos\beta\sin\alpha\sin\delta - \cos\delta\sin\alpha\sin\beta\sin\lambda}\right)^2\right)^{-1/2} \tag{19}$$

2.3. Final impulses

In the third step, after the swing-by is completed, two impulses are used to return the satellite to its initial orbit (with a_0 and e_0), but keeping the modified inclination. First, it is calculated the semi-major axis of the satellite after the swing-by (a_2):

$$a_2 = \frac{\mu_T}{2\mu_T - V_o^2} \tag{20}$$

Next, it is calculated the semi-major axis of the transfer orbit (a_3) and the first impulse (ΔV_2):

$$a_3 = \frac{a_2(1+e_2) + a_0(1-e_0)}{2} \tag{21}$$

$$\Delta V_2 = \left(\frac{2\mu_T}{a_2(1+e_2)} - \frac{\mu_T}{a_2}\right)^{1/2} - \left(\frac{2\mu_T}{a_2(1+e_2)} - \frac{\mu_T}{a_3}\right)^{1/2} \tag{22}$$

When the satellite reaches the perigee of the transfer orbit one last impulse is used to circularize the orbit (ΔV_3):

$$\Delta V_3 = \left(\frac{2\mu_T}{a_0(1+e_0)} - \frac{\mu_T}{a_3}\right)^{1/2} - \left(\frac{2\mu_T}{a_0(1+e_0)} - \frac{\mu_T}{a_0}\right)^{1/2} \tag{23}$$

After the complete maneuver, all impulses are added:

$$\Delta V_t = \Delta V_1 + \Delta V_2 + \Delta V_3 \tag{24}$$

2.4. Classical Maneuver

To test the swing-by as an alternative to the more conventional types of plane change maneuver, this technique is compared to a mono-impulsive maneuver, given by [10]:

$$\Delta V_m = 2\left(\frac{2\mu_T}{a_0(1+e_0)} - \frac{\mu_T}{a_0}\right)^{1/2} \sin\frac{i}{2} \tag{25}$$

Journal of Physics: Conference Series **641** (2015) 012014 doi:10.1088/1742-6596/641/1/012014

3. Results

According to the initial objectives of the present paper and using the proposed technique, two simulations are made for each value of the semi-major axis of the transfer orbit: a_1 (minimum semi-major axis) and $5a_1$. For both simulations, the initial eccentricity was $e_0 = 0$, but in the first simulation (simulation 1) the initial semi-major axis was $a_0 = 0.017$ and the periapsis distance $r_{ap} = 0.0046$ [2]. In the second simulation (simulation 2) the initial value of the semi-major axis was $a_0 = 0.069$ and for the periapsis distance it was used again $r_{ap} = 0.0046$. The initial semi-major axis and the periapsis distance are shown in canonical units (the Earth-Moon distance). The simulations are made for β ranging from 0 to π and the results are shown in figures 3 to 10, displaying the final inclination (i_o) in the vertical axis as a function of β in the horizontal axis, both in radians. Some graphics show ΔV_t - ΔV_m, the difference between the total impulse used by the maneuver that included the swing-by and the magnitude of the mono-impulsive maneuver, given in canonical units of velocity (1 canonical unit of velocity is approximately 1.023 km/s), also as a function of β in the horizontal axis.

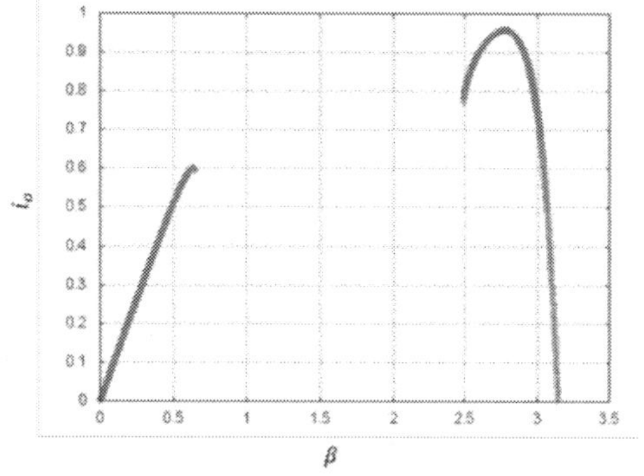

Figure 3. i_o vs. β, for simulation 1 considering the semi-major axis a_1.

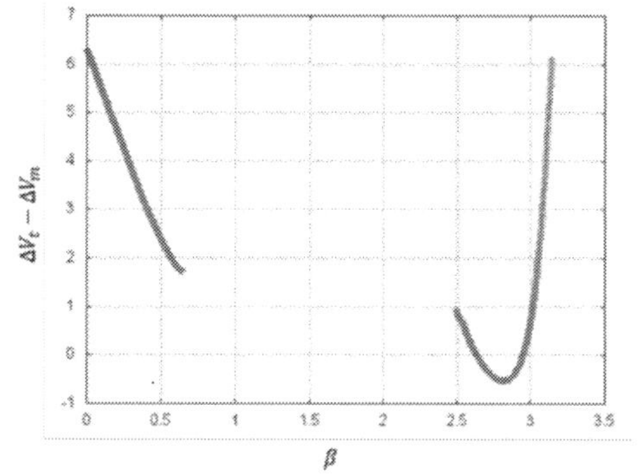

Figure 4. ΔV_t - ΔV_m vs. β, for simulation 1 considering the semi-major axis a_1.

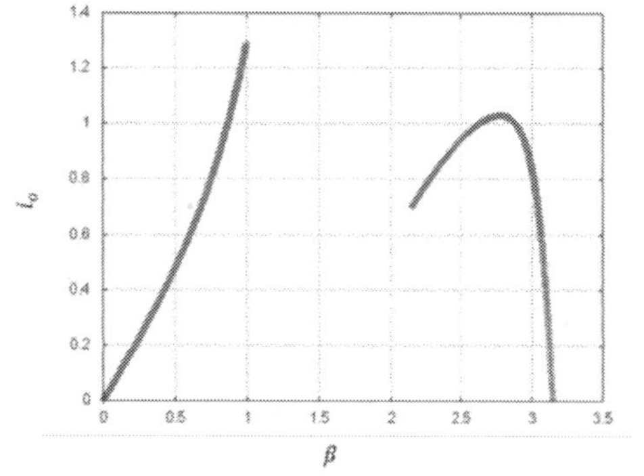

Figure 5. i_o vs. β, for simulation 1 considering the semi-major axis $5a_1$.

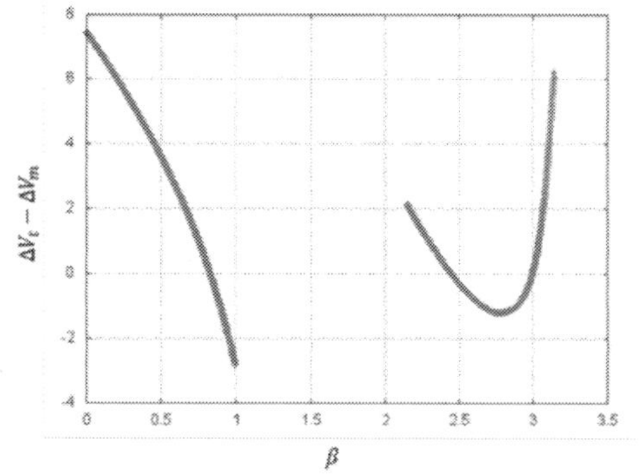

Figure 6. ΔV_t - ΔV_m vs. β, for simulation 1 considering the semi-major axis $5a_1$

The results for the solution 1, considering the semi-major axis a_1, shown in figures 3 and 4, show that the use of the swing-by is justified only for the final inclinations between 0.91 rad and 0.95 rad, which corresponds to β between 2.63 rad and 2.94 rad. Inside this range the savings in the magnitude of the impulses reaches 0.52 canonical units. However, the results given by Solution 1 considering the semi-major axis $5a_1$, as shown in figures 5 and 6, indicates that the use of the swing-is justified by the first family of solutions for the final inclinations between 0.91 rad and 1.29 rad. For values of β between 0.82 rad and 0.99 rad, the savings in the magnitude of the impulse reaches values up to 2.86.

Now, for the second family of solutions, the use of the swing-by is only justified for values of β between 2.44 rad and 2.99 rad, with final inclinations between 0.91 rad and 1.3 rad, with the savings in the magnitude of the impulses up to 1.2 canonical units.

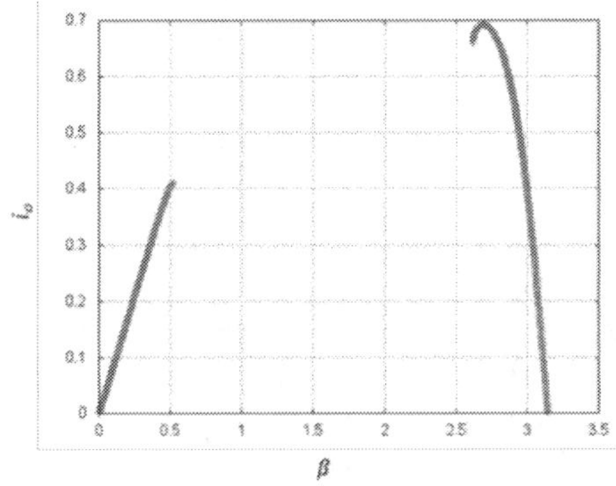

Figure 7. i_o vs. β, for simulation 2 considering the semi-major axis a_1.

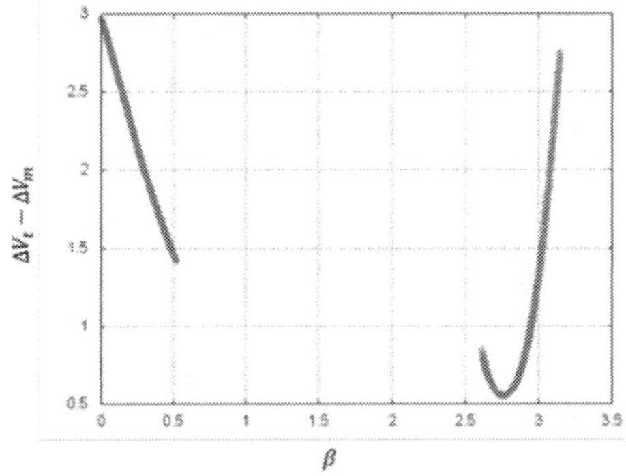

Figure 8. ΔV_t - ΔV_m vs. β, for simulation 2 considering the semi-major axis a_1.

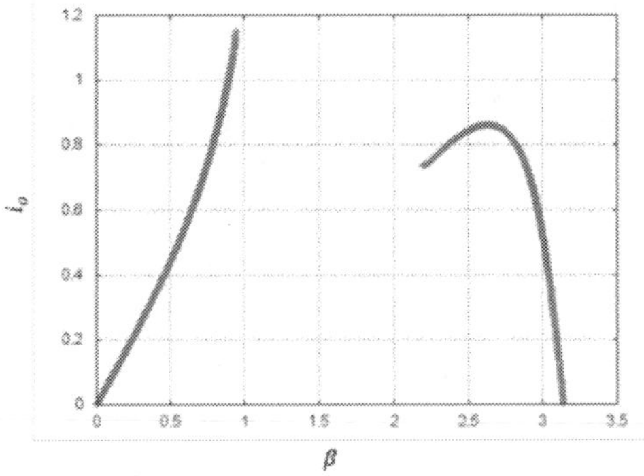

Figure 9. i_o by β, simulation 2 for $5a_1$.

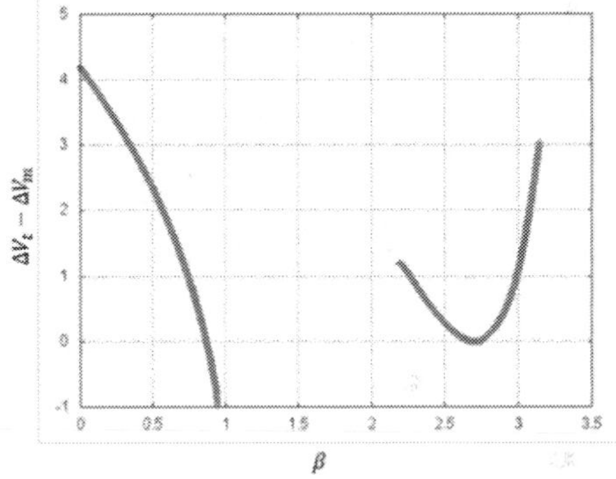

Figure 10. ΔV_t - ΔV_m by β, simulation 2 for $5a_1$

Figures 7 and 8 show the results for the solution 2 considering the semi-major axis as a_1. In this case the use of the swing-by is not justified. The results of the solution 2 for a semi-major axis of $5a_1$ are shown in figures 9 and 10. It indicates that the use of the swing-by is only justified for the first family of solutions and for the final inclinations between 0.93 rad and 1.15 rad, which corresponds to β between 0.86 rad and 0.94 rad. The economy in the magnitude of the impulse can reach up to 0.98 canonical units.

In general, the results are according to the expected for this type of maneuver. There is a cost to send the spacecraft to the Moon and back, even in planar maneuvers. To take advantage of this maneuver, it is necessary to have savings in the out-of-plane maneuver that compensates this extra consumption. This is the physical reasons of the limitations in the region of advantages of the proposed technique. It can be applied when in the situations where the inclination change is larger, so the savings in this part of the maneuver compensates the extra costs involved. Although expected, the results shown in the present paper quantifies the exact locations of the regions where the maneuver has gains over the standard ones and how much are the gains. In particular, the present paper extends studies made before by considering the effects of the semi-major of the transfer orbit in the savings obtained. This is a free important parameter when designing a mission.

Journal of Physics: Conference Series **641** (2015) 012014 doi:10.1088/1742-6596/641/1/012014

4. Conclusion

After analyzing the results, it is concluded that, for some final inclinations desired for the orbits, the use of the swing-by maneuver as an alternative to the classical maneuvers can be a very good option, generating savings even greater than 1 canonical units of velocity. Regarding the analysis of the influence of the semi-major axis of the transfer orbit, the result show that larger semi-major axis allows the existence of new families of solutions, with higher final inclinations. This type of analysis can help mission designers to design missions that can benefit from the technique shown here.

5. References

[1] Prado A F B A and Rios-Neto A 1993 Um estudo bibliográfico sobre o problema de transferências de órbitas *Revista brasileira de ciências mecânicas* **15** pp 65-78

[2] Torre K S and Prado A F B A 2006 Changing inclination of earth satellites using the gravity of the moon *Mat. Prob. Eng.* **2006** pp 1-13

[3] Gomes V M and Prado A F B A 2010 A study of the impact of the initial energy in a close approach of a cloud of particles *WSEAS Trans. Mat.* **09** pp 811-820

[4] Prado A F B A and Broucke R A 1995 Effects of atmospheric drag in swing-by trajectory *Acta Astr.* **36** pp 285-290

[5] Gomes V M and Prado A F B A 2008 Swing-by maneuvers for a cloud of particles with planets of the solar system *WSEAS Trans. Ap. Theor. Mec.* **03** pp 869-878

[6] Prado A F B A and Broucke R A 1995 A classification of swing-by trajectories using the moon *App. Mec. Rev.* **48** pp 138-142

[7] Kohlhase C E and Penzo P A 1977 Voyager mission description *Sp. Sc. Rev.* **21** pp77-101

[8] Wenzel K P, Marsden R G, Page D E and Smith E J 1992 The ULYSSES mission *Astr. and Astrop. Suppl.* **92** pp 207-219

[9] Prado A F B A 2000 An analytical description of the close approach maneuver in three dimensions *Proc. Int. Astr. Congress* (Rio de Janeiro)

[10] Roy A E 1988 Orbital motion (Bristol)

Journal of Physics: Conference Series **641** (2015) 012015 doi:10.1088/1742-6596/641/1/012015

If there is dissipation the particle can gain energy

R Egydio de Carvalho

Universidade Estadual Paulista-UNESP – Rio Claro-SP, Brazil

E-mail: regydio@rc.unesp.br

Abstract. In this work, we summarize two different mechanisms to gain energy from the presence of dissipation in a time-dependent non-linear system. The particles can gain energy, in the average, from two different scenarios: i) for very week dissipation with the creation of an attractor with high velocity, and ii) in the opposite limit, for very strong dissipation, the particles can also gain energy from a boundary crisis. From the thermodynamic viewpoint both results are totally acceptable.

1. Introduction

Dissipation, generically, corresponds to a process that consumes energy in such way the particles of the system lose energy until stabilizing in a stable configuration. A typical stable configuration for a dissipative system corresponds to an attractor. However, depending on the structures of the phase space the results can be different of the common sense.

In a chaotic region of a Hamiltonian system [1], a particle can experiment different velocities in the phase space. When this chaotic region is limited by invariant tori, a half of the height of the column of chaos gives approximately the mean velocity of the particles moving in the chaotic sea. If in the superior limit of the column of chaos there is an island of resonance, then the elliptic fixed point from the inside of the island will play a key role in the limit of weak dissipation. On the other hand, for strong dissipation, the coexistence of a point attractor with a chaotic attractor may lead to the crossing of the stable and unstable manifolds of the point attractor, as the parameter of dissipation is varied. This defines a boundary crisis [2] which destroys the chaotic attractor and drives the trajectories, which were in the chaotic attractor, toward the point attractor. These mechanisms are quite different but both allow the particle to receive part of the energy that would be lost and, in the average, an ensemble of particles can attain higher energies than they had without dissipation. The paper is organized as follows, in section 2 we present the model, the results are shown in section 3 and in section 4 the reader will find the conclusion.

2. The breathing annular billiard

The model we consider is the annular billiard with periodic time-dependent boundaries. However, before introducing it we present the main aspects of the static annular billiard. This is defined by the annulus between two circumferences, which may be concentric or eccentric. The annulus is free of potential and is the only accessible space for the particles. The outer radius is set equal to 1; the inner one we call by r_0 and the distance between both centers, the eccentricity, we call by δ.

For the static case, the particle energy is conserved and the dynamics of a particle moving in the annulus is completely described by the coordinates of a collision with the external boundary. These coordinates are: i) the incidence angle, α , that the trajectory makes with the normal of the external circle and ii) the angle, θ, that the normal makes with the abscissa axis. The corresponding phase space is defined by the variables, S=sin(α) and L=$\theta/2\pi$ in such way that S $\in$ [-1, 1] and L $\in$ [- ½, ½]. We call this plane as geometric phase space. There is an auxiliary circumference, with radius (r + δ) called caustic, that allows us distinguish a class of trajectories. The trajectories that never cross the caustic are called whispering gallery orbits (WGO) and they correspond to preserving angular momentum motions. The dynamics is governed by two maps: i) one called A-motion, corresponding to the case in which between two consecutive hits with the external boundary, the trajectory does not hit the internal one, and ii) another called B-motion, in which the trajectory collides with the internal circumference before hitting the external one. For the static and concentric geometry, besides the energy, the angular momentum is always conserved, in such way the system is globally integrable. However, for the eccentric geometry, the B-motion implies in no conservation of the angular momentum anymore and chaos appears and grows with the eccentricity. δ plays the role of perturbation parameter, as it increases the destruction of invariant tori in the geometric phase space also increases, up the numerical limit $\delta \approx 3r$. A more detailed discussion on the static annular billiard can be found in [3-4].

In order to study how a particle can vary its energy we must introduce a time-dependent perturbation. However, to observe confined motions, we need to preserve a global constant of motion; in this case the constant is the angular momentum. So, we will consider the concentric billiard but with both boundaries periodically breathing in time. We define the ratio between both frequencies of oscillation by ω. The amplitudes of oscillation and the phases are (ε_R, φ) and (ε_r, ϕ) for the outer and inner boundaries respectively. During a collision, the boundaries are not necessarily in the same positions they were in the static case, they have new radii given by $R(t) = 1 + \varepsilon_R \cos(t + \varphi)$ and $r(t) = r_0 + \varepsilon_r \cos(\omega t + \phi)$ in such way there are two collision zones in which the particles can suffer collisions. The collision zone associated with the external circumference is the ring $[1 - \varepsilon_R, 1 + \varepsilon_R]$ while with the internal one it is $[r_0 - \varepsilon_r, r_0 + \varepsilon_r]$. Depending on its velocity the particle entries in a collision zone and depending also on the phase of the moving boundary, it may suffer successive collisions before leaving the collision zone. This problem of successive collisions may receive two different approaches. One called complete problem which consider all possibilities of collisions in the algebraic formulation and along the numeric calculations, and another one called simplified problem which keeps the boundaries algebraically static but the momentum of the particle is changed in the instant of the collision as the boundaries were oscillating. Even though there are small differences between both approaches, essentially for low particle velocities, the dynamics are reasonably similar. Hence, to avoid so long cpu-times, the calculations have been done with the simplified description. In this problem, dissipation is introduced through inelastic collisions between the particle and the boundaries. As the breathing corresponds to a radial motion, the tangential component of the particle velocity, $V_{(n+1)\tau} = V_{(n)\tau}$, does not change in a collision. On the other hand, the radial component of the particle velocity V_η is given by $V_{(n+1)\eta} = -\left| -e\, V_{(n)\eta} + (1+e)\, U \right|$, where (n+1) stands for the next collision after the nth collision, $e \in [0, 1]$ is the restitution coefficient and U is the velocity of the boundary which is the first derivative of the respective radius, $\dot{R}(t)$ or $\dot{r}(t)$, depending on which boundary the collision occurred. If e = 1 there is no dissipation. A collision is obtained through the intersection between the straight lines of the trajectories with the circumferences with time-dependent radii. Hence, for each collision, with any boundary, the particle changes its radial velocity and we get an iteration of the maps. However, we store only the collisions with the external boundary. Along this work, the circumferences will be kept concentric. The variation of the energy due to the time-dependent perturbation defines another plane of phase, which we call energy phase plane, defined by the phase of the external boundary φ and the radial particle velocity v_η. The details concerning the algebraic derivation of the non-dissipative maps and the ideas of the dynamics can be found in [5,6] while the introduction of dissipation is detailed in [7,8].

3. Discussion and Results

A schematic view of the time-dependent annular billiard is shown in figure (1) in which we see a straight line trajectory suffering a single collision, figure (1a) and successive collisions, figure (1b), in the external collision zone, while in figure (1c) the trajectory goes into the internal boundary before hitting the external one. The particle can also suffer successive collisions in the internal collision zone. Throughout this work we kept $\delta = 0$ and consider only the concentric case. *Figures (1a), (1b) and (1c) are authorized to be reproduced from Phys. Rev. E73 (2006) 066229, whose DOI is 10.1103/PhysRevE.73.066229, by the American Physical Society through the license number 3607670344906.*

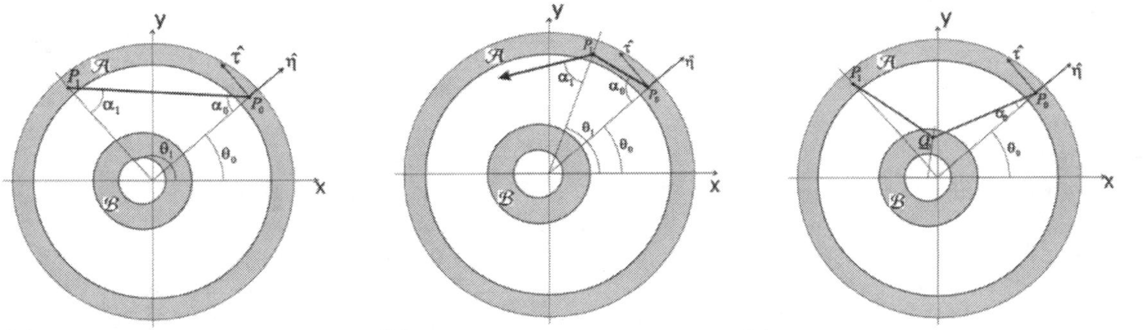

(a) (b) (c)

Figure 1. Schematic illustrations of the time-dependent annular billiard. The collision zones are the regions denoted by the stylized letters A and B. a) and b) show collisions only with the external collision zone, while in c) it is shown the possibility of also happening collisions in the internal collision zone.

In figure (2) it is presented the energy phase plane for the non-dissipative concentric billiard breathing periodically in time. The set of parameters we used were: $r_0 = 0.4$, $R = 1$, $\delta = 0$, $e = 1$, $\omega = 0.5$, $\in_R = \in_r = 0.01$ and $\phi = \pi/4$. We can observe in figure (2a) that there is a chaotic sea for low velocities and for the velocity ~0.20 there is an elliptic fixed point inside the resonance island. Even though there is a thin layer of chaos above the island, we can expect that the mean velocity of an ensemble of particles, starting in the chaotic sea, be approximately $< v > \sim 0.10$, ~a half of the column of chaos. That is confirmed in figure (2b) where we plotted the mean velocity of an ensemble of 1000 initial conditions starting with $-v_\eta = 0.02$ and $\varphi \in [0, 4\pi]$. After 10^6 iterations $< v >$ saturated at ~ 0.10. *Figures (2a), (2b), (3a), (3b), (3c) and (3d) are authorized to be reproduced from Phys. Rev. E77 (2008) 036204, whose DOI is 10.1103/PhysRevE.77.036204, by the American Physical Society through the license number 3607661418103.*

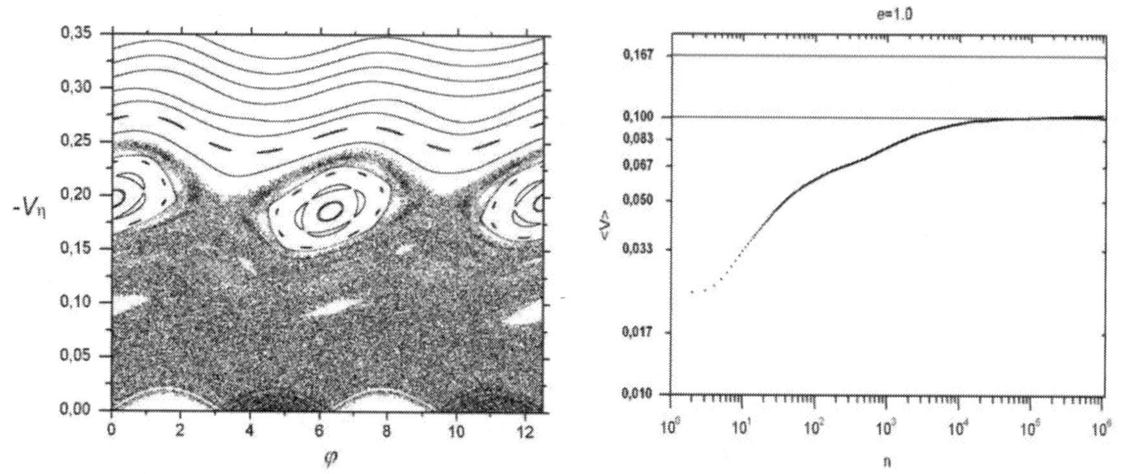

(a) (b)

Figure 2. Non-dissipative case. In (a) we see the energy phase plane with invariant tori limiting the growth of the particle velocities, chaos for low velocities and an elliptic fixed point at velocity ~0.20. In (b) we see the behavior of the mean velocity of an ensemble of particles starting in the chaotic sea. It saturates at velocity ~0.10.

In order to analyze the effect of the inelasticity of the boundaries we set a very weak dissipation with e = 0.9999. In figure (3a) we observe that the invariant tori and the resonance structure have been destroyed and the elliptic fixed point became a point attractor. Figure (3b) shows the amplification of the neighborhood of this attractor. We started the same ensemble of initial conditions used in figures (2) and we observed that the dynamics is driven toward this attractor, which has velocity ~ 0.20. So, it is expected that the mean velocity of the ensemble of initial conditions, started with low velocities, be higher than the non-dissipative counterpart. Figure (3c) confirms that prediction, the mean velocity saturated at ~ 0.167, what implies in a gain of energy of approximately 67% in comparison with the non-dissipative case. This value of the restitution coefficient is not the only possibility to observe this energy gain. Figure (3d) presents a plot for a range of weak dissipation $e \in [1.0, 0.999]$, for which this is observed, however the extreme is at e ~ 0.9999.

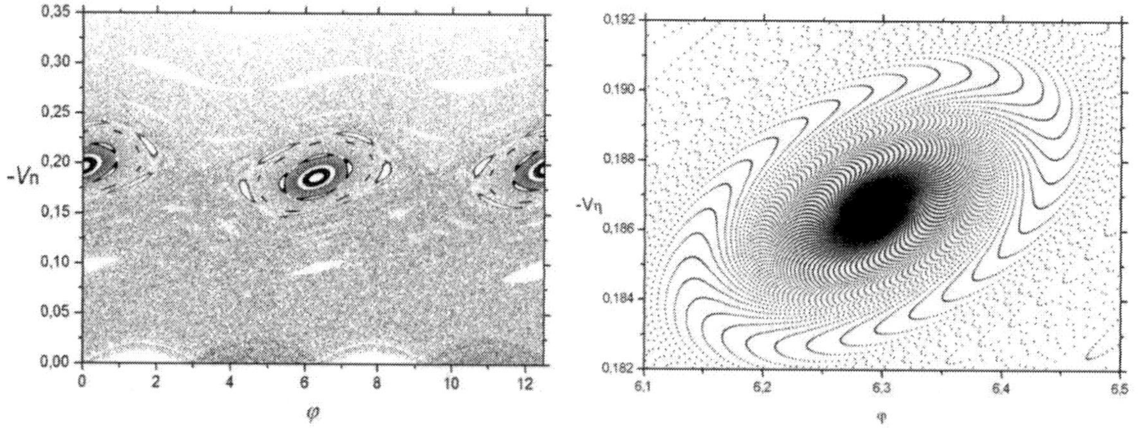

(a) (b)

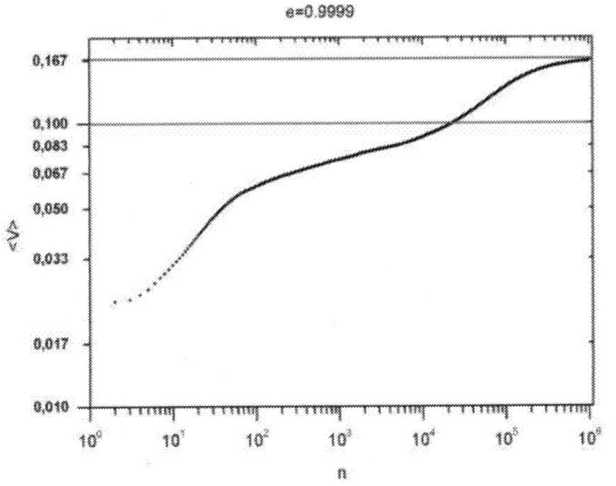

(c)

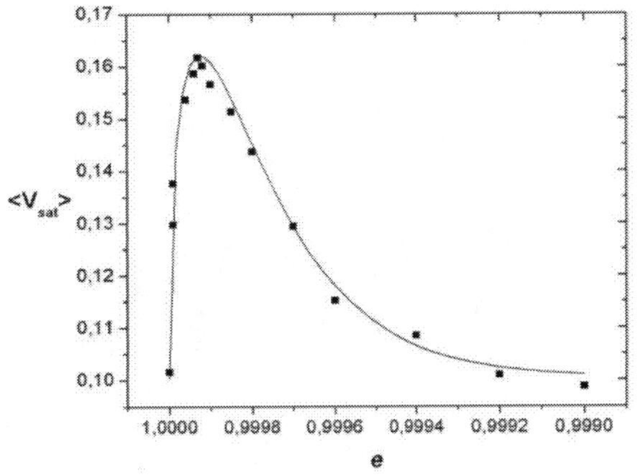

(d)

Figure 3. Dissipative case, in (a), (b) and (c) e=0.9999. In a) it is presented the energy phase plane with a point attractor at velocity ~ 0.20. In b) we see an enlargement around this point attractor. In c) the mean velocity of the same ensemble used in figure (2) converges to the velocity ~ 0.167 (blue line). The red line corresponds to the non-dissipative case. In d) it is shown a range of restitution values for which the particle can gain energy. The peak is at ~0.9999.

We consider now the limit of very strong dissipation. We also set $\omega = 1$ and $\phi = 0$. Figure (4) shows the results we have obtained. In figure (4a) we plot the energy phase plane for the non-dissipative case, which has some similarity with the one of figure (2a), however now we will keep our attention to the elliptic fixed point at $-V\eta \approx 0.065$ and $\varphi \approx 0.50$. In figure (4b) we set $e = 0.8671$ and we observe the coexistence of a chaotic attractor and a point attractor originated from the elliptic fixed point cited above. We plot the stable (black) and the unstable (red) manifolds of the saddle point (S) originated from the unstable fixed point, partner of the stable one in observation. The upper unstable manifold follows toward the attractive fixed point while the bottom one defines the chaotic attractor. The branches of the stable manifold generate the boundaries of the attractive fixed point, splitting the initial conditions which follow toward the stable focus, from those one heading for the chaotic attractor. We call this regime as *before the crisis*. We decrease slightly the restitution coefficient to $\gamma =$ 0.8651, what implies in the collision of the stable manifold, which defines the border of the chaotic attractor, with the unstable manifold. These intersections are represented by arrows in figure 4(c). From this value of the restitution coefficient, the basin of attraction of the chaotic attractor, and the chaotic attractor, do not exist any longer, because a boundary crisis occurred. We observe in the place of the chaotic attractor a chaotic transient, which corresponds to the interval of time the initial conditions, departing from the region where there was the chaotic attractor, are captured by the attractive fixed point. We call this regime as *after the crisis*. In figure (4d) we plot the average velocities of an ensemble the 500 initial conditions started with $-v_\eta = 0.02$ and $\varphi \in [0, 4\pi]$ for the following three cases: i) the non-dissipative case, with $\gamma = 1$; ii) before the crisis, with $\gamma = 0.8671$ and iii) after the crisis, with $\gamma = 0.8651$. The values of the mean energies after 5×10^6 iterations are 5.69×10^{-2}, 2.51×10^{-2} and 6.55×10^{-2} respectively for the case ii) and iii). *Figures (4a), (4b), (4c) and (4d) are authorized to be reproduced from Phys. Rev. E84 (2011) 036204, whose DOI is 0.1103/PhysRevE.84.036204, by the American Physical Society through the license number 3607680086928.*

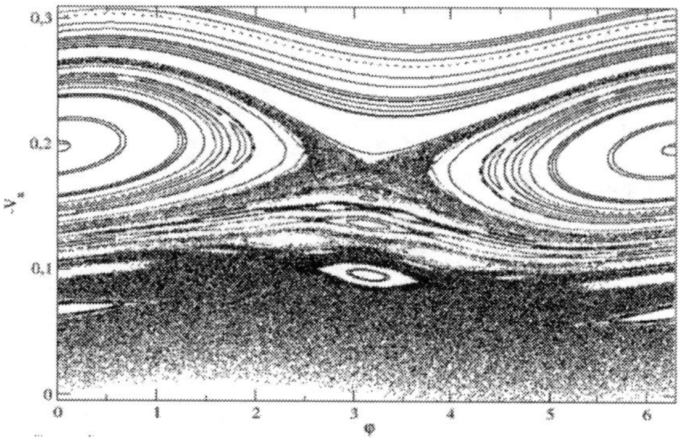

(a)

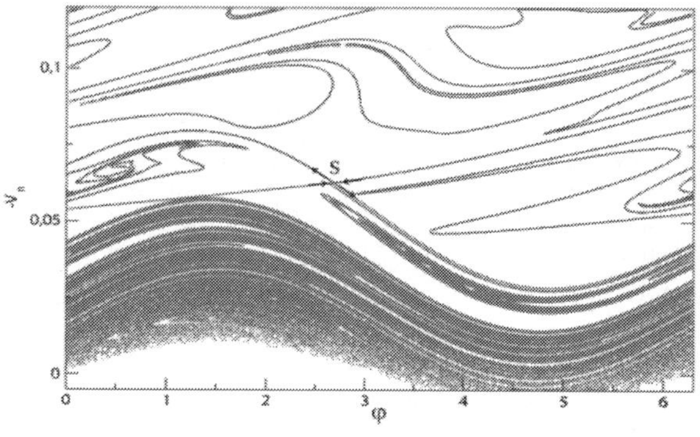

(b)

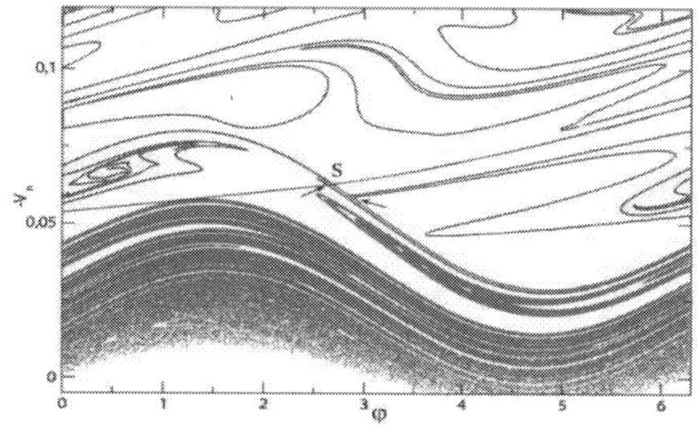

(c)

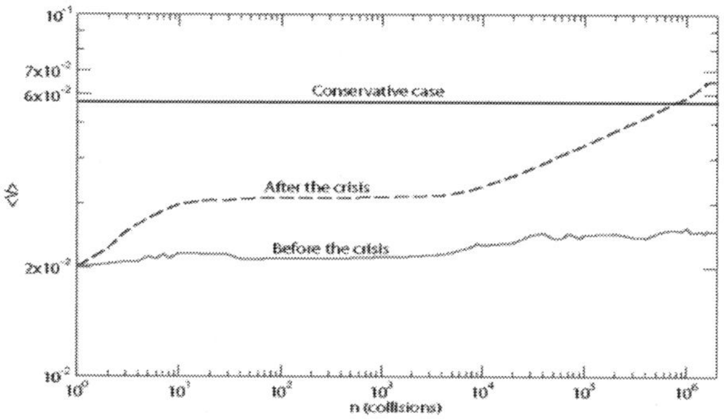

(d)

Figure4. a) Energy phase plane for the non-dissipative scenario. Strong dissipation: b) e=0.8671, before the crisis and c) e=0.8651, after the crisis; d) the mean velocities for the cases: non-dissipative (black), before (red) and after (blue) the crisis.

We observe that even in the regime of strong dissipation, the particles that started with low velocities attained values of energy that are higher than the ones of the non-dissipative configuration. The responsible mechanism for that energy growth is a boundary crisis that destructed the chaotic attractor and has driven the particles toward the point attractor, which has higher velocity than the chaotic attractor.

4. Conclusion

In this work we summarize two mechanisms for particles gain energy from the dissipation in two opposite regimes of dissipation intensity. In the limit of weak dissipation, a point attractor is created in the energy phase space with velocity higher than the average velocity of the non-dissipative dynamics. So, an ensemble of particles with low initial velocities is driven toward the point attractor and in the average these particles gain energy after the introduction of dissipation. In the limit of strong dissipation, there are two attractors, a chaotic and a point one. The velocity of the point attractor is higher than the velocity of the chaotic attractor. A boundary crisis occurs in the system and destroys the chaotic attractor leading to a chaotic transient. The trajectories that were travelling in the chaotic attractor are driven to the point attractor and again they gain energy from the increase of the dissipation intensity. These results have been obtained from a specific model, the breathing annular billiard, however there is no loss of generality in this approach and the ideas can be applied in any system that fulfils the conditions here explained. It is worth noting that all particles that start with initial velocities higher than the velocity of the point attractor, will lose energy.

Acknowledgement
I thank the Brazilian scientific agency, São Paulo Research Foundation-Fapesp for financial support through the grant 2014/00334-9.

References

[1] Ozorio de Almeida A M 1990 *Hamiltonian Systems Chaos and Quantization* (Cambridge: Cambridge Univertisy Preess)
[2] Grebogi C, Ott E and Yorke J A 1982 *Phys. Rev.Lett.* **48** 1507
[3] Saitô N, Hirroka H, Ford J, Vivaldi F and Wilker G H 1982 *Physica D* **5** 273
[4] Bohigas O, Boosé D, Egydio de Carvalho R and Marvulle V 1993 *Nucl. Phys. A* **560** 197
[5] Egydio de Carvalho R, Souza F C and Leonel E D 2006 *Phys. Rev. E* **73** 066229
[6] Egydio de Carvalho R, Souza F C and Leonel E D 2006 *J. Phys A* **39** 3561
[7] Egydio de Carvalho R, Abud C V and Souza F C 2008 *Phys. Rev. E* **77** 036204
[8] Egydio de Carvalho R and Abud C V 2011 *Phys. Rev. E* **84** 036204

Journal of Physics: Conference Series **641** (2015) 012016 doi:10.1088/1742-6596/641/1/012016

Permanent magnet Hall Thrusters development and applications on future brazilian space missions

Jose Leonardo Ferreira[1], Alexandre A. Martins[1], Rodrigo Miranda[1,2], Adriane Schelling[1], Lais de Souza Alves[1], Ernesto Gonçalves Costa[1], Helbert de Oliveira Coelho Junior[1], Artur Castelo Branco[1] and Felipe Nathan de Oliveira Lopes[1]

[1] Plasma Physics Laboratory, Unviersity of Brasilia, 70910-900, Brasilia, DF, Brazil
[2] Aerospace Department, University of Brasilia, 72444-260, Gama, DF, Brazil

E-mail: jleonardoferreira@uol.com.br

Abstract. The Plasma Physics Laboratory (PPLUnB) has been developing a Permanent Magnet Hall Thruster (PHALL) for the Space Research Program for Universities (UNIESPAÇO), part of the Brazilian Space Activities Program (PNAE) since 2004. The PHALL project consists on a plasma source design, construction and characterization of the Hall type that will function as a plasma propulsion engine and characterized by several plasma diagnostics sensors. PHALL is based on a plasma source in which a Hall current is generated inside a cylindrical annular channel with an axial electric field produced by a ring anode and a radial magnetic field produced by permanent magnets. In this work it is shown a brief description of the plasma propulsion engine, its diagnostics instrumentation and possible applications of PHALL on orbit transfer maneuvering for future Brazilian geostationary satellite space missions.

1. Introduction

The Plasma Physics Laboratory of the University of Braslia (PPLUnB) has been developing a Permanent Magnet Hall Thruster (PHALL) for the Brazilian Space Program since 2004 [1, 2]. Our project consists in designing a plasma source and construction/characterization of plasma propulsion engines based on the Hall current ionizing mechanism generated by the perpendicular electric and magnetic fields produced by the vertical electric fields of a bottom ring anode and the perpendicular radial magnetic fields produced by permanent magnets on the exit of a cylindrical annular channel. This thruster has to be studied in vacuum mimicking conditions and therefore we have to perform the tests inside a long and cylindrical vacuum chamber where we have attached a whole paraphernalia of plasma diagnostics and electric power systems for thruster excitation and characterization (Figure 1).

Electric propulsion is now becoming a very successful method for primary and secondary propulsion systems. It is essential for deep space long duration missions and for station keeping of geosynchronous satellites, where the thrusting system can be designed to be used on orbit maneuvering or on satellite attitude control in long term space missions.

One of the main advantages of our PHALL thruster is the production of a steady state magnetic field by permanent magnets providing electron trapping and electron Hall current generation with a significant decrease on the power consumed from the electric power supply. This advantage turns the PHALL thruster into a specially good option when it comes to space

Journal of Physics: Conference Series **641** (2015) 012016 doi:10.1088/1742-6596/641/1/012016

Figure 1. General experimental setup with vacuum chamber and diagnostics for Hall thruster tests.

usage for longer and deep space missions, where solar panels and electric energy storage on batteries is a limiting factor. Two prototype models, PHALL I and II, were already developed and tested with different types of permanent magnets (Figure 2).

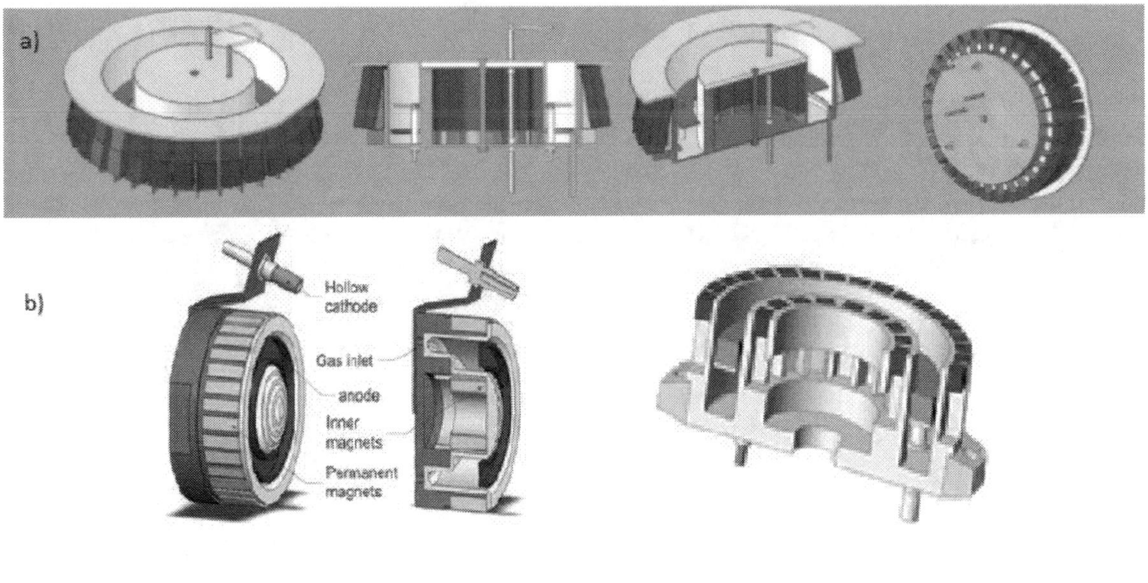

Figure 2. Hall thrusters developed at PPLUnB. a) PHALL I (External diameter of 30 cm); b) PHALL II (External diameter of 15 cm).

Journal of Physics: Conference Series **641** (2015) 012016 doi:10.1088/1742-6596/641/1/012016

2. Physical characteristics of PHALL

Now we will describe briefly the general construction and characteristics of our Hall plasma source. The first prototype, the PHALL I (Figure 2.a) had long magnets that accompanied the length of the whole circular annular channel and were disposed at an angle at the inside and outside of the channel and used a tungsten wire at the top to initiate the discharge by thermionic electron emission. Later on this thruster was refined into PHALL II by using shorter magnets parallel to the channel walls thereby focusing the Hall current only at the channel exit and increasing the ion accelerating electric field between the anode and the virtual cathode (Figure 2.b and Figure 3). In PHALL II the tungsten wire was also replaced with a hollow cathode for electron emission, thereby improving efficiency and performance.

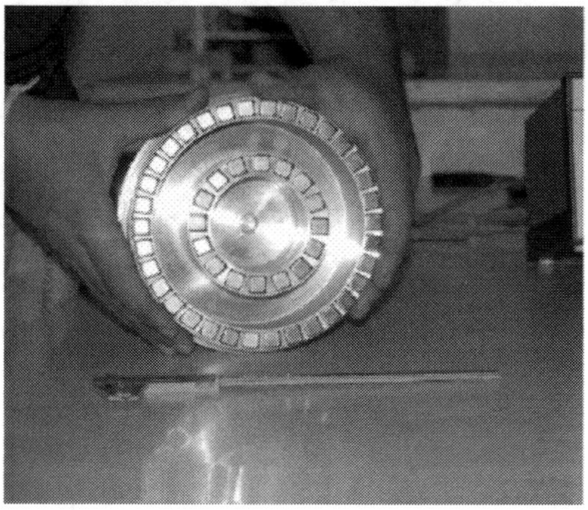

Figure 3. PHALL II. On the left: with top hat dielectric protection , and on the right: showing the magnets without top protection.

3. Diagnostics

Our Integrated Plasma Diagnostic System will now be detailed (Figure 4). This system contains several Langmuir probes that are used for plasma density and temperature measurements. We also use a Faraday Cup, Ion probes and a Spectrograph (Andor SR-750-B2, within 435nm to 700nm - line broadening measurements) to measure ion temperature and transport parameters inside the Hall current channel and the ejected plasma plume. In order to control the argon fuel purity a mass spectrometer is also planned to be used in the future. We are developing at this time a setup for detailed thrust and specific impulse measurements. We also research any relevant plasma physics phenomena that may significantly interfere on PHALL thruster performance due to the occurrence of instabilities inside and outside of the Hall current channel. In order to better understand the turbulence and plasma oscillations that occur during the thruster operation, we have a wide frequency range instability detection system based on a RF detection probe connected to a Spectrum Analyzer (Agilent CSA 100 khz-6 Ghz). Instabilities on the PHALL discharge current are monitored in a real time data acquisition system, based on a PCI-DAS 1602/12 board containing 16 analogic inputs, 24 digital channels operating within a 330 KHz sampling rate.

4. Applications and developments for future Brazilian Space Missions

Future developments of our program will include an experimental PHALL lifetime test system assembly in a vacuum system with a much bigger volume and pumping speed capability. We will

Journal of Physics: Conference Series **641** (2015) 012016 doi:10.1088/1742-6596/641/1/012016

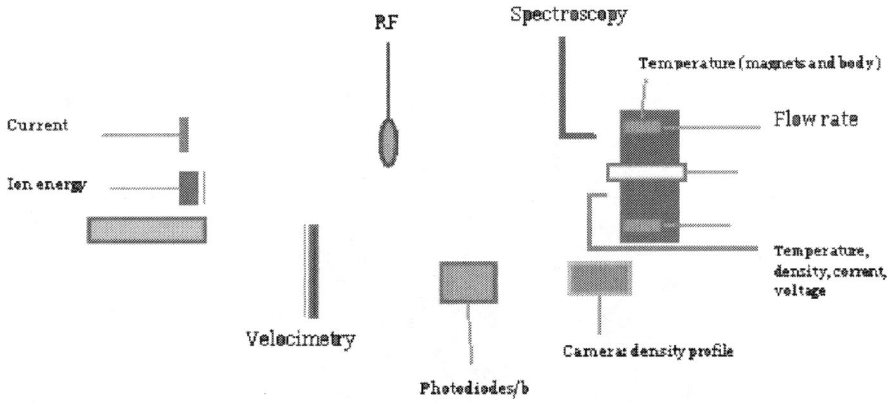

Figure 4. Representation of our Integrated Plasma Diagnostic System.

compare these results with correspondent preliminary simulation results. We will also develop for the bigger chamber a new direct thrust and specific impulse measurement instrumentation system in order to obtain detailed performance characteristics of our PHALL thruster. We are also in the process of concluding a completely automated system for plasma diagnostic measurement which will provide all plasma characteristics and parameters in real time on the spot.

Finally, based on the measured parameters of the accelerated plasma, we are able to present an aptitude figure (see table 1below) for PHALL I and PHALL II. It contains all working parameters such as the specific impulse, total thrust, propellant flow rate, electric power, efficiency and all parameters needed to foresee what type of space mission will be suitable for the thrusters we are developing.

Table 1. Working parameters of PHALL I and PHALL II.

	PHALL-I	**PHALL-II**
Thrust (mN)	26.5 - 126.0	23.0 - 150.0
Thrust power (N/m^2)	1.5	2.4 - 8.7
Specific Impulse (s)	803.5 - 1607.0	1500.0 - 2000.0
Power (W)	1250.0	500.0 - 1000.0
Mass flow rate (mg/s)	5.0	2.0
Electrical efficiency (%)	67.2	80.0
Total efficiency (%)	8.4	45.0 - 60.0

We show numerical simulations results for satellite orbit raising from an altitude of 700 km to 36000 km using a PHALL II operating in the 100mN 500mN thrust range for two situations: in the first case, the satellite is in a circular orbit and is subjected to a continuous thrust tangent to the orbit. In the second case, the satellite is on an elliptic orbit with pericenter of 700 km and apocenter tangent to the geosynchronous orbit. However, in this case we have applied a continuously pulsed thrust with direction tangent to the orbit at the apocenter. The second situation resulted in a faster orbit transfer. In order to perform these calculations, integration techniques of spacecraft trajectory were used. The main simulation parameters

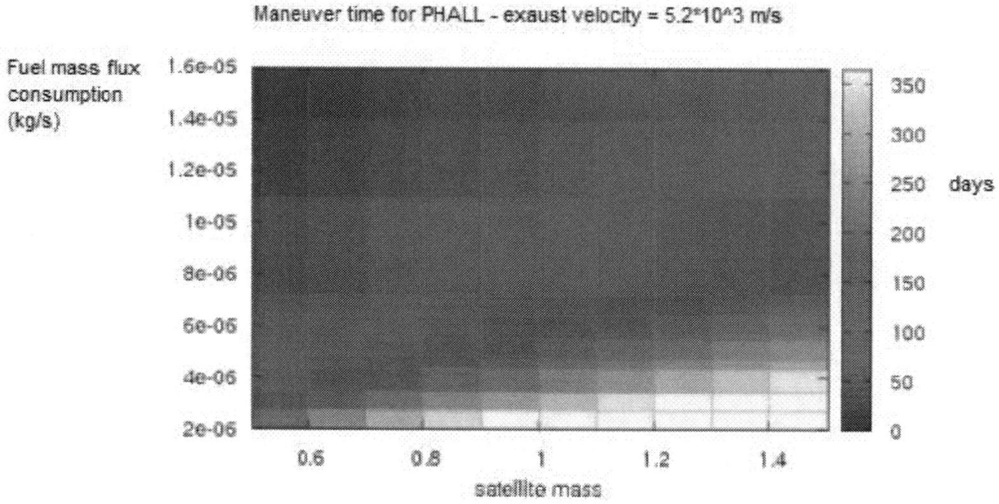

Figure 5. Maneuver time to send a spacecraft to geostationary orbit, from 700 km altitude circular orbit, using continuous thrust as a function of propellant mass flux (measured in laboratory for PHALL II) and of satellite mass.

were: orbit raising time, propellant mass, satellite mass, thrust specific impulse and exhaust velocity. Figure 5 shows the maneuver time to send a spacecraft to geostationary orbit, from 700 km altitude circular orbit, using continuous thrust as a function of propellant mass flux (measured in laboratory for PHALL II) and of satellite mass.

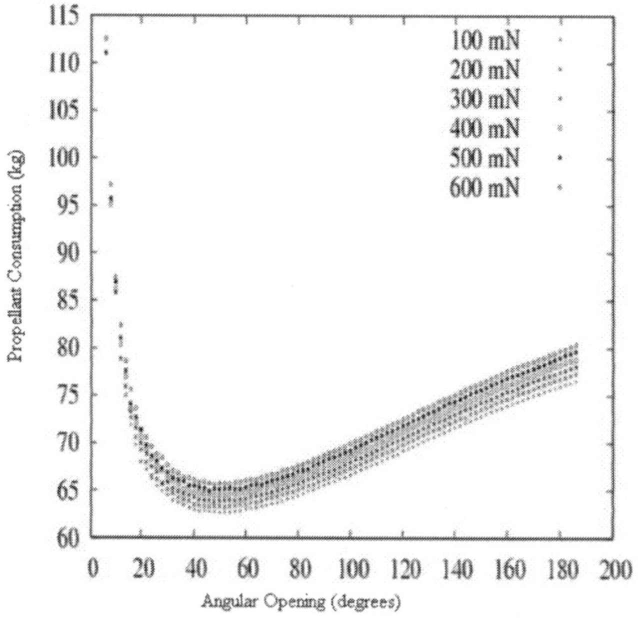

Figure 6. Propellant mass consumption as a function of the apocentral angular opening for various thrusts.

The numerical calculation of propellant consumption for orbital transfer maneuvering are shown in Figure 6. The propellant mass is calculated for pulsed thrust in the correspondent apocentral orbit angle opening. The time necessary for circularized orbit using pulsed thruster, supposing satellite geostationary orbit crossing apocenter, with pericenter at 700 km altitude, for several values of PHALL II total thrust is shown in Figure 7.

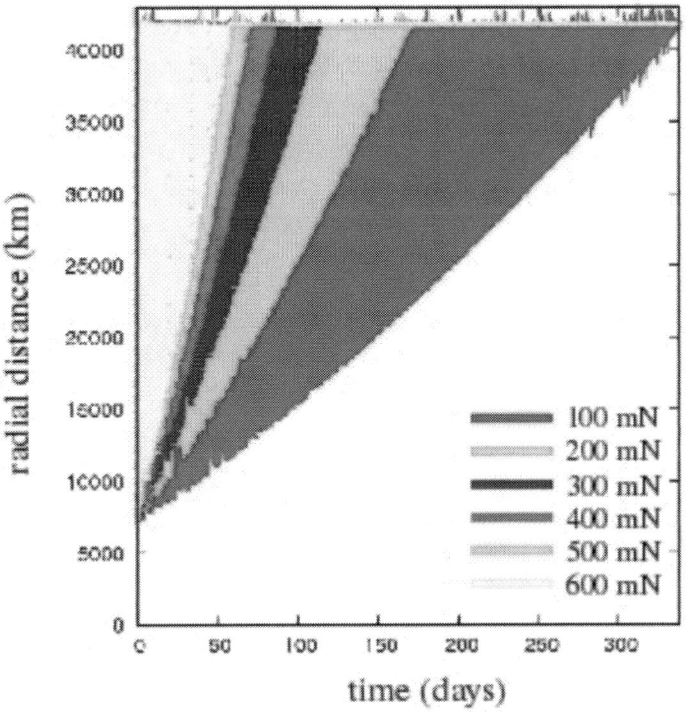

Figure 7. Time necessary for circularized orbit using pulsed thruster, supposing satellite geostationary orbit crossing apocenter, with pericenter at 700 km altitude, for several values of PHALL total thrust.

5. Conclusion

We have come a long way in our study of plasma propulsion of the Hall current type and have progressed from PHALL I to II and are preparing even a third further improvement which will be disclosed soon. We are developing contacts in order to try and send one of these to the outer space for real life testing. Possible applications of PHALL on future Brazilian space missions dedicated to geostationary satellite orbit maneuvering from low earth to geosynchronous orbit is foreseen. PHALL experimental results and computer calculations have shown that Electric propulsion with Hall thrusters can improve and give important contributions to future Brazilian space program missions.

6. References

[1] B. S. Moraes, J. L. Ferreira, D. C. Mouro, O. C. Winter, I. S. Ferreira. "Journal of Physics Conferences Series, vol. 1, pp. 223-254, 2011.

[2] J. L. Ferreira, I. S. Ferreira, B. S. Moraes, J. C. Santos, R. Miranda, P. Gessini, G. Possa, L. T. C. Habl. "Proc. of the 33rd International Electric Propulsion Conference IEPC 2013", pg. 418, Washington D.C. USA, 2013.

Journal of Physics: Conference Series **641** (2015) 012017 doi:10.1088/1742-6596/641/1/012017

Ball's motion, sliding friction, and internal load distribution in a high-speed ball bearing subjected to a combined radial, thrust, and moment load, applied to the inner ring's center of mass: *Mathematical model*

Mário César Ricci
Space Mechanics and Control Division, National Institute for Space Research, Av. dos Astronautas, 1758, 12227-010, São José dos Campos, SP, Brazil

E-mail: mariocesarricci@uol.com.br

Abstract. A set of non-linear algebraic equations, which must to be solved using a numerical procedure, for ball's motion, sliding friction and internal loading distribution computation in a high-speed, single-row, angular-contact ball bearing, subjected to a known combined radial, thrust and moment load, which must be applied to the inner ring's centre of mass, is introduced. For each step of the procedure it is required the iterative solution of $9Z + 3$ simultaneous non-linear equations – where Z is the number of the balls – to yield exact solution for contact angles, ball attitude angles, rolling radii, normal contact deformations and axial, radial, and angular deflections of the inner ring with respect the outer ring. While the focus of this work is obtaining the steady state forces and moments equilibrium conditions on the balls, under the selected loading, the numerical aspects of the procedure are treated in a companion paper. The numerical results derived from the described procedure shall be published later.

1. Introduction

Ball and roller bearings, generically called *rolling bearings*, are commonly used machine elements. They are employed to permit rotary motions of, or about, shafts in simple commercial devices such as bicycles, roller skates, and electric motors. They are also used in complex engineering mechanisms such as aircraft gas turbines, rolling mills, dental drills, gyroscopes, reaction and momentum wheels, and power transmissions.

The standardized forms of ball or roller bearings permit rotary motion between two machine elements and always include a complement of ball or rollers that maintain the shaft and a usually stationary supporting structure, frequently called *housing*, in a radially or axially spaced-apart relationship. Usually, a bearing may be obtained as a unit, which includes two steel rings each of which has a hardened raceway on which hardened balls or rollers roll. The balls or rollers, also called *rolling elements*, are usually held in an angularly spaced relationship by a *cage*, also called a *separator* or *retainer*.

There are many different kinds of rolling bearings. This work is concerned with *single-row angular-contact ball bearings* – see figure 1 – which are designed to support combined radial and thrust loads or heavy thrust loads depending on the *contact angle* magnitude. The bearings having large contact angle can support heavier thrust loads. The figure 1 shows bearings having small and

Journal of Physics: Conference Series **641** (2015) 012017 doi:10.1088/1742-6596/641/1/012017

large contact angles. The bearings generally have groove curvature radii in the range of 52-53% of the ball diameter. The contact angle does not usually exceed 40°.

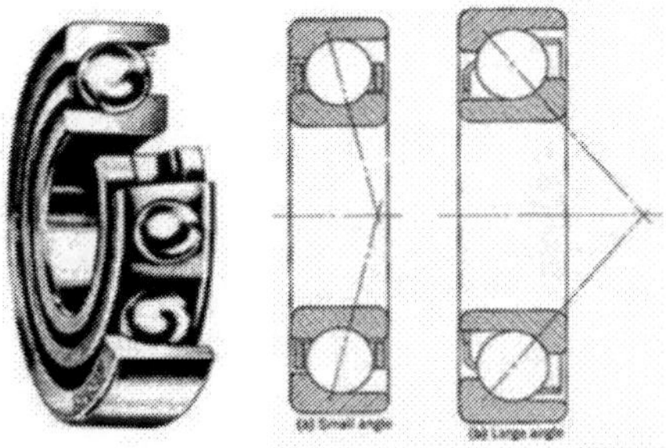

Figure 1. Angular-contact ball bearing.

This work is devoted to study of internal load distribution in a *high-speed* angular-contact ball bearing. Several researchers have studied the subject of internal load distribution in a *statically loaded* angular-contact ball bearing (see [1]-[10]). The methods developed by them to calculate distribution of load among the balls and rollers of rolling bearings can be used in most bearing applications because rotational speeds are usually slow to moderate. Under these speed conditions, the effects of rolling element centrifugal forces and gyroscopic moments are negligible. At high speeds of rotation these body forces become significant, tending to alter contact angles and clearance. Thus, they can affect the static load distribution to a great extension.

Harris [11] described methods for internal loading distribution in statically loaded bearings addressing pure radial; pure thrust (centric and eccentric loads); combined radial and thrust load, which uses radial and thrust integrals introduced by Sjoväll; and for ball bearings under combined radial, thrust, and moment load, initially due to Jones.

The first great contribution to the study of ball motion, sliding friction and internal load distribution in a *high-speed* angular-contact ball bearing must be credited to A B Jones [12]-[13]. Harris describes the orbital, pivotal and spinning ball's motions and load distribution in ball bearings, in general reproducing the Jones's developments. In this work the Jones's works is revisited and differences are introduced under the yoke of critical analysis, which will be detailed. Then, particularly, in this work, a set of non-linear algebraic equations, which must to be solved using a numerical procedure, for ball's motion, sliding friction and internal loading distribution computation in a high-speed, single-row, angular-contact ball bearing, subjected to a known combined radial, thrust and moment load, which must be applied to the inner ring center of mass, is introduced. For each step of the procedure it is required the iterative solution of $9Z + 3$ simultaneous non-linear equations – where Z is the number of the balls – to yield exact solution for contact angles, ball attitude angles, rolling radii, normal contact deformations and axial, radial, and angular deflections of the inner ring with respect the outer ring. While the focus of this work is obtaining the steady state forces and moments equilibrium conditions on the balls, under the selected external loading, the numerical aspects of the procedure are treated in a companion paper. The numerical results derived from the described procedure shall be published later.

2. Mathematical model

Having defined in other works analytical expressions for bearing geometry and the contact stress and deformations for a given ball or roller-raceway contact (point or line loading) in terms of load (see, e.g., [12]) it is possible to consider how the bearing load is distributed among the rolling elements. In this section a specific internal loading distribution resulting from a combined radial, thrust, and moment external load, which must be applied to the center of mass of the inner ring of a high speed ball bearing, is considered.

The figure 2 shows the displacements of an inner ring related to the outer ring due to a generalized loading system including radial, axial, and moment loads. The figure 3 shows the relative angular position of each ball in the bearing.

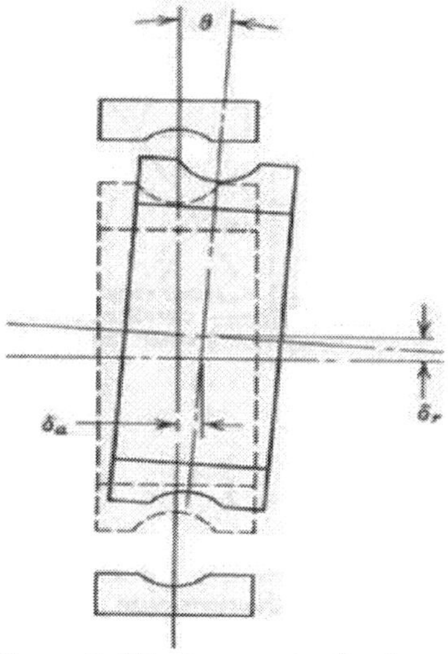

Figure 2. Displacements of an inner ring (outer ring fixed) due to a combined radial, axial, and moment external loading.

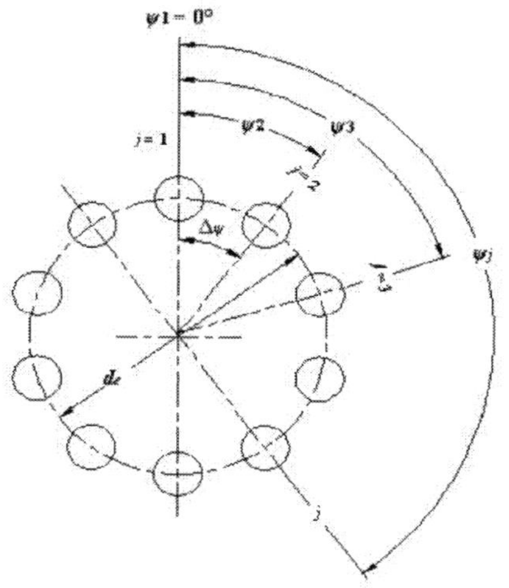

Figure 3. Ball angular positions in the radial plane that is perpendicular to the bearing's axis of rotation, $\Delta\psi = 2\pi/Z$, $\psi_j = 2\pi(j-1)/Z$, $j = 1 \ldots Z$, in which Z is the number of balls.

Let a ball bearing with Z balls, each with diameter D, symmetrically distributed about a pitch circle according to figure 3, to be subjected to a combined radial, thrust, and moment load applied to the inner ring's center of mass. Then, a *relative axial displacement*, δ_a, a *relative angular displacement*, θ, and a *relative radial displacement*, δ_r, between the inner and outer ring raceways may be expected according figure 2. Let $\psi = 0$ to be the angular position of the maximum loaded ball.

Under zero load the centers of raceway groove curvature radii are separated by a distance A given by

$$A = (f_o + f_i - 1)D, \tag{1}$$

in which f_o, f_i are the conformities for outer and inner raceways, respectively.

Under an applied static load, the distance s between centers will increase from A to A plus the amount of the contact deformation δ_i plus δ_o, as show by figure 4. The line of action between centers is collinear with A. If, however, a centrifugal force acts on the ball, then because the inner and outer raceway contact angles are dissimilar, the line of action between raceway groove curvature radii centers is not collinear with A, but is discontinuous as indicated by figure 5. It is assumed in figure 5 that the outer raceway groove curvature center is fixed in space and the inner raceway groove curvature center moves relative to that fixed center. Moreover, the ball center shifts by virtue of the dissimilar contact angles.

The figure 5 when compared with similar figures in [11] and [13] shows minor differences. The inner contact angle must be $\beta_{ij} + \theta\cos\psi_j$ rather than β_{ij}, to take into account the tilting of the rigid inner ring with respect the rigid outer ring, during the external loading application. Furthermore, since the problem is to be solved numerically, no makes sense to linearize the distances between the final and initial inner raceway groove curvature center positions, as done in previous works.

In accordance with figure 5 the distance between the fixed outer raceway groove curvature center and the final position of the ball center at any ball location j is

Journal of Physics: Conference Series **641** (2015) 012017 doi:10.1088/1742-6596/641/1/012017

$$\Delta_{oj} = r_o - \frac{D}{2} + \delta_{oj}. \tag{2}$$

Since $r_o = f_o D$,

$$\Delta_{oj} = (f_o - 0.5)D + \delta_{oj}. \tag{3}$$

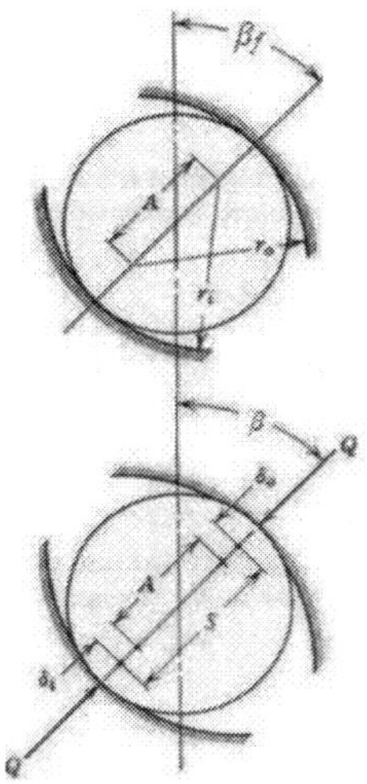

Figure 4. (a) Ball-raceway contact before loading; (b) Ball-raceway contact under load.

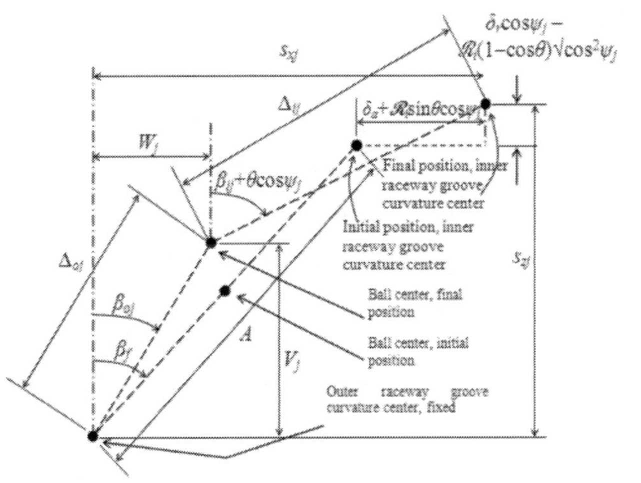

Figure 5. Positions of ball center and raceway groove curvature centers at angular position ψ_j with and without applied load.

Similarly, the distance between the moving inner raceway groove curvature center and the final position of the ball center at any ball location j is

$$\Delta_{ij} = (f_i - 0.5)D + \delta_{ij}, \tag{4}$$

in which δ_{oj} and δ_{ij} are the normal contact deformations at the outer and inner raceway contacts, respectively.

In accordance with the relative axial displacement between inner and outer rings mass centers, δ_a, and the relative angular displacement θ, the axial distance between inner and outer raceway groove curvature centers at ball position j is

$$s_{xj} = A\sin\beta_f + \delta_a + \mathscr{R}_i\sin\theta\cos\psi_j, \tag{5}$$

in which

$$\mathscr{R}_i = ½d_e + (f_i - ½)D\cos\beta_f \tag{6}$$

is the radius to locus of inner raceway groove curvature centers, d_e is the unloaded pitch diameter, and β_f is the unloaded contact angle. Further, in accordance with the relative radial displacement between inner and outer rings mass centers, δ_r, and the relative angular displacement θ, the radial distance between inner and outer groove curvature centers at each ball location j is

$$s_{zj} = A\cos\beta_f + \delta_r\cos\psi_j - \mathscr{R}_i(1 - \cos\theta)\sqrt{\cos^2\psi_j}. \tag{7}$$

Since the iterative techniques of the Newton-Raphson method will be used to solve the associated nonlinear equations, the angles β_{oj} and β_{ij} are best stated in terms of the co-ordinates V and W, in figure 5. Then

$$\sin\beta_{oj} = \frac{W_j}{(f_o - 0.5)D + \delta_{oj}}, \tag{8}$$

$$\cos\beta_{oj} = \frac{V_j}{(f_o - 0.5)D + \delta_{oj}}, \tag{9}$$

$$\sin(\beta_{ij}+\theta\cos\psi_j)=\frac{s_{xj}-W_j}{(f_i-0.5)D+\delta_{ij}}, \tag{10}$$

$$\cos(\beta_{ij}+\theta\cos\psi_j)=\frac{s_{zj}-V_j}{(f_i-0.5)D+\delta_{ij}}. \tag{11}$$

Similarly, the ball angular speed about its own center pitch and yaw angles, α_j and α'_j, are best stated in terms of the ball angular velocity components: $\omega_{x'j}$, $\omega_{y'j}$, and $\omega_{z'j}$; in which x', y', and z' are the axes of the coordinate frame whose origin is at the ball center; x' is parallel to the longitudinal axis of the bearing around which the balls circulate in its orbital motion, and z' is the radial axis. Then

$$\sin\alpha_j=\frac{\omega_{z'j}}{\sqrt{\omega_{x'j}^2+\omega_{y'j}^2+\omega_{z'j}^2}}, \tag{12}$$

$$\cos\alpha_j=\frac{\sqrt{\omega_{x'j}^2+\omega_{y'j}^2}}{\sqrt{\omega_{x'j}^2+\omega_{y'j}^2+\omega_{z'j}^2}}, \tag{13}$$

$$\sin\alpha'_j=\frac{\omega_{y'j}}{\sqrt{\omega_{x'j}^2+\omega_{y'j}^2}}, \tag{14}$$

$$\cos\alpha'_j=\frac{\omega_{x'j}}{\sqrt{\omega_{x'j}^2+\omega_{y'j}^2}}. \tag{15}$$

Using the Pythagorean Theorem, it can be seen from figure 5 that

$$(s_{zj}-V_j)^2+(s_{xj}-W_j)^2-[(f_i-0.5)D+\delta_{ij}]^2=0=\epsilon_j, \tag{16}$$

$$V_j^2+W_j^2-[(f_o-0.5)D+\delta_{oj}]^2=0=\epsilon_{j+Z}. \tag{17}$$

From (12)-(15)

$$\omega_{x'j}^2+\omega_{y'j}^2+\omega_{z'j}^2-\omega_{Rj}^2=0=\epsilon_{j+2Z}, \qquad \omega_{Rj}=\sqrt{\omega_{x'j}^2+\omega_{y'j}^2+\omega_{z'j}^2}. \tag{18}$$

For steady state operation of a ball bearing at high speed, the forces and moments acting on each ball are as shown by figure 6.

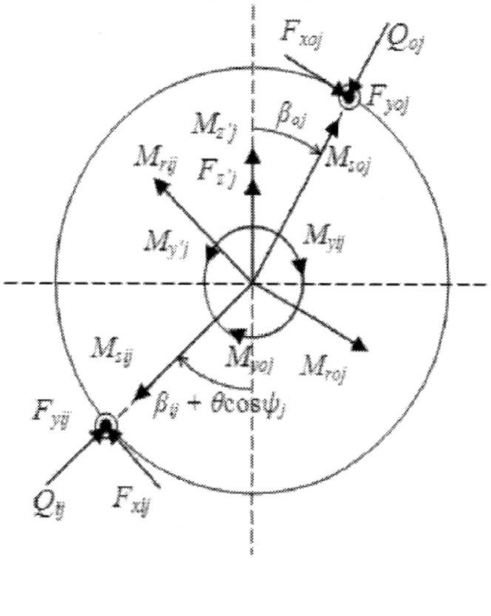

Figure 6. Ball loading at angular position ψ_j.

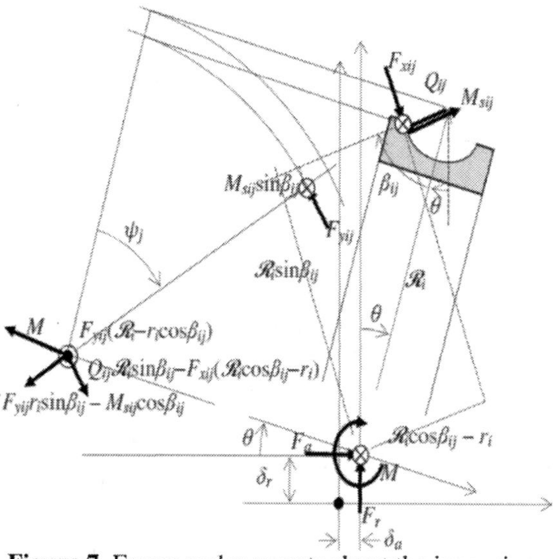

Figure 7. Forces and moments about the inner ring center of mass.

Journal of Physics: Conference Series **641** (2015) 012017 doi:10.1088/1742-6596/641/1/012017

The normal ball loads are related to normal contact deformations by

$$Q_{oj} = K_{oj}\delta_{oj}^{1.5}, \qquad Q_{ij} = K_{ij}\delta_{ij}^{1.5}, \tag{19}$$

in which K_{oj} and K_{ij} are functions of contact angles [11].

From figure 6 considering the three axes equilibrium forces:

$$Q_{ij}\sin(\beta_{ij} + \theta\cos\psi_j) - Q_{oj}\sin\beta_{oj} - F_{xij}\cos(\beta_{ij} + \theta\cos\psi_j) + F_{xoj}\cos\beta_{oj} = 0, \tag{20}$$

$$Q_{ij}\cos(\beta_{ij} + \theta\cos\psi_j) - Q_{oj}\cos\beta_{oj} + F_{xij}\sin(\beta_{ij} + \theta\cos\psi_j) - F_{xoj}\sin\beta_{oj} + F_{z'j} = 0, \tag{21}$$

$$F_{yoj} + F_{yij} = 0 = \epsilon_{j+5Z}, \tag{22}$$

Substituting (8)-(11) and (19) into (20)-(21) yields

$$\frac{F_{xoj}V_j - K_{oj}\delta_{oj}^{1.5}W_j}{(f_o-0.5)D+\delta_{oj}} + \frac{K_{ij}\delta_{ij}^{1.5}(s_{xj}-W_j) - F_{xij}(s_{zj}-V_j)}{(f_i-0.5)D+\delta_{ij}} = 0 = \epsilon_{j+3Z}, \tag{23}$$

$$\frac{K_{oj}\delta_{oj}^{1.5}V_j + F_{xoj}W_j}{(f_o-0.5)D+\delta_{oj}} - \frac{K_{ij}\delta_{ij}^{1.5}(s_{zj}-V_j) + F_{xij}(s_{xj}-W_j)}{(f_i-0.5)D+\delta_{ij}} - F_{z'j} = 0 = \epsilon_{j+4Z}. \tag{24}$$

From figure 6 considering the three axes equilibrium moments:

$$-M_{sij}\sin(\beta_{ij} + \theta\cos\psi_j) + M_{soj}\sin\beta_{oj} - M_{Rij}\cos(\beta_{ij} + \theta\cos\psi_j) + M_{Roj}\cos\beta_{oj} = 0, \tag{25}$$

$$-M_{sij}\cos(\beta_{ij} + \theta\cos\psi_j) + M_{soj}\cos\beta_{oj} + M_{Rij}\sin(\beta_{ij} + \theta\cos\psi_j) - M_{Roj}\sin\beta_{oj} + M_{z'j} = 0, \tag{26}$$

$$M_{y'j} - M_{yij} - M_{yoj} = 0 = \epsilon_{j+8Z}. \tag{27}$$

Substituting (8)-(11) into (25)-(26) yields

$$\frac{M_{Roj}V_j + M_{soj}W_j}{(f_o-0.5)D+\delta_{oj}} - \frac{M_{sij}(s_{xj}-W_j) + M_{Rij}(s_{zj}-V_j)}{(f_i-0.5)D+\delta_{ij}} = 0 = \epsilon_{j+6Z}, \tag{28}$$

$$\frac{M_{soj}V_j - M_{Roj}W_j}{(f_o-0.5)D+\delta_{oj}} - \frac{M_{sij}(s_{zj}-V_j) - M_{Rij}(s_{xj}-W_j)}{(f_i-0.5)D+\delta_{ij}} + M_{z'j} = 0 = \epsilon_{j+7Z}. \tag{29}$$

The centrifugal force acting on the ball at angular position ψ_j is given by [11]

$$F_{z'j} = \frac{1}{2} m d_{mj}\omega_{mj}^2, \tag{30}$$

in which m is the mass of ball,

$$d_{mj} = d_e + 2[V_j - (f_o - ½)D\cos\beta_f] \tag{31}$$

is the operational ball's pitch diameter at position j, and ω_{mj} is the absolute orbital speed of the ball about of the bearing axis.

Substituting the identity $\omega_{mj}^2 = (\omega_{mj}/\omega)^2\omega^2$ in (30) gives

$$F_{z'j} = \frac{1}{2} m\omega^2 d_{mj}\left(\frac{\omega_{mj}}{\omega}\right)^2, \tag{32}$$

in which ω is the absolute angular velocity of the rotating ring.

For the outer race to be stationary $\omega_{mj} = -\omega_{oj}$, $\omega = \omega_{ij} + \omega_{mj}$,

$$\frac{\omega_{mj}}{\omega} = \frac{1}{1 + \frac{r'_{ij}\left\{\frac{d_{mj}}{2}[(f_o-0.5)D+\delta_{oj}] + r'_{oj}V_j\right\}[\omega_{x'j}(s_{zj}-V_j) + \omega_{z'j}(s_{xj}-W_j)]}{r'_{oj}\left\{\frac{d_{mj}}{2}[(f_i-0.5)D+\delta_{ij}] - r'_{ij}(s_{zj}-V_j)\right\}(\omega_{x'j}V_j + \omega_{z'j}W_j)}}, \tag{33}$$

and

$$\frac{\omega_{Rj}}{\omega} = \frac{-\sqrt{\omega_{x'j}^2 + \omega_{y'j}^2 + \omega_{z'j}^2}}{\frac{r'_{oj}(\omega_{x'j}V_j + \omega_{z'j}W_j)}{\frac{d_{mj}}{2}[(f_o-0.5)D+\delta_{oj}] + r'_{oj}V_j} + \frac{r'_{ij}[\omega_{x'j}(s_{zj}-V_j) + \omega_{z'j}(s_{xj}-W_j)]}{\frac{d_{mj}}{2}[(f_i-0.5)D+\delta_{ij}] - r'_{ij}(s_{zj}-V_j)}}, \tag{34}$$

Journal of Physics: Conference Series **641** (2015) 012017 doi:10.1088/1742-6596/641/1/012017

in which ω_{ij}, ω_{oj} are the angular velocities about the bearing axis of the inner and outer rings with respect to the ball at position j, and r'_{ij}, r'_{oj} are the inner and outer rolling radii [11].

Likewise, for the inner race to be stationary $\omega_{mj} = -\omega_{ij}$, $\omega = \omega_{oj} + \omega_{mj}$,

$$\frac{\omega_{mj}}{\omega} = \frac{1}{1+\frac{r'_{oj}\left\{\frac{d_{mj}}{2}\left[(f_i-0.5)D+\delta_{ij}\right]-r'_{ij}(s_{zj}-V_j)\right\}\left(\omega_{x'j}V_j+\omega_{z'j}W_j\right)}{r'_{ij}\left\{\frac{d_{mj}}{2}\left[(f_o-0.5)D+\delta_{oj}\right]+r'_{oj}V_j\right\}\left[\omega_{x'j}(s_{zj}-V_j)+\omega_{z'j}(s_{xj}-W_j)\right]}} \tag{35}$$

and ω_{Rj}/ω is given by (34) with opposite sign.

Similarly, the gyroscopic moments acting on the ball at angular position ψ_j are given by [11]

$$M_{y'j} = J\omega^2\left(\frac{\omega_{Rj}}{\omega}\right)\left(\frac{\omega_{mj}}{\omega}\right)\frac{\omega_{z'j}}{\sqrt{\omega^2_{x'j}+\omega^2_{y'j}+\omega^2_{z'j}}}, \tag{36}$$

and

$$M_{z'j} = -J\omega^2\left(\frac{\omega_{Rj}}{\omega}\right)\left(\frac{\omega_{mj}}{\omega}\right)\frac{\omega_{y'j}}{\sqrt{\omega^2_{x'j}+\omega^2_{y'j}+\omega^2_{z'j}}}, \tag{37}$$

in which J is the ball's mass moment of inertia.

The friction forces due to sliding in the x and y-directions of inner and outer ball-raceway elliptical contact areas are given by [11]

$$F_{xij} = \frac{3\mu K_{ij}\delta_{ij}^{1.5}}{2\pi a_{ij}b_{ij}}\int_{-a_{ij}}^{a_{ij}}\int_{-b_{ij}\sqrt{1-\frac{x_{ij}^2}{a_{ij}^2}}}^{b_{ij}\sqrt{1-\frac{x_{ij}^2}{a_{ij}^2}}}\sqrt{1-\frac{x_{ij}^2}{a_{ij}^2}-\frac{y_{ij}^2}{b_{ij}^2}}\sin\gamma_{ij}dy_{ij}\,dx_{ij}, \tag{38}$$

$$F_{xoj} = \frac{3\mu K_{oj}\delta_{oj}^{1.5}}{2\pi a_{oj}b_{oj}}\int_{-a_{oj}}^{a_{oj}}\int_{-b_{oj}\sqrt{1-\frac{x_{oj}^2}{a_{oj}^2}}}^{b_{oj}\sqrt{1-\frac{x_{oj}^2}{a_{oj}^2}}}\sqrt{1-\frac{x_{oj}^2}{a_{oj}^2}-\frac{y_{oj}^2}{b_{oj}^2}}\sin\gamma_{oj}dy_{oj}\,dx_{oj}, \tag{39}$$

$$F_{yij} = \frac{3\mu K_{ij}\delta_{ij}^{1.5}}{2\pi a_{ij}b_{ij}}\int_{-a_{ij}}^{a_{ij}}\int_{-b_{ij}\sqrt{1-\frac{x_{ij}^2}{a_{ij}^2}}}^{b_{ij}\sqrt{1-\frac{x_{ij}^2}{a_{ij}^2}}}\sqrt{1-\frac{x_{ij}^2}{a_{ij}^2}-\frac{y_{ij}^2}{b_{ij}^2}}\cos\gamma_{ij}dy_{ij}\,dx_{ij}, \tag{40}$$

$$F_{yoj} = \frac{3\mu K_{oj}\delta_{oj}^{1.5}}{2\pi a_{oj}b_{oj}}\int_{-a_{oj}}^{a_{oj}}\int_{-b_{oj}\sqrt{1-\frac{x_{oj}^2}{a_{oj}^2}}}^{b_{oj}\sqrt{1-\frac{x_{oj}^2}{a_{oj}^2}}}\sqrt{1-\frac{x_{oj}^2}{a_{oj}^2}-\frac{y_{oj}^2}{b_{oj}^2}}\cos\gamma_{oj}dy_{oj}\,dx_{oj}, \tag{41}$$

in which μ is the friction coefficient; a_{ij}, b_{ij}, a_{oj}, and b_{oj} are semimajor and semiminor-axes of inner and outer pressure ellipses; x_{ij}, y_{ij}, x_{oj}, y_{oj} are the co-ordinates of an element of area, $dydx$, inside the contact ellipse, which has a resultant velocity of slip V of the race on the ball acting at the angle γ with respect to the y-direction, which are given by

$$\gamma_{ij} = \tan^{-1}\frac{y_{ij}-\frac{V_{xij}}{\omega_{sij}}}{x_{ij}+\frac{V_{yij}}{\omega_{sij}}}, \qquad \gamma_{oj} = \tan^{-1}\frac{y_{oj}-\frac{V_{xoj}}{\omega_{soj}}}{x_{oj}+\frac{V_{yoj}}{\omega_{soj}}}. \tag{42}$$

V_{xij}, V_{xoj}, V_{yij}, V_{yoj}, ω_{sij}, and ω_{soj} are the relative linear and angular slip velocities of inner and outer races with respect to the ball located at position j. The terms involving these velocities for use in (42) are given by [11]

$$\frac{V_{xij}}{\omega_{sij}} = \frac{\left(\sqrt{R_i^2 - x_{ij}^2} - \sqrt{R_i^2 - a_{ij}^2} + \sqrt{\left(\frac{D}{2}\right)^2 - a_{ij}^2}\right)\left[(f_i - 0.5)D + \delta_{ij} - \frac{r'_{ij}}{\frac{d_{mj}}{2}}(s_{zj} - V_j)\right]\omega_{y'j}}{\omega_{x'j}(s_{xj} - W_j) - \omega_{z'j}\left[s_{zj} - V_j - \frac{(f_i - 0.5)D + \delta_{ij}}{\frac{d_{mj}}{2}} r'_{ij}\right]}, \tag{43}$$

$$\frac{V_{yij}}{\omega_{sij}} = \frac{\left(\sqrt{R_i^2 - x_{ij}^2} - \sqrt{R_i^2 - a_{ij}^2} + \sqrt{\left(\frac{D}{2}\right)^2 - a_{ij}^2} - r'_{ij}\right)\left[\omega_{x'j}(s_{zj} - V_j) + \omega_{z'j}(s_{xj} - W_j)\right]}{\omega_{x'j}(s_{xj} - W_j) - \omega_{z'j}\left[s_{zj} - V_j - \frac{(f_i - 0.5)D + \delta_{ij}}{\frac{d_{mj}}{2}} r'_{ij}\right]}, \tag{44}$$

$$\frac{V_{xoj}}{\omega_{soj}} = \frac{-\left(\sqrt{R_o^2 - x_{oj}^2} - \sqrt{R_o^2 - a_{oj}^2} + \sqrt{\left(\frac{D}{2}\right)^2 - a_{oj}^2}\right)\left[(f_o - 0.5)D + \delta_{oj} + \frac{r'_{oj}}{\frac{d_{mj}}{2}} V_j\right]\omega_{y'j}}{\omega_{x'j} W_j - \omega_{z'j}\left[V_j + \frac{(f_o - 0.5)D + \delta_{oj}}{\frac{d_{mj}}{2}} r'_{oj}\right]}, \tag{45}$$

$$\frac{V_{yoj}}{\omega_{soj}} = \frac{\left(\sqrt{R_o^2 - x_{oj}^2} - \sqrt{R_o^2 - a_{oj}^2} + \sqrt{\left(\frac{D}{2}\right)^2 - a_{oj}^2} - r'_{oj}\right)\left(\omega_{x'j} V_j + \omega_{z'j} W_j\right)}{\omega_{x'j} W_j - \omega_{z'j}\left[V_j + \frac{(f_o - 0.5)D + \delta_{oj}}{\frac{d_{mj}}{2}} r'_{oj}\right]}, \tag{46}$$

in which R_i and R_o are the curvature radii of deformed surfaces, given by

$$R_i = \frac{2f_i D}{2f_i + 1}, \qquad R_o = \frac{2f_o D}{2f_o + 1}. \tag{47}$$

The total frictional moments of the friction forces about the normal at the center of the contact ellipse are [11]

$$M_{sij} = \frac{3\mu K_{ij}\delta_{ij}^{1.5}}{2\pi a_{ij} b_{ij}} \int_{-a_{ij}}^{a_{ij}} \int_{-b_{ij}\sqrt{1-\frac{x_{ij}^2}{a_{ij}^2}}}^{b_{ij}\sqrt{1-\frac{x_{ij}^2}{a_{ij}^2}}} \sqrt{x_{ij}^2 + y_{ij}^2}\sqrt{1 - \frac{x_{ij}^2}{a_{ij}^2} - \frac{y_{ij}^2}{b_{ij}^2}} \cos\left(\gamma_{ij} - \tan^{-1}\frac{y_{ij}}{x_{ij}}\right) dy_{ij}\, dx_{ij}, \tag{48}$$

$$M_{soj} = \frac{3\mu K_{oj}\delta_{oj}^{1.5}}{2\pi a_{oj} b_{oj}} \int_{-a_{oj}}^{a_{oj}} \int_{-b_{oj}\sqrt{1-\frac{x_{oj}^2}{a_{oj}^2}}}^{b_{oj}\sqrt{1-\frac{x_{oj}^2}{a_{oj}^2}}} \sqrt{x_{oj}^2 + y_{oj}^2}\sqrt{1 - \frac{x_{oj}^2}{a_{oj}^2} - \frac{y_{oj}^2}{b_{oj}^2}} \cos\left(\gamma_{oj} - \tan^{-1}\frac{y_{oj}}{x_{oj}}\right) dy_{oj}\, dx_{oj}. \tag{49}$$

The moments of the friction forces about the y'-axis are [11]

$$M_{yij} = \frac{3\mu K_{ij}\delta_{ij}^{1.5}}{2\pi a_{ij} b_{ij}} \int_{-a_{ij}}^{a_{ij}} \int_{-b_{ij}\sqrt{1-\frac{x_{ij}^2}{a_{ij}^2}}}^{b_{ij}\sqrt{1-\frac{x_{ij}^2}{a_{ij}^2}}} \left(\sqrt{R_i^2 - x_{ij}^2} - \sqrt{R_i^2 - a_{ij}^2} + \sqrt{\left(\frac{D}{2}\right)^2 - a_{ij}^2}\right)\sqrt{1 - \frac{x_{ij}^2}{a_{ij}^2} - \frac{y_{ij}^2}{b_{ij}^2}} \sin\gamma_{ij} dy_{ij}\, dx_{ij}, \tag{50}$$

$$M_{yoj} = \frac{3\mu K_{oj}\delta_{oj}^{1.5}}{2\pi a_{oj} b_{oj}} \int_{-a_{oj}}^{a_{oj}} \int_{-b_{oj}\sqrt{1-\frac{x_{oj}^2}{a_{oj}^2}}}^{b_{oj}\sqrt{1-\frac{x_{oj}^2}{a_{oj}^2}}} \left(\sqrt{R_o^2 - x_{oj}^2} - \sqrt{R_o^2 - a_{oj}^2} + \sqrt{\left(\frac{D}{2}\right)^2 - a_{oj}^2}\right)\sqrt{1 - \frac{x_{oj}^2}{a_{oj}^2} - \frac{y_{oj}^2}{b_{oj}^2}} \sin\gamma_{oj} dy_{oj}\, dx_{oj}. \tag{51}$$

The frictional moments about an axis through the ball center perpendicular to the line defining the contact angle, which line lies in the $x'z'$-plane, are [11]

$$M_{Rij} = \frac{3\mu K_{ij}\delta_{ij}^{1.5}}{2\pi a_{ij} b_{ij}} \int_{-a_{ij}}^{a_{ij}} \int_{-b_{ij}\sqrt{1-\frac{x_{ij}^2}{a_{ij}^2}}}^{b_{ij}\sqrt{1-\frac{x_{ij}^2}{a_{ij}^2}}} \left(\sqrt{R_i^2 - x_{ij}^2} - \sqrt{R_i^2 - a_{ij}^2} + \sqrt{\left(\frac{D}{2}\right)^2 - a_{ij}^2}\right) \sqrt{1 - \frac{x_{ij}^2}{a_{ij}^2} - \frac{y_{ij}^2}{b_{ij}^2}} \cos\gamma_{ij} dy_{ij}\, dx_{ij}, \quad (52)$$

$$M_{Roj} = \frac{3\mu K_{oj}\delta_{oj}^{1.5}}{2\pi a_{oj} b_{oj}} \int_{-a_{oj}}^{a_{oj}} \int_{-b_{oj}\sqrt{1-\frac{x_{oj}^2}{a_{oj}^2}}}^{b_{oj}\sqrt{1-\frac{x_{oj}^2}{a_{oj}^2}}} \left(\sqrt{R_o^2 - x_{oj}^2} - \sqrt{R_o^2 - a_{oj}^2} + \sqrt{\left(\frac{D}{2}\right)^2 - a_{oj}^2}\right) \sqrt{1 - \frac{x_{oj}^2}{a_{oj}^2} - \frac{y_{oj}^2}{b_{oj}^2}} \cos\gamma_{oj} dy_{oj}\, dx_{oj}. \quad (53)$$

Equations (16)-(18), (22)-(24) and (27)-(29) may be solved simultaneously for V_j, W_j, δ_{oj}, δ_{ij}, r'_{oj}, r'_{ij}, $\omega_{x'j}$, $\omega_{y'j}$, and $\omega_{z'j}$ at each ball angular location once values for δ_a, δ_r, and θ are assumed.

An iterative procedure is to be used to solve the equations simultaneously. Since K_{oj} and K_{ij} are functions of contact angle, equations (8)-(11) may be used to establish K_{oj} and K_{ij} values iteratively.

To find the values of δ_a, δ_r, and θ, it remains to establish the equilibrium conditions of forces and moments about the inner ring center of mass, as shown by figure 7, which are

$$F_a - \sum_{j=1}^{Z} \left[\frac{K_{ij}\delta_{ij}^{1.5}(s_{xj} - W_j) - F_{xij}(s_{zj} - V_j)}{(f_i - 0.5)D + \delta_{ij}}\right] = 0 = \epsilon_{9Z+1}, \quad (54)$$

$$F_r - \sum_{j=1}^{Z} \left[\frac{K_{ij}\delta_{ij}^{1.5}(s_{zj} - V_j) + F_{xij}(s_{xj} - W_j)}{(f_i - 0.5)D + \delta_{ij}}\right] \cos\psi_j = 0 = \epsilon_{9Z+2}, \quad (55)$$

$$M - \sum_{j=1}^{Z} \{\mathscr{R}_i[K_{ij}\delta_{ij}^{1.5}\sin\beta_{ij} - F_{xij}(\cos\beta_{ij} - r_i/\mathscr{R}_i)]\cos\psi_j - (F_{yij} r_i \sin\beta_{ij} - M_{sij}\cos\beta_{ij})\sin\psi_j\} = 0 = \epsilon_{9Z+3}, \quad (56)$$

in which F_a, F_r, and M are external forces and moment applied to the inner ring center of mass.

Having computed values for V_j, W_j, δ_{oj}, δ_{ij}, r'_{oj}, r'_{ij}, $\omega_{x'j}$, $\omega_{y'j}$, and $\omega_{z'j}$ at each angular position and knowing F_a, F_r, and M as input conditions the values of δ_a, δ_r, and θ may be computed by equations (54)-(56). After obtaining the primary unknown quantities δ_a, δ_r, and θ, it is necessary to repeat the calculation of V_j, W_j, δ_{oj}, δ_{ij}, r'_{oj}, r'_{ij}, $\omega_{x'j}$, $\omega_{y'j}$, and $\omega_{z'j}$, until compatible values of primary unknown quantities δ_a, δ_r, and θ are obtained.

3. References

[1] Stribeck R 1907 *Trans. ASME* **29** 420-63

[2] Sjoväll H 1933 *Teknisk Tidskrift, Mek* **h.9**

[3] Jones A 1946 *Analysis of Stresses and Deflections* (Bristol, CT: New Departure Engineering Data)

[4] Rumbarger J 1962 *Mach. Des.*

[5] Ricci M C 2009 *Proc. 8th Brazilian Conf. on Dynamics, Control and Appl.* (Bauru, Brazil, 18-22 May)

[6] Ricci M C 2009 *1st Int. Conf. on Comp. Contact Mech., Program and Abstracts* (Lecce, Italy, 16-18 Sept) p. 21-2

[7] Ricci M C 2009 *Proc. 6th Int. Congress of Croatian Soc. Mech.* (Dubrovnik, Croatia Sept. 30 to Oct. 2)

[8] Ricci M C 2009 *Proc. 60th Int. Astron. Congress* (Daejeon, South Korea, October, 12-16)

[9] Ricci, M C 2009 *Math. Probl. Eng.*

[10] Ricci, M C 2010 *Proc. 11th Pan-American Congress of Appl. Mech.*, January, 04-10, Foz do Iguaçu, Brazil

[11] Harris T 2001 *Rolling Bearing Analysis*, 4th ed (New York: John Wiley & Sons)

[12] Jones A B 1959 *ASME J. Basic Eng.* **3** 1-12

[13] Jones A B 1960 *J. Fluids Eng.* **82(2)** 309-20

Journal of Physics: Conference Series **641** (2015) 012018 doi:10.1088/1742-6596/641/1/012018

Ball's motion, sliding friction, and internal load distribution in a high-speed ball bearing subjected to a combined radial, thrust, and moment load, applied to the inner ring's center of mass: *Numerical procedure*

Mário César Ricci
Space Mechanics and Control Division, National Institute for Space Research, Av. dos Astronautas, 1758, 12227-010, São José dos Campos, SP, Brazil

E-mail: mariocesarricci@uol.com.br

Abstract. In a companion paper of this was introduced a set of non-linear algebraic equations for ball's motion, sliding friction and internal loading distribution computation in a high-speed, single-row, angular-contact ball bearing, subjected to a known combined radial, thrust and moment load, which must be applied to the inner ring's center of mass. It was shown there that it is required the iterative solution of $9Z + 3$ simultaneous non-linear equations – where Z is the number of balls – to yield exact solution for contact angles, ball attitude angles, rolling radii, normal contact deformations and axial, radial, and angular deflections of the inner ring with respect the outer ring. The Newton-Rhapson method is to be used to solve the problem. This paper deals with the numerical procedure description. The numerical results derived from the described procedure shall be published later.

1. Introduction

When an external load is applied to one of the rings of a rolling bearing it is transmitted through rolling elements to the other ring. Because the internal load distribution on the rolling elements is an important operating characteristic of a bearing a great number of authors have addressing the problem. A literature review on the subject can be found in [1] and [2], in which a mathematical model for necessary radial displacement between rings, and a mathematical model for external radial load, so that the q-th rolling element passes to participate in the load transfer were presented.

In [3] a model was developed, which enables a very simple determination of the number of active rolling elements participating in an external load transfer, depending on the bearing type and internal radial clearance.

In [4] the theoretical analysis of a single-row radial bearing with radial clearance under constant external radial load was presented. The analysis was focused on finding the rolling element deflection that allows determining the number of active rolling elements that participate in the load transfer. Taking into account the bearing internal geometry, a mathematical model to calculate the rolling elements deflections during the bearing rotation has been derived.

In [5], [6] and [7] capacitive probes were inserted into the fixed ring of the bearing such that forms with the raceway a capacitor with variable gap that depends on the transmitted load by the rolling element. A numerical model of this capacitor's capacitance as a function of transmitted load by the rolling element has been established. An experimental prototype has been established in order to precisely measure the probe's capacitance. Finally, this technique has been generalized with a capacitive probe in front of each rolling element. Thus, knowing the load transmitted by each of the rolling elements, the external load on the bearing of the rotating machine can be easily reconstructed.

The evaluation of change in contact angle due to applied load is vital in order to study the load carrying capacity of large diameter bearings. Analytical and numerical procedures have been developed to calculate various design factors such as contact angle, contact stress and deformation. In [8] the change in contact angle of balls was determined by using FEA. The change in contact angle was compared with analytical, FEA and the Newton-Raphson method. The results show a good agreement with the values calculated using Hertz's relations for deformation. The FEA method was used to get the nodal solution of contact angle, contact Stress and deflection for various loading conditions.

In [9] the dynamic modeling of a centrally supported symmetrical disk-shaft bearing system has been analyzed using Timoshenko beam elements. Intermittent ball bearing contact forces and Muszynska's force [10] at seal-disk interface were considered in the model to simulate a real-time system. Results show that there was a marked effect of each type of nonlinear excitation on the overall system response.

In [11] a wheel bearing life prediction method, which considers the bearing dynamics characteristics, was proposed. The results were compared with existing formulas and static analyses results from structural dynamics commercial software.

In [12] a unidirectional compression spring was used to model the contact between a rolling element and the raceway of a heavy-duty slewing bearing accounting for the supporting structure flexibility and the plastic deformation of the bearing. The spring constant was determined by the load against elastic-plastic deformation relationship of a single rolling element, which was obtained by finite element contact method. The difference between the traditional Hertz contact results and the FEM results is very obvious for the slewing bearings with plastic deformation, such as contact deflection of the rolling elements and the raceway, load distribution on the rolling elements, stress in the raceway and contact pressure between the rolling elements and the raceway. Therefore, the method based on the Hertz contact mechanics theory is not applicable for the performance analysis of the heavy-duty slewing bearing.

The focus of this paper is to describe the numerical procedure for ball's motion, sliding friction and internal load distribution computation in a *high-speed*, single-row, angular contact ball bearing, subjected to a known combined radial, thrust and moment load, which must be applied to the inner ring's center of mass. In a companion paper of this [13] a set of non-linear algebraic equations was introduced in order to accomplish this task. It was shown there that it is required the iterative solution of $9Z + 3$ simultaneous non-linear equations – where Z is the number of balls – to yield exact solution for contact angles, ball attitude angles, rolling radii, normal contact deformations and axial, radial, and angular deflections of the inner ring with respect the outer ring. The Newton-Rhapson method is to be used to solve the problem.

2. Symbols

a Semimajor axis of the projected contact, m; elements of square matrix
b Semiminor axis of the projected contact, m
c Short for cos
d Bearing pitch diameter, m
D Ball's diameter, m
f r/D
F External, friction or centrifugal forces, N
J Ball's mass moment of inertia, kgm^2
K Load-deflection factor, $N/m^{3/2}$
m Ball's mass, kg
M External, gyroscope or frictional moments, Nm
r Raceway groove curvature radius, m
r' Rolling radius, m
s,s Distance between loci of inner and outer raceway groove curvature centers, m; short for sin
R Curvature radius of deformed surfaces, m
$\mathcal{R}$ Radius to locus of raceway groove curvature centers, m
V Radial projection of distance between ball center and outer raceway groove curvature center, m; slip velocity of the race on the ball, m/sec

Journal of Physics: Conference Series **641** (2015) 012018 doi:10.1088/1742-6596/641/1/012018

- W Axial projection of distance between ball center and outer raceway groove curvature center, m
- Z Number of rolling elements
- β Contact angle, rad, °
- γ Angle of the race on the ball slip velocity with respect to the semiminor axes of the pressure ellipses, rad, °
- δ Normal contact deformation, m; relative displacement between inner and outer rings, m; vector of systems variables
- ϵ Error function
- θ Relative angular displacement between inner and outer rings, rad, °
- μ Friction coefficient
- ψ Azimuth angle, rad, °
- ω rotational speed, rad/sec

Superscripts:

- 0,' Refers to initial and improved estimates of variables values

Subscripts:

- a Refers to axial direction
- g,h index numbers
- i Refers to inner raceway
- j Refers to rolling element position
- k index numbers
- m Refers to pitch diameter
- o Refers to outer raceway
- r Refers to radial direction
- R Refers to the axis through the ball center perpendicular to the line defining the contact angle; refers to ball angular speed axis about its own center
- s Refers to the normal at the center of the contact ellipse
- x Refers to x-direction
- x' Refers to x'-direction
- y Refers to y-direction
- y' Refers to y'-direction
- z Refers to z-direction
- Z Refers to number of rolling elements
- z' Refers to z'-direction

3. Numerical procedure

Equations (16)-(18), (22)-(24), (27)-(29), and (54)-(56) of [13] may be written as

$$\epsilon_g(\delta_h) = 0, \qquad g, h = 1, \ldots, 9Z + 3, \tag{1}$$

in which $\delta_1 = V_1, \ldots, \delta_Z = V_Z, \delta_{Z+1} = W_1, \ldots, \delta_{2Z} = W_Z, \delta_{2Z+1} = \delta_{o1}, \ldots, \delta_{3Z} = \delta_{oZ}, \delta_{3Z+1} = \delta_{i1}, \ldots, \delta_{4Z} = \delta_{iZ}, \delta_{4Z+1} = r'_{o1}, \ldots, \delta_{5Z} = r'_{oZ}, \delta_{5Z+1} = r'_{i1}, \ldots, \delta_{6Z} = r'_{iZ}, \delta_{6Z+1} = \omega_{x'1}, \ldots, \delta_{7Z} = \omega_{x'Z}, \delta_{7Z+1} = \omega_{y'1}, \ldots, \delta_{8Z} = \omega_{y'Z}, \delta_{8Z+1} = \omega_{z'1}, \ldots, \delta_{9Z} = \omega_{z'Z}, \delta_{9Z+1} = \delta_a, \delta_{9Z+2} = \delta_r, \delta_{9Z+3} = \theta$.

The first 9Z equations from (1) must be solved simultaneously for $\delta_1, \ldots, \delta_{9Z}$ once values for $\delta_{9Z+1}, \ldots, \delta_{9Z+3}$ are assumed. If δ_h^0, $h = 1, \ldots, 9Z$, is a 9Z-dimensional vector with the initial estimates of the variables $\delta_1, \ldots, \delta_{9Z}$, improved values are given by

$$\delta'_h = \delta_h^0 - \left[a_{gh}\right]^{-1}\left\{\epsilon_g\right\}, \tag{2}$$

in which $\left\{\epsilon_g\right\}$, $g = 1, \ldots, 9Z$, is the 9Z-dimensional vector with the first 9Z errors functions from (1). The elements of the square 9Z×9Z-matrix $\left[a_{gh}\right]$ are

$$a_{jh} = -2\left(s_{zj} - \delta_j\right)\frac{\partial \delta_j}{\partial \delta_h} - 2\left(s_{xj} - \delta_{Z+j}\right)\frac{\partial \delta_{Z+j}}{\partial \delta_h} - 2\left[(f_i - 0.5)D + \delta_{3Z+j}\right]\frac{\partial \delta_{3Z+j}}{\partial \delta_h}, \tag{3}$$

$$a_{(j+Z)h} = 2\delta_j\frac{\partial \delta_j}{\partial \delta_h} + 2\delta_{Z+j}\frac{\partial \delta_{Z+j}}{\partial \delta_h} - 2\left[(f_o - 0.5)D + \delta_{2Z+j}\right]\frac{\partial \delta_{2Z+j}}{\partial \delta_h}, \tag{4}$$

$$a_{(j+2Z)h} = 2\delta_{6Z+j}\frac{\partial \delta_{6Z+j}}{\partial \delta_h} + 2\delta_{7Z+j}\frac{\partial \delta_{7Z+j}}{\partial \delta_h} + 2\delta_{8Z+j}\frac{\partial \delta_{8Z+j}}{\partial \delta_h}, \tag{5}$$

$$a_{(j+3Z)h} = \frac{\left\{\begin{array}{c}\left(F_{xoj}\frac{\partial \delta_j}{\partial \delta_h} - K_{oj}\delta_{2Z+j}^{1.5}\frac{\partial \delta_{Z+j}}{\partial \delta_h} - 1.5K_{oj}\delta_{2Z+j}^{0.5}\delta_{Z+j}\frac{\partial \delta_{2Z+j}}{\partial \delta_h} + \delta_j\frac{\partial F_{xoj}}{\partial \delta_h}\right)\times \\ \left[(f_o - 0.5)D + \delta_{2Z+j}\right] - \left(F_{xoj}\delta_j - K_{oj}\delta_{2Z+j}^{1.5}\delta_{Z+j}\right)\frac{\partial \delta_{2Z+j}}{\partial \delta_h}\end{array}\right\}}{\left[(f_o - 0.5)D + \delta_{2Z+j}\right]^2} + \frac{\left\{\begin{array}{c}\left[F_{xij}\frac{\partial \delta_j}{\partial \delta_h} - K_{ij}\delta_{3Z+j}^{1.5}\frac{\partial \delta_{Z+j}}{\partial \delta_h} + 1.5K_{ij}\delta_{3Z+j}^{0.5}\left(s_{xj} - \delta_{Z+j}\right)\frac{\partial \delta_{3Z+j}}{\partial \delta_h} - \left(s_{zj} - \delta_j\right)\frac{\partial F_{xij}}{\partial \delta_h}\right]\times \\ \left[(f_i - 0.5)D + \delta_{3Z+j}\right] - \left[K_{ij}\delta_{3Z+j}^{1.5}\left(s_{xj} - \delta_{Z+j}\right) - F_{xij}\left(s_{zj} - \delta_j\right)\right]\frac{\partial \delta_{3Z+j}}{\partial \delta_h}\end{array}\right\}}{\left[(f_i - 0.5)D + \delta_{3Z+j}\right]^2}, \tag{6}$$

$$a_{(j+4Z)h} = \frac{\left\{\begin{array}{c}\left(K_{oj}\delta_{2Z+j}^{1.5}\frac{\partial \delta_j}{\partial \delta_h} + F_{xoj}\frac{\partial \delta_{Z+j}}{\partial \delta_h} + 1.5K_{oj}\delta_{2Z+j}^{0.5}\delta_j\frac{\partial \delta_{2Z+j}}{\partial \delta_h} + \delta_{Z+j}\frac{\partial F_{xoj}}{\partial \delta_h}\right)\times \\ \left[(f_o - 0.5)D + \delta_{2Z+j}\right] - \left(K_{oj}\delta_{2Z+j}^{1.5}\delta_j + F_{xoj}\delta_{Z+j}\right)\frac{\partial \delta_{2Z+j}}{\partial \delta_h}\end{array}\right\}}{\left[(f_o - 0.5)D + \delta_{2Z+j}\right]^2} + \frac{\left\{\begin{array}{c}\left[K_{ij}\delta_{3Z+j}^{1.5}\frac{\partial \delta_j}{\partial \delta_h} + F_{xij}\frac{\partial \delta_{Z+j}}{\partial \delta_h} - 1.5K_{ij}\delta_{3Z+j}^{0.5}\left(s_{zj} - \delta_j\right)\frac{\partial \delta_{3Z+j}}{\partial \delta_h} - \left(s_{xj} - \delta_{Z+j}\right)\frac{\partial F_{xij}}{\partial \delta_h}\right]\times \\ \left[(f_i - 0.5)D + \delta_{2Z+j}\right] + \left[K_{ij}\delta_{3Z+j}^{1.5}\left(s_{zj} - \delta_j\right) + F_{xij}\left(s_{xj} - \delta_{Z+j}\right)\right]\frac{\partial \delta_{3Z+j}}{\partial \delta_h}\end{array}\right\}}{\left[(f_i - 0.5)D + \delta_{3Z+j}\right]^2} - \frac{\partial F_{z'j}}{\partial \delta_h}, \tag{7}$$

$$a_{(j+5Z)h} = \frac{\partial F_{yoj}}{\partial \delta_h} + \frac{\partial F_{yij}}{\partial \delta_h}, \tag{8}$$

$$a_{(j+6Z)h} = \frac{\left\{\begin{array}{l}\left(M_{Roj}\frac{\partial\delta_j}{\partial\delta_h}+M_{soj}\frac{\partial\delta_{Z+j}}{\partial\delta_h}+\delta_j\frac{\partial M_{Roj}}{\partial\delta_h}+\delta_{Z+j}\frac{\partial M_{soj}}{\partial\delta_h}\right)\times\\ \left[(f_o-0.5)D+\delta_{2Z+j}\right]-\left(M_{Roj}\delta_j+M_{soj}\delta_{Z+j}\right)\frac{\partial\delta_{2Z+j}}{\partial\delta_h}\end{array}\right\}}{\left[(f_o-0.5)D+\delta_{2Z+j}\right]^2} + \frac{\left\{\begin{array}{l}\left[M_{Rij}\frac{\partial\delta_j}{\partial\delta_h}+M_{sij}\frac{\partial\delta_{Z+j}}{\partial\delta_h}-(s_{xj}-\delta_{Z+j})\frac{\partial M_{sij}}{\partial\delta_h}-(s_{zj}-\delta_j)\frac{\partial M_{Rij}}{\partial\delta_h}\right]\times\\ \left[(f_i-0.5)D+\delta_{3Z+j}\right]+\left[M_{sij}(s_{xj}-\delta_{Z+j})+M_{Rij}(s_{zj}-\delta_j)\right]\frac{\partial\delta_{3Z+j}}{\partial\delta_h}\end{array}\right\}}{\left[(f_i-0.5)D+\delta_{3Z+j}\right]^2}, \quad (9)$$

$$a_{(j+7Z)h} = \frac{\left\{\begin{array}{l}\left(M_{soj}\frac{\partial\delta_j}{\partial\delta_h}-M_{Roj}\frac{\partial\delta_{Z+j}}{\partial\delta_h}+\delta_j\frac{\partial M_{soj}}{\partial\delta_h}-\delta_{Z+j}\frac{\partial M_{Roj}}{\partial\delta_h}\right)\times\\ \left[(f_o-0.5)D+\delta_{2Z+j}\right]-\left(M_{soj}\delta_j-M_{Roj}\delta_{Z+j}\right)\frac{\partial\delta_{2Z+j}}{\partial\delta_h}\end{array}\right\}}{\left[(f_o-0.5)D+\delta_{2Z+j}\right]^2} + \frac{\left\{\begin{array}{l}\left[M_{sij}\frac{\partial\delta_j}{\partial\delta_h}-M_{Rij}\frac{\partial\delta_{Z+j}}{\partial\delta_h}-(s_{zj}-\delta_j)\frac{\partial M_{sij}}{\partial\delta_h}+(s_{xj}-\delta_{Z+j})\frac{\partial M_{Rij}}{\partial\delta_h}\right]\times\\ \left[(f_i-0.5)D+\delta_{3Z+j}\right]+\left[M_{sij}(s_{zj}-\delta_j)-M_{Rij}(s_{xj}-\delta_{Z+j})\right]\frac{\partial\delta_{3Z+j}}{\partial\delta_h}\end{array}\right\}}{\left[(f_i-0.5)D+\delta_{3Z+j}\right]^2} + \frac{\partial M_{z'j}}{\partial\delta_h}, \quad (10)$$

$$a_{(j+8Z)h} = \frac{\partial M_{y'j}}{\partial\delta_h} - \frac{\partial M_{yij}}{\partial\delta_h} - \frac{\partial M_{yoj}}{\partial\delta_h}. \quad (11)$$

The forces $F_{z'j}$, F_{xij} and F_{xoj} to be used in (6) and (7) are given by (32), (38) and (39) of the companion paper of this [13] and their differentiation with respect to δ_h, $h = 1, \ldots, 9Z$, yields

$$\frac{\partial F_{z'j}}{\partial\delta_h} = m\omega^2\left[\left(\frac{\omega_{mj}}{\omega}\right)^2\frac{\partial\delta_j}{\partial\delta_h} + d_{mj}\left(\frac{\omega_{mj}}{\omega}\right)\frac{\partial\left(\frac{\omega_{mj}}{\omega}\right)}{\partial\delta_h}\right], \quad (12)$$

$$\frac{\partial F_{xij}}{\partial\delta_h} = \frac{3\mu K_{ij}\delta_{3Z+j}^{1.5}}{2\pi a_{ij}b_{ij}}\int_{-a_{ij}}^{a_{ij}}\int_{-b_{ij}\sqrt{1-\frac{x_{ij}^2}{a_{ij}^2}}}^{b_{ij}\sqrt{1-\frac{x_{ij}^2}{a_{ij}^2}}}\sqrt{1-\frac{x_{ij}^2}{a_{ij}^2}-\frac{y_{ij}^2}{b_{ij}^2}}\cos\gamma_{ij}\frac{\partial\gamma_{ij}}{\partial\delta_h}dy_{ij}\,dx_{ij} + \frac{3}{2}\frac{F_{xij}}{\delta_{3Z+j}}\frac{\partial\delta_{3Z+j}}{\partial\delta_h}, \quad (13)$$

$$\frac{\partial F_{xoj}}{\partial\delta_h} = \frac{3\mu K_{oj}\delta_{2Z+j}^{1.5}}{2\pi a_{oj}b_{oj}}\int_{-a_{oj}}^{a_{oj}}\int_{-b_{oj}\sqrt{1-\frac{x_{oj}^2}{a_{oj}^2}}}^{b_{oj}\sqrt{1-\frac{x_{oj}^2}{a_{oj}^2}}}\sqrt{1-\frac{x_{oj}^2}{a_{oj}^2}-\frac{y_{oj}^2}{b_{oj}^2}}\cos\gamma_{oj}\frac{\partial\gamma_{oj}}{\partial\delta_h}dy_{oj}\,dx_{oj} + \frac{3}{2}\frac{F_{xoj}}{\delta_{2Z+j}}\frac{\partial\delta_{2Z+j}}{\partial\delta_h}. \quad (14)$$

The forces F_{yij} and F_{yoj} to be used in (8) are given by (40) and (41) of the companion paper of this [13] and their differentiation with respect to δ_h, $h = 1, \ldots, 9Z$, yields

$$\frac{\partial F_{yij}}{\partial\delta_h} = -\frac{3\mu K_{ij}\delta_{3Z+j}^{1.5}}{2\pi a_{ij}b_{ij}}\int_{-a_{ij}}^{a_{ij}}\int_{-b_{ij}\sqrt{1-\frac{x_{ij}^2}{a_{ij}^2}}}^{b_{ij}\sqrt{1-\frac{x_{ij}^2}{a_{ij}^2}}}\sqrt{1-\frac{x_{ij}^2}{a_{ij}^2}-\frac{y_{ij}^2}{b_{ij}^2}}\sin\gamma_{ij}\frac{\partial\gamma_{ij}}{\partial\delta_h}dy_{ij}\,dx_{ij} + \frac{3}{2}\frac{F_{yij}}{\delta_{3Z+j}}\frac{\partial\delta_{3Z+j}}{\partial\delta_h}, \quad (15)$$

$$\frac{\partial F_{yoj}}{\partial\delta_h} = -\frac{3\mu K_{oj}\delta_{2Z+j}^{1.5}}{2\pi a_{oj}b_{oj}}\int_{-a_{oj}}^{a_{oj}}\int_{-b_{oj}\sqrt{1-\frac{x_{oj}^2}{a_{oj}^2}}}^{b_{oj}\sqrt{1-\frac{x_{oj}^2}{a_{oj}^2}}}\sqrt{1-\frac{x_{oj}^2}{a_{oj}^2}-\frac{y_{oj}^2}{b_{oj}^2}}\sin\gamma_{oj}\frac{\partial\gamma_{oj}}{\partial\delta_h}dy_{oj}\,dx_{oj} + \frac{3}{2}\frac{F_{yoj}}{\delta_{2Z+j}}\frac{\partial\delta_{2Z+j}}{\partial\delta_h}. \quad (16)$$

The moments M_{sij}, M_{soj}, M_{Rij}, M_{Roj} and $M_{z'j}$ to be used in (9) and (10) are given by (48), (49), (52), (53) and (37) of [13] and their differentiation with respect to δ_h, $h = 1, \ldots, 9Z$, yields

$$\frac{\partial M_{sij}}{\partial\delta_h} = \frac{-3\mu K_{ij}\delta_{3Z+j}^{1.5}}{2\pi a_{ij}b_{ij}}\int_{-a_{ij}}^{a_{ij}}\int_{-b_{ij}\sqrt{1-\frac{x_{ij}^2}{a_{ij}^2}}}^{b_{ij}\sqrt{1-\frac{x_{ij}^2}{a_{ij}^2}}}\sqrt{x_{ij}^2+y_{ij}^2}\sqrt{1-\frac{x_{ij}^2}{a_{ij}^2}-\frac{y_{ij}^2}{b_{ij}^2}}\sin\left(\gamma_{ij}-\tan^{-1}\frac{y_{ij}}{x_{ij}}\right)\frac{\partial\gamma_{ij}}{\partial\delta_h}dy_{ij}\,dx_{ij} + \frac{3}{2}\frac{M_{sij}}{\delta_{3Z+j}}\frac{\partial\delta_{3Z+j}}{\partial\delta_h}, \quad (17)$$

$$\frac{\partial M_{soj}}{\partial\delta_h} = \frac{-3\mu K_{oj}\delta_{2Z+j}^{1.5}}{2\pi a_{oj}b_{oj}}\int_{-a_{oj}}^{a_{oj}}\int_{-b_{oj}\sqrt{1-\frac{x_{oj}^2}{a_{oj}^2}}}^{b_{oj}\sqrt{1-\frac{x_{oj}^2}{a_{oj}^2}}}\sqrt{x_{oj}^2+y_{oj}^2}\sqrt{1-\frac{x_{oj}^2}{a_{oj}^2}-\frac{y_{oj}^2}{b_{oj}^2}}\sin\left(\gamma_{oj}-\tan^{-1}\frac{y_{oj}}{x_{oj}}\right)\frac{\partial\gamma_{oj}}{\partial\delta_h}dy_{oj}\,dx_{oj} + \frac{3}{2}\frac{M_{soj}}{\delta_{2Z+j}}\frac{\partial\delta_{2Z+j}}{\partial\delta_h}, \quad (18)$$

$$\frac{\partial M_{Rij}}{\partial\delta_h} = \frac{-3\mu K_{ij}\delta_{2Z+j}^{1.5}}{2\pi a_{ij}b_{ij}}\int_{-a_{ij}}^{a_{ij}}\int_{-b_{ij}\sqrt{1-\frac{x_{ij}^2}{a_{ij}^2}}}^{b_{ij}\sqrt{1-\frac{x_{ij}^2}{a_{ij}^2}}}\left(\sqrt{R_i^2-x_{ij}^2}-\sqrt{R_i^2-a_{ij}^2}+\sqrt{\left(\frac{D}{2}\right)^2-a_{ij}^2}\right)\sqrt{1-\frac{x_{ij}^2}{a_{ij}^2}-\frac{y_{ij}^2}{b_{ij}^2}}\sin\gamma_{ij}\frac{\partial\gamma_{ij}}{\partial\delta_h}dy_{ij}\,dx_{ij} + \frac{3}{2}\frac{M_{Rij}}{\delta_{3Z+j}}\frac{\partial\delta_{3Z+j}}{\partial\delta_h}, \quad (19)$$

$$\frac{\partial M_{Roj}}{\partial\delta_h}=\frac{-3\mu K_{oj}\delta_{2Z+j}^{1.5}}{2\pi a_{oj}b_{oj}}\int_{-a_{oj}}^{a_{oj}}\int_{-b_{oj}\sqrt{1-\frac{x_{oj}^2}{a_{oj}^2}}}^{b_{oj}\sqrt{1-\frac{x_{oj}^2}{a_{oj}^2}}}\left(\sqrt{R_o^2-x_{oj}^2}-\sqrt{R_o^2-a_{oj}^2}+\sqrt{\left(\frac{D}{2}\right)^2-a_{oj}^2}\right)\sqrt{1-\frac{x_{oj}^2}{a_{oj}^2}-\frac{y_{oj}^2}{b_{oj}^2}}\sin\gamma_{oj}\frac{\partial\gamma_{oj}}{\partial\delta_h}dy_{oj}\,dx_{oj}+\frac{3}{2}\frac{M_{Roj}}{\delta_{2Z+j}}\frac{\partial\delta_{2Z+j}}{\partial\delta_h},\tag{20}$$

$$\frac{\partial M_{z'j}}{\partial\delta_h}=-J\omega^2\left\{\frac{\delta_{7Z+j}}{\sqrt{\delta_{6Z+j}^2+\delta_{7Z+j}^2+\delta_{8Z+j}^2}}\left[\left(\frac{\omega_{Rj}}{\omega}\right)\frac{\partial\left(\frac{\omega_{mj}}{\omega}\right)}{\partial\delta_h}+\frac{\partial\left(\frac{\omega_{Rj}}{\omega}\right)}{\partial\delta_h}\left(\frac{\omega_{mj}}{\omega}\right)\right]-\frac{\delta_{6Z+j}\delta_{7Z+j}\frac{\partial\delta_{6Z+j}}{\partial\delta_h}-\left(\delta_{6Z+j}^2+\delta_{8Z+j}^2\right)\frac{\partial\delta_{7Z+j}}{\partial\delta_h}+\delta_{7Z+j}\delta_{8Z+j}\frac{\partial\delta_{8Z+j}}{\partial\delta_h}}{\left(\delta_{6Z+j}^2+\delta_{7Z+j}^2+\delta_{8Z+j}^2\right)^{\frac{3}{2}}}\left(\frac{\omega_{Rj}}{\omega}\right)\left(\frac{\omega_{mj}}{\omega}\right)\right\}.\tag{21}$$

The moments M_{yij}, M_{yoj} and $M_{y'j}$ to be used in (11) are given by (50), (51) and (36) of the companion paper of this [13] and their differentiation with respect to δ_h, $h = 1, \ldots, 9Z$, yields

$$\frac{\partial M_{yij}}{\partial\delta_h}=\frac{3\mu K_{ij}\delta_{3Z+j}^{1.5}}{2\pi a_{ij}b_{ij}}\int_{-a_{ij}}^{a_{ij}}\int_{-b_{ij}\sqrt{1-\frac{x_{ij}^2}{a_{ij}^2}}}^{b_{ij}\sqrt{1-\frac{x_{ij}^2}{a_{ij}^2}}}\left(\sqrt{R_i^2-x_{ij}^2}-\sqrt{R_i^2-a_{ij}^2}+\sqrt{\left(\frac{D}{2}\right)^2-a_{ij}^2}\right)\sqrt{1-\frac{x_{ij}^2}{a_{ij}^2}-\frac{y_{ij}^2}{b_{ij}^2}}\cos\gamma_{ij}\frac{\partial\gamma_{ij}}{\partial\delta_h}dy_{ij}\,dx_{ij}+\frac{3}{2}\frac{M_{yij}}{\delta_{3Z+j}}\frac{\partial\delta_{3Z+j}}{\partial\delta_h},\tag{22}$$

$$\frac{\partial M_{yoj}}{\partial\delta_h}=\frac{3\mu K_{oj}\delta_{2Z+j}^{1.5}}{2\pi a_{oj}b_{oj}}\int_{-a_{oj}}^{a_{oj}}\int_{-b_{oj}\sqrt{1-\frac{x_{oj}^2}{a_{oj}^2}}}^{b_{oj}\sqrt{1-\frac{x_{oj}^2}{a_{oj}^2}}}\left(\sqrt{R_o^2-x_{oj}^2}-\sqrt{R_o^2-a_{oj}^2}+\sqrt{\left(\frac{D}{2}\right)^2-a_{oj}^2}\right)\sqrt{1-\frac{x_{oj}^2}{a_{oj}^2}-\frac{y_{oj}^2}{b_{oj}^2}}\cos\gamma_{oj}\frac{\partial\gamma_{oj}}{\partial\delta_h}dy_{oj}\,dx_{oj}+\frac{3}{2}\frac{M_{yoj}}{\delta_{2Z+j}}\frac{\partial\delta_{2Z+j}}{\partial\delta_h},\tag{23}$$

$$\frac{\partial M_{y'j}}{\partial\delta_h}=J\omega^2\left\{\frac{\delta_{8Z+j}}{\sqrt{\delta_{6Z+j}^2+\delta_{7Z+j}^2+\delta_{8Z+j}^2}}\left[\left(\frac{\omega_{Rj}}{\omega}\right)\frac{\partial\left(\frac{\omega_{mj}}{\omega}\right)}{\partial\delta_h}+\frac{\partial\left(\frac{\omega_{Rj}}{\omega}\right)}{\partial\delta_h}\left(\frac{\omega_{mj}}{\omega}\right)\right]-\frac{\delta_{6Z+j}\delta_{8Z+j}\frac{\partial\delta_{6Z+j}}{\partial\delta_h}+\delta_{7Z+j}\delta_{8Z+j}\frac{\partial\delta_{7Z+j}}{\partial\delta_h}-\left(\delta_{6Z+j}^2+\delta_{7Z+j}^2\right)\frac{\partial\delta_{8Z+j}}{\partial\delta_h}}{\left(\delta_{6Z+j}^2+\delta_{7Z+j}^2+\delta_{8Z+j}^2\right)^{\frac{3}{2}}}\left(\frac{\omega_{Rj}}{\omega}\right)\left(\frac{\omega_{mj}}{\omega}\right)\right\}.\tag{24}$$

As in (38)-(41) and (48)-(53) of [13], also in (13)-(20), (22)-(23) γ_{ij}, γ_{oj} are given by (42) of [13]. The derivatives of γ_{ij}, γ_{oj} with respect δ_h, $h = 1, \ldots, 9Z$, to be used in (13)-(20), (22)-(23) are given by

$$\frac{\partial\gamma_{ij}}{\partial\delta_h}=\frac{-\left(x_{ij}+\frac{V_{yij}}{\omega_{sij}}\right)\frac{\partial\left(\frac{V_{xij}}{\omega_{sij}}\right)}{\partial\delta_h}-\left(y_{ij}-\frac{V_{xij}}{\omega_{sij}}\right)\frac{\partial\left(\frac{V_{yij}}{\omega_{sij}}\right)}{\partial\delta_h}}{\left(y_{ij}-\frac{V_{xij}}{\omega_{sij}}\right)^2+\left(x_{ij}+\frac{V_{yij}}{\omega_{sij}}\right)^2},\qquad\frac{\partial\gamma_{oj}}{\partial\delta_h}=\frac{-\left(x_{oj}+\frac{V_{yoj}}{\omega_{soj}}\right)\frac{\partial\left(\frac{V_{xoj}}{\omega_{soj}}\right)}{\partial\delta_h}-\left(y_{oj}-\frac{V_{xoj}}{\omega_{soj}}\right)\frac{\partial\left(\frac{V_{yoj}}{\omega_{soj}}\right)}{\partial\delta_h}}{\left(y_{oj}-\frac{V_{xoj}}{\omega_{soj}}\right)^2+\left(x_{oj}+\frac{V_{yoj}}{\omega_{soj}}\right)^2},\tag{25}$$

in which V_{xij}/ω_{sij}, V_{yij}/ω_{sij}, V_{xoj}/ω_{soj}, V_{yoj}/ω_{soj} are given by (43)-(46) of [13]. The derivatives of V_{xij}/ω_{sij}, V_{yij}/ω_{sij}, V_{xoj}/ω_{soj} and V_{yoj}/ω_{soj} with respect δ_h, $h = 1, \ldots, 9Z$, to be used in (25) are given by

$$\begin{aligned}\frac{\partial\left(\frac{V_{xij}}{\omega_{sij}}\right)}{\partial\delta_h}&=\left(\sqrt{R_i^2-x_{ij}^2}-\sqrt{R_i^2-a_{ij}^2}+\sqrt{\left(\frac{D}{2}\right)^2-a_{ij}^2}\right)\times\\&\frac{\left\{\begin{aligned}&\left\{\left\{\frac{\delta_{5Z+j}}{\frac{d_{mj}}{2}}\left[\left(1+\frac{s_{zj}-\delta_j}{\frac{d_{mj}}{2}}\right)\frac{\partial\delta_j}{\partial\delta_h}-c\psi_j\frac{\partial\delta_{9Z+2}}{\partial\delta_h}+\mathcal{R}|c\psi_j|s\delta_{9Z+3}\frac{\partial\delta_{9Z+3}}{\partial\delta_h}\right]+\frac{\partial\delta_{3Z+j}}{\partial\delta_h}-\frac{s_{zj}-\delta_j}{\frac{d_{mj}}{2}}\frac{\partial\delta_{5Z+j}}{\partial\delta_h}\right\}\delta_{7Z+j}+\left[(f_i-0.5)D+\delta_{3Z+j}-\frac{s_{zj}-\delta_j}{\frac{d_{mj}}{2}}\delta_{5Z+j}\right]\frac{\partial\delta_{7Z+j}}{\partial\delta_h}\right\}\left\{\delta_{6Z+j}(s_{xj}-\delta_{Z+j})\right.\\&\left.-\delta_{8Z+j}\left[s_{zj}-\delta_j-\frac{(f_i-0.5)D+\delta_{3Z+j}}{\frac{d_{mj}}{2}}\delta_{5Z+j}\right]\right\}-\left[(f_i-0.5)D+\delta_{3Z+j}-\frac{s_{zj}-\delta_j}{\frac{d_{mj}}{2}}\delta_{5Z+j}\right]\delta_{7Z+j}\left\{\delta_{6Z+j}\left(\frac{\partial\delta_{9Z+1}}{\partial\delta_h}\mathcal{R}c\psi_j c\delta_{9Z+3}\frac{\partial\delta_{9Z+3}}{\partial\delta_h}-\frac{\partial\delta_{Z+j}}{\partial\delta_h}\right)+(s_{xj}-\delta_{Z+j})\frac{\partial\delta_{6Z+j}}{\partial\delta_h}\right.\\&\left.-\delta_{8Z+j}\left\{c\psi_j\frac{\partial\delta_{9Z+2}}{\partial\delta_h}-\mathcal{R}|c\psi_j|s\delta_{9Z+3}\frac{\partial\delta_{9Z+3}}{\partial\delta_h}+\left[\frac{\delta_{5Z+j}}{\frac{d_{mj}}{2}}\frac{(f_i-0.5)D+\delta_{3Z+j}}{\frac{d_{mj}}{2}}-1\right]\frac{\partial\delta_j}{\partial\delta_h}-\frac{\delta_{5Z+j}}{\frac{d_{mj}}{2}}\frac{\partial\delta_{3Z+j}}{\partial\delta_h}-\frac{(f_i-0.5)D+\delta_{3Z+j}}{\frac{d_{mj}}{2}}\frac{\partial\delta_{5Z+j}}{\partial\delta_h}\right\}-\left[s_{zj}-\delta_j-\frac{(f_i-0.5)D+\delta_{3Z+j}}{\frac{d_{mj}}{2}}\delta_{5Z+j}\right]\frac{\partial\delta_{8Z+j}}{\partial\delta_h}\right\}\end{aligned}\right\}}{\left\{\delta_{6Z+j}(s_{xj}-\delta_{Z+j})-\delta_{8Z+j}\left[s_{zj}-\delta_j-\frac{(f_i-0.5)D+\delta_{3Z+j}}{\frac{d_{mj}}{2}}\delta_{5Z+j}\right]\right\}^2},\end{aligned}\tag{26}$$

$$\frac{\partial\left(\frac{V_{yij}}{\omega_{sij}}\right)}{\partial\delta_h}=\frac{\left\{\begin{aligned}&\left\{\left(\sqrt{R_i^2-x_{ij}^2}-\sqrt{R_i^2-a_{ij}^2}+\sqrt{\left(\frac{D}{2}\right)^2-a_{ij}^2}-\delta_{5Z+j}\right)\left[\delta_{8Z+j}\left(c\psi_j\frac{\partial\delta_{9Z+2}}{\partial\delta_h}-\mathcal{R}|c\psi_j|s\delta_{9Z+3}\frac{\partial\delta_{9Z+3}}{\partial\delta_h}-\frac{\partial\delta_j}{\partial\delta_h}\right)+(s_{zj}-\delta_j)\frac{\partial\delta_{8Z+j}}{\partial\delta_h}+\delta_{6Z+j}\left(\frac{\partial\delta_{9Z+1}}{\partial\delta_h}\mathcal{R}c\psi_j c\delta_{9Z+3}\times\right.\right.\right.\\&\left.\left.\left.\frac{\partial\delta_{9Z+3}}{\partial\delta_h}-\frac{\partial\delta_{Z+j}}{\partial\delta_h}\right)+(s_{xj}-\delta_{Z+j})\frac{\partial\delta_{6Z+j}}{\partial\delta_h}\right]-\left[\delta_{6Z+j}(s_{zj}-\delta_j)+\delta_{8Z+j}(s_{xj}-\delta_{Z+j})\right]\frac{\partial\delta_{5Z+j}}{\partial\delta_h}\right\}\left\{\delta_{6Z+j}(s_{xj}-\delta_{Z+j})-\delta_{8Z+j}\left[s_{zj}-\delta_j-\frac{(f_i-0.5)D+\delta_{3Z+j}}{\frac{d_{mj}}{2}}\delta_{5Z+j}\right]\right\}\\&-\left(\sqrt{R_i^2-x_{ij}^2}-\sqrt{R_i^2-a_{ij}^2}+\sqrt{\left(\frac{D}{2}\right)^2-a_{ij}^2}-\delta_{5Z+j}\right)\left[\delta_{6Z+j}(s_{zj}-\delta_j)+\delta_{8Z+j}(s_{xj}-\delta_{Z+j})\right]\left\{\delta_{6Z+j}\left(\frac{\partial\delta_{9Z+1}}{\partial\delta_h}\mathcal{R}c\psi_j c\delta_{9Z+3}\frac{\partial\delta_{9Z+3}}{\partial\delta_h}-\frac{\partial\delta_{Z+j}}{\partial\delta_h}\right)+(s_{xj}-\delta_{Z+j})\frac{\partial\delta_{6Z+j}}{\partial\delta_h}-\delta_{8Z+j}\right.\\&\left.\times\left\{c\psi_j\frac{\partial\delta_{9Z+2}}{\partial\delta_h}-\mathcal{R}|c\psi_j|s\delta_{9Z+3}\frac{\partial\delta_{9Z+3}}{\partial\delta_h}+\left[\frac{(f_i-0.5)D+\delta_{3Z+j}}{\frac{d_{mj}}{2}}\frac{\delta_{5Z+j}}{\frac{d_{mj}}{2}}-1\right]\frac{\partial\delta_j}{\partial\delta_h}-\frac{\delta_{5Z+j}}{\frac{d_{mj}}{2}}\frac{\partial\delta_{3Z+j}}{\partial\delta_h}-\frac{(f_i-0.5)D+\delta_{3Z+j}}{\frac{d_{mj}}{2}}\frac{\partial\delta_{5Z+j}}{\partial\delta_h}\right\}-\left[s_{zj}-\delta_j-\frac{(f_i-0.5)D+\delta_{3Z+j}}{\frac{d_{mj}}{2}}\delta_{5Z+j}\right]\frac{\partial\delta_{8Z+j}}{\partial\delta_h}\right\}\end{aligned}\right\}}{\left\{\delta_{6Z+j}(s_{xj}-\delta_{Z+j})-\delta_{8Z+j}\left[s_{zj}-\delta_j-\frac{(f_i-0.5)D+\delta_{3Z+j}}{\frac{d_{mj}}{2}}\delta_{5Z+j}\right]\right\}^2},\tag{27}$$

Journal of Physics: Conference Series **641** (2015) 012018 doi:10.1088/1742-6596/641/1/012018

$$
\begin{aligned}
\frac{\partial\left(\frac{V_{xoj}}{\omega_{soj}}\right)}{\partial\delta_h} = -\left(\sqrt{R_o^2-x_{oj}^2}-\sqrt{R_o^2-a_{oj}^2}+\sqrt{\left(\frac{D}{2}\right)^2-a_{oj}^2}\right)\times \\
\frac{\begin{gathered}\Big\{\Big\{\Big[\frac{\delta_{4Z+j}}{\frac{d_{mj}}{2}}\Big(1-\frac{\delta_j}{\frac{d_{mj}}{2}}\Big)\frac{\partial\delta_j}{\partial\delta_h}+\frac{\partial\delta_{2Z+j}}{\partial\delta_h}+\frac{\delta_j}{\frac{d_{mj}}{2}}\frac{\partial\delta_{4Z+j}}{\partial\delta_h}\Big]\delta_{7Z+j}+\Big[(f_o-0.5)D+\delta_{2Z+j}+\frac{\delta_j}{\frac{d_{mj}}{2}}\delta_{4Z+j}\Big]\frac{\partial\delta_{7Z+j}}{\partial\delta_h}\Big\}\Big\{\delta_{6Z+j}\delta_{Z+j}-\delta_{8Z+j}\Big[\delta_j+\frac{(f_o-0.5)D+\delta_{2Z+j}}{\frac{d_{mj}}{2}}\delta_{4Z+j}\Big]\Big\}-\Big[(f_o-0.5)D+\delta_{2Z+j}+ \\
\frac{\delta_j}{\frac{d_{mj}}{2}}\delta_{4Z+j}\Big]\delta_{7Z+j}\Big\{\delta_{6Z+j}\frac{\partial\delta_{Z+j}}{\partial\delta_h}+\delta_{Z+j}\frac{\partial\delta_{6Z+j}}{\partial\delta_h}-\delta_{8Z+j}\Big\{\Big[1-\frac{\delta_{4Z+j}}{\frac{d_{mj}}{2}}\frac{(f_o-0.5)D+\delta_{2Z+j}}{\frac{d_{mj}}{2}}\Big]\frac{\partial\delta_j}{\partial\delta_h}+\frac{\delta_{4Z+j}}{\frac{d_{mj}}{2}}\frac{\partial\delta_{2Z+j}}{\partial\delta_h}+\frac{(f_o-0.5)D+\delta_{2Z+j}}{\frac{d_{mj}}{2}}\frac{\partial\delta_{4Z+j}}{\partial\delta_h}\Big\}-\Big[\delta_j+\frac{(f_o-0.5)D+\delta_{2Z+j}}{\frac{d_{mj}}{2}}\delta_{4Z+j}\Big]\frac{\partial\delta_{8Z+j}}{\partial\delta_h}\Big\}\Big\}\end{gathered}}{\Big\{\delta_{6Z+j}\delta_{Z+j}-\delta_{8Z+j}\Big[\delta_j+\frac{(f_o-0.5)D+\delta_{2Z+j}}{\frac{d_{mj}}{2}}\delta_{4Z+j}\Big]\Big\}^2},
\end{aligned}
\tag{28}
$$

$$
\frac{\partial\left(\frac{V_{yoj}}{\omega_{soj}}\right)}{\partial\delta_h} = \frac{\begin{gathered}\Big\{\Big[\Big(\sqrt{R_o^2-x_{oj}^2}-\sqrt{R_o^2-a_{oj}^2}+\sqrt{\left(\frac{D}{2}\right)^2-a_{oj}^2}-\delta_{4Z+j}\Big)\Big(\delta_{6Z+j}\frac{\partial\delta_j}{\partial\delta_h}+\delta_j\frac{\partial\delta_{6Z+j}}{\partial\delta_h}+\delta_{8Z+j}\frac{\partial\delta_{Z+j}}{\partial\delta_h}+\delta_{Z+j}\frac{\partial\delta_{8Z+j}}{\partial\delta_h}\Big)-(\delta_{6Z+j}\delta_j+\delta_{8Z+j}\delta_{Z+j})\frac{\partial\delta_{4Z+j}}{\partial\delta_h}\Big]\Big\{\delta_{6Z+j}\delta_{Z+j} \\
-\delta_{8Z+j}\Big[\delta_j+\frac{(f_o-0.5)D+\delta_{2Z+j}}{\frac{d_{mj}}{2}}\delta_{4Z+j}\Big]\Big\}-\Big(\sqrt{R_o^2-x_{oj}^2}-\sqrt{R_o^2-a_{oj}^2}+\sqrt{\left(\frac{D}{2}\right)^2-a_{oj}^2}-\delta_{4Z+j}\Big)(\delta_{6Z+j}\delta_j+\delta_{8Z+j}\delta_{Z+j})\Big\{\delta_{6Z+j}\frac{\partial\delta_{Z+j}}{\partial\delta_h}+\delta_{Z+j}\frac{\partial\delta_{6Z+j}}{\partial\delta_h}- \\
\delta_{8Z+j}\Big\{\Big[1-\frac{\delta_{4Z+j}}{\frac{d_{mj}}{2}}\frac{(f_o-0.5)D+\delta_{2Z+j}}{\frac{d_{mj}}{2}}\Big]\frac{\partial\delta_j}{\partial\delta_h}+\frac{\delta_{4Z+j}}{\frac{d_{mj}}{2}}\frac{\partial\delta_{2Z+j}}{\partial\delta_h}+\frac{(f_o-0.5)D+\delta_{2Z+j}}{\frac{d_{mj}}{2}}\frac{\partial\delta_{4Z+j}}{\partial\delta_h}\Big\}-\Big[\delta_j+\frac{(f_o-0.5)D+\delta_{2Z+j}}{\frac{d_{mj}}{2}}\delta_{4Z+j}\Big]\frac{\partial\delta_{8Z+j}}{\partial\delta_h}\Big\}\Big\}\end{gathered}}{\Big\{\delta_{6Z+j}\delta_{Z+j}-\delta_{8Z+j}\Big[\delta_j+\frac{(f_o-0.5)D+\delta_{2Z+j}}{\frac{d_{mj}}{2}}\delta_{4Z+j}\Big]\Big\}^2}.
\tag{29}
$$

For the outer race to be stationary ω_{mj}/ω and ω_{Rj}/ω are given by (33)-(34) of [13]. The derivatives of (33) and (34) of [13] with respect δ_h, $h = 1, \ldots, 9Z$, to be used in (12), (21) and (24) are given by

$$
\frac{\partial\left(\frac{\omega_{mj}}{\omega}\right)}{\partial\delta_h} = \frac{\begin{gathered}\Big\{\Big\{\Big[\delta_{4Z+j}\Big\{\frac{d_{mj}}{2}\frac{\partial\delta_{3Z+j}}{\partial\delta_h}+[(f_i-0.5)D+\delta_{3Z+j}]\frac{\partial\delta_j}{\partial\delta_h}-\delta_{5Z+j}\Big(c\psi_j\frac{\partial\delta_{9Z+2}}{\partial\delta_h}-\mathcal{R}|c\psi_j|s\delta_{9Z+3}\frac{\partial\delta_{9Z+3}}{\partial\delta_h}\frac{\partial\delta_j}{\partial\delta_h}\Big)-(s_{zj}-\delta_j)\frac{\partial\delta_{5Z+j}}{\partial\delta_h}\Big\}+\Big\{\frac{d_{mj}}{2}[(f_i-0.5)D+\delta_{3Z+j}]-\delta_{5Z+j}(s_{zj}-\delta_j)\Big\}\frac{\partial\delta_{4Z+j}}{\partial\delta_h}\Big\}\times \\
(\delta_{6Z+j}\delta_j+\delta_{8Z+j}\delta_{Z+j})+\delta_{4Z+j}\Big\{\frac{d_{mj}}{2}[(f_i-0.5)D+\delta_{3Z+j}]-\delta_{5Z+j}(s_{zj}-\delta_j)\Big\}\Big(\delta_{6Z+j}\frac{\partial\delta_j}{\partial\delta_h}+\delta_{8Z+j}\frac{\partial\delta_{Z+j}}{\partial\delta_h}+\delta_j\frac{\partial\delta_{6Z+j}}{\partial\delta_h}+\delta_{Z+j}\frac{\partial\delta_{8Z+j}}{\partial\delta_h}\Big)\Big]\Big\{\delta_{4Z+j}\Big\{\frac{d_{mj}}{2}[(f_i-0.5)D+\delta_{3Z+j}]-\delta_{5Z+j}\times \\
(s_{zj}-\delta_j)\Big\}(\delta_{6Z+j}\delta_j+\delta_{8Z+j}\delta_{Z+j})+\delta_{5Z+j}\Big\{\frac{d_{mj}}{2}[(f_o-0.5)D+\delta_{2Z+j}]+\delta_{4Z+j}\delta_j\Big\}[\delta_{6Z+j}(s_{zj}-\delta_j)+\delta_{8Z+j}(s_{xj}-\delta_{Z+j})]\Big\}-\delta_{4Z+j}\Big\{\frac{d_{mj}}{2}[(f_i-0.5)D+\delta_{3Z+j}]-\delta_{5Z+j}(s_{zj}-\delta_j)\Big\}\times \\
(\delta_{6Z+j}\delta_j+\delta_{8Z+j}\delta_{Z+j})\Big\{\Big\{\delta_{4Z+j}\Big\{\frac{d_{mj}}{2}\frac{\partial\delta_{3Z+j}}{\partial\delta_h}+[(f_i-0.5)D+\delta_{3Z+j}]\frac{\partial\delta_j}{\partial\delta_h}-\delta_{5Z+j}\Big(c\psi_j\frac{\partial\delta_{9Z+2}}{\partial\delta_h}-\mathcal{R}|c\psi_j|s\delta_{9Z+3}\frac{\partial\delta_{9Z+3}}{\partial\delta_h}\frac{\partial\delta_j}{\partial\delta_h}\Big)-(s_{zj}-\delta_j)\frac{\partial\delta_{5Z+j}}{\partial\delta_h}\Big\}+\Big\{\frac{d_{mj}}{2}[(f_i-0.5)D+\delta_{3Z+j}]- \\
\delta_{5Z+j}(s_{zj}-\delta_j)\Big\}\frac{\partial\delta_{4Z+j}}{\partial\delta_h}\Big\}(\delta_{6Z+j}\delta_j+\delta_{8Z+j}\delta_{Z+j})+\delta_{4Z+j}\Big\{\frac{d_{mj}}{2}[(f_i-0.5)D+\delta_{3Z+j}]-\delta_{5Z+j}(s_{zj}-\delta_j)\Big\}\Big(\delta_{6Z+j}\frac{\partial\delta_j}{\partial\delta_h}+\delta_{8Z+j}\frac{\partial\delta_{Z+j}}{\partial\delta_h}+\delta_j\frac{\partial\delta_{6Z+j}}{\partial\delta_h}+\delta_{Z+j}\frac{\partial\delta_{8Z+j}}{\partial\delta_h}\Big)+\Big\{\delta_{5Z+j}\Big\{\frac{d_{mj}}{2}\times \\
\frac{\partial\delta_{2Z+j}}{\partial\delta_h}+[(f_o-0.5)D+\delta_{2Z+j}+\delta_{4Z+j}]\frac{\partial\delta_j}{\partial\delta_h}+\delta_j\frac{\partial\delta_{4Z+j}}{\partial\delta_h}\Big\}+\Big\{\frac{d_{mj}}{2}[(f_o-0.5)D+\delta_{2Z+j}]+\delta_{4Z+j}\delta_j\Big\}\frac{\partial\delta_{5Z+j}}{\partial\delta_h}\Big\}[\delta_{6Z+j}(s_{zj}-\delta_j)+\delta_{8Z+j}(s_{xj}-\delta_{Z+j})]+\delta_{5Z+j}\Big\{\frac{d_{mj}}{2}[(f_o-0.5)D+\delta_{2Z+j}] \\
+\delta_{4Z+j}\delta_j\Big\}\Big[\delta_{6Z+j}\Big(c\psi_j\frac{\partial\delta_{9Z+2}}{\partial\delta_h}-\mathcal{R}|c\psi_j|s\delta_{9Z+3}\frac{\partial\delta_{9Z+3}}{\partial\delta_h}\frac{\partial\delta_j}{\partial\delta_h}\Big)+\delta_{8Z+j}\Big(\frac{\partial\delta_{9Z+1}}{\partial\delta_h}+\mathcal{R}c\psi_j c\delta_{9Z+3}\frac{\partial\delta_{9Z+3}}{\partial\delta_h}\frac{\partial\delta_{Z+j}}{\partial\delta_h}\Big)+(s_{zj}-\delta_j)\frac{\partial\delta_{6Z+j}}{\partial\delta_h}+(s_{xj}-\delta_{Z+j})\frac{\partial\delta_{8Z+j}}{\partial\delta_h}\Big]\Big\}\Big\}\end{gathered}}{\begin{Bmatrix}\delta_{4Z+j}\Big\{\frac{d_{mj}}{2}[(f_i-0.5)D+\delta_{3Z+j}]-\delta_{5Z+j}(s_{zj}-\delta_j)\Big\}(\delta_{6Z+j}\delta_j+\delta_{8Z+j}\delta_{Z+j})+ \\ \delta_{5Z+j}\Big\{\frac{d_{mj}}{2}[(f_o-0.5)D+\delta_{2Z+j}]+\delta_{4Z+j}\delta_j\Big\}[\delta_{6Z+j}(s_{zj}-\delta_j)+\delta_{8Z+j}(s_{xj}-\delta_{Z+j})]\end{Bmatrix}^2},
\tag{30}
$$

$$
\frac{\partial\left(\frac{\omega_{Rj}}{\omega}\right)}{\partial\delta_h} = \frac{\begin{gathered}-\Big\{\Big\{\Big[\Big\{\frac{d_{mj}}{2}[(f_o-0.5)D+\delta_{2Z+j}]+\delta_{4Z+j}\delta_j\Big\}\Big\{\frac{d_{mj}}{2}\frac{\partial\delta_{3Z+j}}{\partial\delta_h}+[(f_i-0.5)D+\delta_{3Z+j}]\frac{\partial\delta_j}{\partial\delta_h}-\delta_{5Z+j}\Big(c\psi_j\frac{\partial\delta_{9Z+2}}{\partial\delta_h}-\mathcal{R}|c\psi_j|s\delta_{9Z+3}\frac{\partial\delta_{9Z+3}}{\partial\delta_h}\frac{\partial\delta_j}{\partial\delta_h}\Big)-(s_{zj}-\delta_j)\frac{\partial\delta_{5Z+j}}{\partial\delta_h}\Big\}+\Big\{\frac{d_{mj}}{2}[(f_i-0.5)D+ \\
\delta_{3Z+j}]-\delta_{5Z+j}(s_{zj}-\delta_j)\Big\}\Big\{\frac{d_{mj}}{2}\frac{\partial\delta_{2Z+j}}{\partial\delta_h}+[(f_o-0.5)D+\delta_{2Z+j}+\delta_{4Z+j}]\frac{\partial\delta_j}{\partial\delta_h}+\delta_j\frac{\partial\delta_{4Z+j}}{\partial\delta_h}\Big\}\Big]\sqrt{\delta_{6Z+j}^2+\delta_{7Z+j}^2+\delta_{8Z+j}^2}+\Big\{\frac{d_{mj}}{2}[(f_o-0.5)D+\delta_{2Z+j}]+\delta_{4Z+j}\delta_j\Big\}\Big\{\frac{d_{mj}}{2}[(f_i-0.5)D+\delta_{3Z+j}]- \\
\delta_{5Z+j}(s_{zj}-\delta_j)\Big\}\frac{\delta_{6Z+j}\frac{\partial\delta_{6Z+j}}{\partial\delta_h}+\delta_{7Z+j}\frac{\partial\delta_{7Z+j}}{\partial\delta_h}+\delta_{8Z+j}\frac{\partial\delta_{8Z+j}}{\partial\delta_h}}{\sqrt{\delta_{6Z+j}^2+\delta_{7Z+j}^2+\delta_{8Z+j}^2}}\Big\}\Big\{\delta_{4Z+j}\Big\{\frac{d_{mj}}{2}[(f_i-0.5)D+\delta_{3Z+j}]-\delta_{5Z+j}(s_{zj}-\delta_j)\Big\}(\delta_{6Z+j}\delta_j+\delta_{8Z+j}\delta_{Z+j})+\delta_{5Z+j}\Big\{\frac{d_{mj}}{2}[(f_o-0.5)D+\delta_{2Z+j}]+ \\
\delta_{4Z+j}\delta_j\Big\}[\delta_{6Z+j}(s_{zj}-\delta_j)+\delta_{8Z+j}(s_{xj}-\delta_{Z+j})]\Big\}+\sqrt{\delta_{6Z+j}^2+\delta_{7Z+j}^2+\delta_{8Z+j}^2}\Big\{\frac{d_{mj}}{2}[(f_o-0.5)D+\delta_{2Z+j}]+\delta_{4Z+j}\delta_j\Big\}\Big\{\frac{d_{mj}}{2}[(f_i-0.5)D+\delta_{3Z+j}]-\delta_{5Z+j}(s_{zj}-\delta_j)\Big\}\Big\{\Big\{\delta_{4Z+j}\Big\{\frac{d_{mj}}{2}\frac{\partial\delta_{3Z+j}}{\partial\delta_h}+ \\
[(f_i-0.5)D+\delta_{3Z+j}]\frac{\partial\delta_j}{\partial\delta_h}-\delta_{5Z+j}\Big(c\psi_j\frac{\partial\delta_{9Z+2}}{\partial\delta_h}-\mathcal{R}|c\psi_j|s\delta_{9Z+3}\frac{\partial\delta_{9Z+3}}{\partial\delta_h}\frac{\partial\delta_j}{\partial\delta_h}\Big)-(s_{zj}-\delta_j)\frac{\partial\delta_{5Z+j}}{\partial\delta_h}\Big\}+\Big\{\frac{d_{mj}}{2}[(f_i-0.5)D+\delta_{3Z+j}]-\delta_{5Z+j}(s_{zj}-\delta_j)\Big\}\frac{\partial\delta_{4Z+j}}{\partial\delta_h}\Big\}(\delta_{6Z+j}\delta_j+\delta_{8Z+j}\delta_{Z+j})+ \\
\delta_{4Z+j}\Big\{\frac{d_{mj}}{2}[(f_i-0.5)D+\delta_{3Z+j}]-\delta_{5Z+j}(s_{zj}-\delta_j)\Big\}\Big(\delta_{6Z+j}\frac{\partial\delta_j}{\partial\delta_h}+\delta_j\frac{\partial\delta_{6Z+j}}{\partial\delta_h}+\delta_{8Z+j}\frac{\partial\delta_{Z+j}}{\partial\delta_h}+\delta_{Z+j}\frac{\partial\delta_{8Z+j}}{\partial\delta_h}\Big)+\Big\{\delta_{5Z+j}\Big\{\frac{d_{mj}}{2}\frac{\partial\delta_{2Z+j}}{\partial\delta_h}+[(f_o-0.5)D+\delta_{2Z+j}+\delta_{4Z+j}]\frac{\partial\delta_j}{\partial\delta_h}+\delta_j\frac{\partial\delta_{4Z+j}}{\partial\delta_h}\Big\}+ \\
\Big\{\frac{d_{mj}}{2}[(f_o-0.5)D+\delta_{2Z+j}]+\delta_{4Z+j}\delta_j\Big\}\frac{\partial\delta_{5Z+j}}{\partial\delta_h}\Big\}[\delta_{6Z+j}(s_{zj}-\delta_j)+\delta_{8Z+j}(s_{xj}-\delta_{Z+j})]+\delta_{5Z+j}\Big\{\frac{d_{mj}}{2}[(f_o-0.5)D+\delta_{2Z+j}]+\delta_{4Z+j}\delta_j\Big\}\times \\
\Big[\delta_{6Z+j}\Big(c\psi_j\frac{\partial\delta_{9Z+2}}{\partial\delta_h}-\mathcal{R}|c\psi_j|s\delta_{9Z+3}\frac{\partial\delta_{9Z+3}}{\partial\delta_h}\frac{\partial\delta_j}{\partial\delta_h}\Big)+(s_{zj}-\delta_j)\frac{\partial\delta_{6Z+j}}{\partial\delta_h}+\delta_{8Z+j}\Big(\frac{\partial\delta_{9Z+1}}{\partial\delta_h}+\mathcal{R}c\psi_j c\delta_{9Z+3}\frac{\partial\delta_{9Z+3}}{\partial\delta_h}\frac{\partial\delta_{Z+j}}{\partial\delta_h}\Big)+(s_{xj}-\delta_{Z+j})\frac{\partial\delta_{8Z+j}}{\partial\delta_h}\Big]\Big\}\Big\}\end{gathered}}{\begin{Bmatrix}\delta_{4Z+j}\Big\{\frac{d_{mj}}{2}[(f_i-0.5)D+\delta_{3Z+j}]-\delta_{5Z+j}(s_{zj}-\delta_j)\Big\}(\delta_{6Z+j}\delta_j+\delta_{8Z+j}\delta_{Z+j})+ \\ \delta_{5Z+j}\Big\{\frac{d_{mj}}{2}[(f_o-0.5)D+\delta_{2Z+j}]+\delta_{4Z+j}\delta_j\Big\}[\delta_{6Z+j}(s_{zj}-\delta_j)+\delta_{8Z+j}(s_{xj}-\delta_{Z+j})]\end{Bmatrix}^2}
\tag{31}
$$

Likewise, for the inner race to be stationary,

$$
\frac{\partial\left(\frac{\omega_{mj}}{\omega}\right)}{\partial\delta_h} = \frac{\begin{gathered}\Big\{\Big\{\delta_{5Z+j}\Big\{\frac{d_{mj}}{2}\frac{\partial\delta_{2Z+j}}{\partial\delta_h}+[(f_o-0.5)D+\delta_{2Z+j}+\delta_{4Z+j}]\frac{\partial\delta_j}{\partial\delta_h}+\delta_j\frac{\partial\delta_{4Z+j}}{\partial\delta_h}\Big\}+\Big\{\frac{d_{mj}}{2}[(f_o-0.5)D+\delta_{2Z+j}]+\delta_{4Z+j}\delta_j\Big\}\frac{\partial\delta_{5Z+j}}{\partial\delta_h}\Big\}[\delta_{6Z+j}(s_{zj}-\delta_j)+\delta_{8Z+j}(s_{xj}-\delta_{Z+j})]+\delta_{5Z+j}\Big\{\frac{d_{mj}}{2}[(f_o-0.5)D \\
+\delta_{2Z+j}]+\delta_{4Z+j}\delta_j\Big\}\Big[\delta_{6Z+j}\Big(c\psi_j\frac{\partial\delta_{9Z+2}}{\partial\delta_h}-\mathcal{R}|c\psi_j|s\delta_{9Z+3}\frac{\partial\delta_{9Z+3}}{\partial\delta_h}\frac{\partial\delta_j}{\partial\delta_h}\Big)+(s_{zj}-\delta_j)\frac{\partial\delta_{6Z+j}}{\partial\delta_h}+\delta_{8Z+j}\Big(\frac{\partial\delta_{9Z+1}}{\partial\delta_h}+\mathcal{R}c\psi_j c\delta_{9Z+3}\frac{\partial\delta_{9Z+3}}{\partial\delta_h}\frac{\partial\delta_{Z+j}}{\partial\delta_h}\Big)+(s_{xj}-\delta_{Z+j})\frac{\partial\delta_{8Z+j}}{\partial\delta_h}\Big]\Big\}\Big\{\delta_{4Z+j}\Big\{\frac{d_{mj}}{2} \\
\times[(f_i-0.5)D+\delta_{3Z+j}]-\delta_{5Z+j}(s_{zj}-\delta_j)\Big\}(\delta_{6Z+j}\delta_j+\delta_{8Z+j}\delta_{Z+j})+\delta_{5Z+j}\Big\{\frac{d_{mj}}{2}[(f_o-0.5)D+\delta_{2Z+j}]+\delta_{4Z+j}\delta_j\Big\}[\delta_{6Z+j}(s_{zj}-\delta_j)+\delta_{8Z+j}(s_{xj}-\delta_{Z+j})]\Big\}-\delta_{5Z+j}\Big\{\frac{d_{mj}}{2}[(f_o-0.5)D+ \\
\delta_{2Z+j}]+\delta_{4Z+j}\delta_j\Big\}[\delta_{6Z+j}(s_{zj}-\delta_j)+\delta_{8Z+j}(s_{xj}-\delta_{Z+j})]\Big\{\Big\{\delta_{4Z+j}\Big\{\frac{d_{mj}}{2}\frac{\partial\delta_{3Z+j}}{\partial\delta_h}+[(f_i-0.5)D+\delta_{3Z+j}]\frac{\partial\delta_j}{\partial\delta_h}-\delta_{5Z+j}\Big(c\psi_j\frac{\partial\delta_{9Z+2}}{\partial\delta_h}-\mathcal{R}|c\psi_j|s\delta_{9Z+3}\frac{\partial\delta_{9Z+3}}{\partial\delta_h}\frac{\partial\delta_j}{\partial\delta_h}\Big)-(s_{zj}-\delta_j)\frac{\partial\delta_{5Z+j}}{\partial\delta_h}\Big\} \\
+\Big\{\frac{d_{mj}}{2}[(f_i-0.5)D+\delta_{3Z+j}]-\delta_{5Z+j}(s_{zj}-\delta_j)\Big\}\frac{\partial\delta_{4Z+j}}{\partial\delta_h}\Big\}(\delta_{6Z+j}\delta_j+\delta_{8Z+j}\delta_{Z+j})+\delta_{4Z+j}\Big\{\frac{d_{mj}}{2}[(f_i-0.5)D+\delta_{3Z+j}]-\delta_{5Z+j}(s_{zj}-\delta_j)\Big\}\Big(\delta_{6Z+j}\frac{\partial\delta_j}{\partial\delta_h}+\delta_j\frac{\partial\delta_{6Z+j}}{\partial\delta_h}+\delta_{8Z+j}\frac{\partial\delta_{Z+j}}{\partial\delta_h}+\delta_{Z+j}\times \\
\frac{\partial\delta_{8Z+j}}{\partial\delta_h}\Big)+\Big\{\delta_{5Z+j}\Big\{\frac{d_{mj}}{2}\frac{\partial\delta_{2Z+j}}{\partial\delta_h}+[(f_o-0.5)D+\delta_{2Z+j}+\delta_{4Z+j}]\frac{\partial\delta_j}{\partial\delta_h}+\delta_j\frac{\partial\delta_{4Z+j}}{\partial\delta_h}\Big\}+\Big\{\frac{d_{mj}}{2}[(f_o-0.5)D+\delta_{2Z+j}]+\delta_{4Z+j}\delta_j\Big\}\frac{\partial\delta_{5Z+j}}{\partial\delta_h}\Big\}[\delta_{6Z+j}(s_{zj}-\delta_j)+\delta_{8Z+j}(s_{xj}-\delta_{Z+j})]+\delta_{5Z+j}\Big\{\frac{d_{mj}}{2} \\
\times[(f_o-0.5)D+\delta_{2Z+j}]+\delta_{4Z+j}\delta_j\Big\}\Big[(s_{zj}-\delta_j)\frac{\partial\delta_{6Z+j}}{\partial\delta_h}+(s_{xj}-\delta_{Z+j})\frac{\partial\delta_{8Z+j}}{\partial\delta_h}+\delta_{6Z+j}\Big(c\psi_j\frac{\partial\delta_{9Z+2}}{\partial\delta_h}-\mathcal{R}|c\psi_j|s\delta_{9Z+3}\frac{\partial\delta_{9Z+3}}{\partial\delta_h}\frac{\partial\delta_j}{\partial\delta_h}\Big)+\delta_{8Z+j}\Big(\frac{\partial\delta_{9Z+1}}{\partial\delta_h}+\mathcal{R}c\psi_j c\delta_{9Z+3}\frac{\partial\delta_{9Z+3}}{\partial\delta_h}\frac{\partial\delta_{Z+j}}{\partial\delta_h}\Big)\Big]\Big\}\Big\}\end{gathered}}{\begin{Bmatrix}\delta_{4Z+j}\Big\{\frac{d_{mj}}{2}[(f_i-0.5)D+\delta_{3Z+j}]-\delta_{5Z+j}(s_{zj}-\delta_j)\Big\}(\delta_{6Z+j}\delta_j+\delta_{8Z+j}\delta_{Z+j})+ \\ \delta_{5Z+j}\Big\{\frac{d_{mj}}{2}[(f_o-0.5)D+\delta_{2Z+j}]+\delta_{4Z+j}\delta_j\Big\}[\delta_{6Z+j}(s_{zj}-\delta_j)+\delta_{8Z+j}(s_{xj}-\delta_{Z+j})]\end{Bmatrix}^2},
\tag{32}
$$

Journal of Physics: Conference Series **641** (2015) 012018 doi:10.1088/1742-6596/641/1/012018

and $\partial(\omega_{Rj}/\omega)/\partial\delta$ is given by (31) with the opposite sign.

The last three equations from (1) must be solved simultaneously for δ_{9Z+1}, …, δ_{9Z+3} after obtaining updated values for: β_{ij}, β_{oj}, K_{ij}, K_{oj}, s_{xj}, s_{zj}, F_{xij}, F_{yij}, F_{xoj}, F_{yoj}, M_{sij}, M_{soj}, M_{Rij}, M_{Roj}, δ_{kZ+j}, $k = 0$, …, 8; $j = 1$, …, Z. If δ_h^0, $h = 9Z+1$, …, $9Z+3$, is a 3-dimensional vector with the initial estimates of the variables δ_{9Z+1}, …, δ_{9Z+3}, in that order, improved values are given by (2), in which $\{\epsilon_g\}$, $g = 9Z+1$, …, $9Z+3$, is the 3-dimensional vector with the errors functions, in that order, from (1). The elements of the 3×3-matrix $[a_{gh}]$ are

$$a_{(9Z+1)h}=\frac{\partial F_a}{\partial\delta_h}-\sum_{j=1}^{Z}\frac{\left\{\begin{matrix}[(f_i-0.5)D+\delta_{3Z+j}]\Big[K_{ij}\delta_{3Z+j}^{1.5}\left(\frac{\partial\delta_{9Z+1}}{\partial\delta_h}+\mathcal{R}_i c\psi_j c\delta_{9Z+3}\frac{\partial\delta_{9Z+3}}{\partial\delta_h}-\frac{\partial\delta_{Z+j}}{\partial\delta_h}\right)+1.5K_{ij}\delta_{3Z+j}^{0.5}(s_{xj}-\delta_{Z+j})\frac{\partial\delta_{3Z+j}}{\partial\delta_h}-\\ F_{xij}\left(c\psi_j\frac{\partial\delta_{9Z+2}}{\partial\delta_h}-\mathcal{R}_i|c\psi_j|s\delta_{9Z+3}\frac{\partial\delta_{9Z+3}}{\partial\delta_h}-\frac{\partial\delta_j}{\partial\delta_h}\right)-(s_{zj}-\delta_j)\frac{\partial F_{xij}}{\partial\delta_h}\Big]-\left[K_{ij}\delta_{3Z+j}^{1.5}(s_{xj}-\delta_{Z+j})-F_{xij}(s_{zj}-\delta_j)\right]\frac{\partial\delta_{3Z+j}}{\partial\delta_h}\end{matrix}\right\}}{[(f_i-0.5)D+\delta_{3Z+j}]^2},\tag{33}$$

$$a_{(9Z+2)h}=\frac{\partial F_r}{\partial\delta_h}-\sum_{j=1}^{Z}\frac{\left\{\begin{matrix}[(f_i-0.5)D+\delta_{3Z+j}]\Big[K_{ij}\delta_{3Z+j}^{1.5}\left(c\psi_j\frac{\partial\delta_{9Z+2}}{\partial\delta_h}-\mathcal{R}_i|c\psi_j|s\delta_{9Z+3}\frac{\partial\delta_{9Z+3}}{\partial\delta_h}-\frac{\partial\delta_j}{\partial\delta_h}\right)+1.5K_{ij}\delta_{3Z+j}^{0.5}(s_{zj}-\delta_j)\frac{\partial\delta_{3Z+j}}{\partial\delta_h}+\\ F_{xij}\left(\frac{\partial\delta_{9Z+1}}{\partial\delta_h}+\mathcal{R}_i c\psi_j c\delta_{9Z+3}\frac{\partial\delta_{9Z+3}}{\partial\delta_h}-\frac{\partial\delta_{Z+j}}{\partial\delta_h}\right)+(s_{xj}-\delta_{Z+j})\frac{\partial F_{xij}}{\partial\delta_h}\Big]-\left[K_{ij}\delta_{3Z+j}^{1.5}(s_{zj}-\delta_j)+F_{xij}(s_{xj}-\delta_{Z+j})\right]\frac{\partial\delta_{3Z+j}}{\partial\delta_h}\end{matrix}\right\}}{[(f_i-0.5)D+\delta_{3Z+j}]^2}c\psi_j,\tag{34}$$

$$\begin{aligned}a_{(9Z+3)h}=\frac{\partial M}{\partial\delta_h}-\sum_{j=1}^{Z}\Bigg\{\mathcal{R}_i\Bigg\{&(K_{ij}\delta_{3Z+j}^{1.5}c\beta_{ij}+F_{xij}s\beta_{ij})\Bigg[\frac{(s_{zj}-\delta_j)\left(\frac{\partial\delta_{9Z+1}}{\partial\delta_h}+\mathcal{R}_i c\psi_j c\delta_{9Z+3}\frac{\partial\delta_{9Z+3}}{\partial\delta_h}-\frac{\partial\delta_{Z+j}}{\partial\delta_h}\right)-(s_{xj}-\delta_{Z+j})\left(c\psi_j\frac{\partial\delta_{9Z+2}}{\partial\delta_h}-\mathcal{R}_i|c\psi_j|s\delta_{9Z+3}\frac{\partial\delta_{9Z+3}}{\partial\delta_h}-\frac{\partial\delta_j}{\partial\delta_h}\right)}{(s_{zj}-\delta_j)^2+(s_{xj}-\delta_{Z+j})^2}\\&-c\psi_j\frac{\partial\delta_{9Z+3}}{\partial\delta_h}\Bigg]+1.5K_{ij}\delta_{3Z+j}^{0.5}s\beta_{ij}\frac{\partial\delta_{3Z+j}}{\partial\delta_h}-(c\beta_{ij}-r_i/\mathcal{R}_i)\frac{\partial F_{xij}}{\partial\delta_h}\Bigg\}c\psi_j-\Bigg\{(F_{yij}r_ic\beta_{ij}+M_{sij}s\beta_{ij})\Bigg[\frac{(s_{zj}-\delta_j)\left(\frac{\partial\delta_{9Z+1}}{\partial\delta_h}+\mathcal{R}_i c\psi_j c\delta_{9Z+3}\frac{\partial\delta_{9Z+3}}{\partial\delta_h}-\frac{\partial\delta_{Z+j}}{\partial\delta_h}\right)}{(s_{zj}-\delta_j)^2+(s_{xj}-\delta_{Z+j})^2}-\\&\frac{(s_{xj}-\delta_{Z+j})\left(c\psi_j\frac{\partial\delta_{9Z+2}}{\partial\delta_h}-\mathcal{R}_i|c\psi_j|s\delta_{9Z+3}\frac{\partial\delta_{9Z+3}}{\partial\delta_h}-\frac{\partial\delta_j}{\partial\delta_h}\right)}{(s_{zj}-\delta_j)^2+(s_{xj}-\delta_{Z+j})^2}-c\psi_j\frac{\partial\delta_{9Z+3}}{\partial\delta_h}\Bigg]+r_is\beta_{ij}\frac{\partial F_{yij}}{\partial\delta_h}-c\beta_{ij}\frac{\partial M_{sij}}{\partial\delta_h}\Bigg\}s\psi_j\Bigg\}\end{aligned}\tag{35}$$

Differentiating (16)-(18), (22)-(24) and (27)-(29) of [13] with respect δ_h, $h = 9Z+1$, …, $9Z+3$, 27Z simultaneous linear equations in $\partial\delta_{kZ+j}/\partial\delta_h$, $k = 0$, …, 8; $j = 1$, …, Z, results, which are

$$\begin{aligned}&(s_{zj}-\delta_j)\frac{\partial\delta_j}{\partial\delta_h}+(s_{xj}-\delta_{Z+j})\frac{\partial\delta_{Z+j}}{\partial\delta_h}+[(f_i-0.5)D+\delta_{3Z+j}]\frac{\partial\delta_{3Z+j}}{\partial\delta_h}=\\&(s_{xj}-\delta_{Z+j})\left(\frac{\partial\delta_{9Z+1}}{\partial\delta_h}+\mathcal{R}_i c\psi_j c\delta_{9Z+3}\frac{\partial\delta_{9Z+3}}{\partial\delta_h}\right)+(s_{zj}-\delta_j)\left(c\psi_j\frac{\partial\delta_{9Z+2}}{\partial\delta_h}-\mathcal{R}_i|c\psi_j|s\delta_{9Z+3}\frac{\partial\delta_{9Z+3}}{\partial\delta_h}\right),\end{aligned}\tag{36}$$

$$\delta_j\frac{\partial\delta_j}{\partial\delta_h}+\delta_{Z+j}\frac{\partial\delta_{Z+j}}{\partial\delta_h}-[(f_o-0.5)D+\delta_{2Z+j}]\frac{\partial\delta_{2Z+j}}{\partial\delta_h}=0,\tag{37}$$

$$\delta_{6Z+j}\frac{\partial\delta_{6Z+j}}{\partial\delta_h}+\delta_{7Z+j}\frac{\partial\delta_{7Z+j}}{\partial\delta_h}+\delta_{8Z+j}\frac{\partial\delta_{8Z+j}}{\partial\delta_h}=0,\tag{38}$$

$$\begin{aligned}&\left[\frac{F_{xoj}}{(f_o-0.5)D+\delta_{2Z+j}}+\frac{F_{xij}}{(f_i-0.5)D+\delta_{3Z+j}}\right]\frac{\partial\delta_j}{\partial\delta_h}+\frac{\delta_j}{(f_o-0.5)D+\delta_{2Z+j}}\frac{\partial F_{xoj}}{\partial\delta_h}-\frac{(s_{zj}-\delta_j)}{(f_i-0.5)D+\delta_{3Z+j}}\frac{\partial F_{xij}}{\partial\delta_h}-\left[\frac{K_{oj}\delta_{2Z+j}^{1.5}}{(f_o-0.5)D+\delta_{2Z+j}}+\frac{K_{ij}\delta_{3Z+j}^{1.5}}{(f_i-0.5)D+\delta_{ij}}\right]\frac{\partial\delta_{Z+j}}{\partial\delta_h}-\\&\left\{\frac{1.5K_{oj}\delta_{2Z+j}^{0.5}\delta_{Z+j}[(f_o-0.5)D+\delta_{2Z+j}]+F_{xoj}\delta_j-K_{oj}\delta_{2Z+j}^{1.5}\delta_{Z+j}}{[(f_o-0.5)D+\delta_{2Z+j}]^2}\right\}\frac{\partial\delta_{2Z+j}}{\partial\delta_h}+\left\{\frac{1.5K_{ij}\delta_{3Z+j}^{0.5}(s_{xj}-\delta_{Z+j})[(f_i-0.5)D+\delta_{3Z+j}]+F_{xij}(s_{zj}-\delta_j)-K_{ij}\delta_{3Z+j}^{1.5}(s_{xj}-\delta_{Z+j})}{[(f_i-0.5)D+\delta_{3Z+j}]^2}\right\}\frac{\partial\delta_{3Z+j}}{\partial\delta_h}=\\&\frac{-K_{ij}\delta_{3Z+j}^{1.5}\left(\frac{\partial\delta_{9Z+1}}{\partial\delta_h}+\mathcal{R}_i c\psi_j c\delta_{9Z+3}\frac{\partial\delta_{9Z+3}}{\partial\delta_h}\right)+F_{xij}\left(c\psi_j\frac{\partial\delta_{9Z+2}}{\partial\delta_h}-\mathcal{R}_i|c\psi_j|s\delta_{9Z+3}\frac{\partial\delta_{9Z+3}}{\partial\delta_h}\right)}{(f_i-0.5)D+\delta_{3Z+j}},\end{aligned}\tag{39}$$

$$\begin{aligned}&\left[\frac{K_{oj}\delta_{2Z+j}^{1.5}}{(f_o-0.5)D+\delta_{2Z+j}}+\frac{K_{ij}\delta_{3Z+j}^{1.5}}{(f_i-0.5)D+\delta_{3Z+j}}\right]\frac{\partial\delta_j}{\partial\delta_h}+\frac{\delta_{Z+j}}{(f_o-0.5)D+\delta_{2Z+j}}\frac{\partial F_{xoj}}{\partial\delta_h}-\frac{(s_{xj}-\delta_{Z+j})}{(f_i-0.5)D+\delta_{3Z+j}}\frac{\partial F_{xij}}{\partial\delta_h}+\left[\frac{F_{xoj}}{(f_o-0.5)D+\delta_{2Z+j}}+\frac{F_{xij}}{(f_i-0.5)D+\delta_{3Z+j}}\right]\frac{\partial\delta_{Z+j}}{\partial\delta_h}+\\&\left\{\frac{1.5K_{oj}\delta_{2Z+j}^{0.5}\delta_j[(f_o-0.5)D+\delta_{2Z+j}]-K_{oj}\delta_{2Z+j}^{1.5}\delta_j-F_{xoj}\delta_{Z+j}}{[(f_o-0.5)D+\delta_{2Z+j}]^2}\right\}\frac{\partial\delta_{2Z+j}}{\partial\delta_h}-\\&\left\{\frac{1.5K_{ij}\delta_{3Z+j}^{0.5}(s_{zj}-\delta_j)[(f_i-0.5)D+\delta_{3Z+j}]-K_{ij}\delta_{3Z+j}^{1.5}(s_{zj}-\delta_j)-F_{xij}(s_{xj}-\delta_{Z+j})}{[(f_i-0.5)D+\delta_{3Z+j}]^2}\right\}\frac{\partial\delta_{3Z+j}}{\partial\delta_h}-\frac{\partial F_{zij}}{\partial\delta_h}=\\&\frac{K_{ij}\delta_{3Z+j}^{1.5}\left(c\psi_j\frac{\partial\delta_{9Z+2}}{\partial\delta_h}-\mathcal{R}_i|c\psi_j|s\delta_{9Z+3}\frac{\partial\delta_{9Z+3}}{\partial\delta_h}\right)+F_{xij}\left(\frac{\partial\delta_{9Z+1}}{\partial\delta_h}+\mathcal{R}_i c\psi_j c\delta_{9Z+3}\frac{\partial\delta_{9Z+3}}{\partial\delta_h}\right)}{(f_i-0.5)D+\delta_{3Z+j}},\end{aligned}\tag{40}$$

$$\frac{\partial F_{yoj}}{\partial\delta_h}+\frac{\partial F_{yij}}{\partial\delta_h}=0,\tag{41}$$

$$\begin{aligned}&\left[\frac{M_{Roj}}{(f_o-0.5)D+\delta_{2Z+j}}+\frac{M_{Rij}}{(f_i-0.5)D+\delta_{3Z+j}}\right]\frac{\partial\delta_j}{\partial\delta_h}+\frac{\delta_j}{(f_o-0.5)D+\delta_{2Z+j}}\frac{\partial M_{Roj}}{\partial\delta_h}+\frac{\delta_{Z+j}}{(f_o-0.5)D+\delta_{2Z+j}}\frac{\partial M_{soj}}{\partial\delta_h}+\\&\left[\frac{M_{soj}}{(f_o-0.5)D+\delta_{2Z+j}}+\frac{M_{sij}}{(f_i-0.5)D+\delta_{3Z+j}}\right]\frac{\partial\delta_{Z+j}}{\partial\delta_h}-\frac{(s_{xj}-\delta_{Z+j})}{(f_i-0.5)D+\delta_{3Z+j}}\frac{\partial M_{sij}}{\partial\delta_h}-\frac{(s_{zj}-\delta_j)}{(f_i-0.5)D+\delta_{3Z+j}}\frac{\partial M_{Rij}}{\partial\delta_h}-\frac{M_{Roj}\delta_j+M_{soj}\delta_{Z+j}}{[(f_o-0.5)D+\delta_{2Z+j}]^2}\frac{\partial\delta_{2Z+j}}{\partial\delta_h}+\\&\frac{M_{sij}(s_{xj}-\delta_{Z+j})+M_{Rij}(s_{zj}-\delta_j)}{[(f_i-0.5)D+\delta_{3Z+j}]^2}\frac{\partial\delta_{3Z+j}}{\partial\delta_h}=\frac{M_{sij}\left(\frac{\partial\delta_{9Z+1}}{\partial\delta_h}+\mathcal{R}_i c\psi_j c\delta_{9Z+3}\frac{\partial\delta_{9Z+3}}{\partial\delta_h}\right)+M_{Rij}\left(c\psi_j\frac{\partial\delta_{9Z+2}}{\partial\delta_h}-\mathcal{R}_i|c\psi_j|s\delta_{9Z+3}\frac{\partial\delta_{9Z+3}}{\partial\delta_h}\right)}{(f_i-0.5)D+\delta_{3Z+j}},\end{aligned}\tag{42}$$

Journal of Physics: Conference Series **641** (2015) 012018 doi:10.1088/1742-6596/641/1/012018

$$\left[\frac{M_{soj}}{(f_o-0.5)D+\delta_{2Z+j}}+\frac{M_{sij}}{(f_i-0.5)D+\delta_{3Z+j}}\right]\frac{\partial\delta_j}{\partial\delta_h}+\frac{\delta_j}{(f_o-0.5)D+\delta_{2Z+j}}\frac{\partial M_{soj}}{\partial\delta_h}-\frac{\delta_{Z+j}}{(f_o-0.5)D+\delta_{2Z+j}}\frac{\partial M_{Roj}}{\partial\delta_h}-$$
$$\left[\frac{M_{Roj}}{(f_o-0.5)D+\delta_{2Z+j}}+\frac{M_{Rij}}{(f_i-0.5)D+\delta_{3Z+j}}\right]\frac{\partial\delta_{Z+j}}{\partial\delta_h}+\frac{(s_{xj}-\delta_{Z+j})}{(f_i-0.5)D+\delta_{3Z+j}}\frac{\partial M_{Rij}}{\partial\delta_h}-\frac{(s_{zj}-\delta_j)}{(f_i-0.5)D+\delta_{3Z+j}}\frac{\partial M_{sij}}{\partial\delta_h}-\frac{M_{soj}\delta_j-M_{Roj}\delta_{Z+j}}{[(f_o-0.5)D+\delta_{2Z+j}]^2}\frac{\partial\delta_{2Z+j}}{\partial\delta_h}+$$
$$\frac{M_{sij}(s_{zj}-\delta_j)-M_{Rij}(s_{xj}-\delta_{Z+j})}{[(f_i-0.5)D+\delta_{3Z+j}]^2}\frac{\partial\delta_{3Z+j}}{\partial\delta_h}+\frac{\partial M_{z'j}}{\partial\delta_h}=\frac{M_{sij}\left(c\psi_j\frac{\partial\delta_{9Z+2}}{\partial\delta_h}-\mathcal{R}_i|c\psi_j|s\delta_{9Z+3}\frac{\partial\delta_{9Z+3}}{\partial\delta_h}\right)-M_{Rij}\left(\frac{\partial\delta_{9Z+1}}{\partial\delta_h}+\mathcal{R}_i c\psi_j c\delta_{9Z+3}\frac{\partial\delta_{9Z+3}}{\partial\delta_h}\right)}{(f_i-0.5)D+\delta_{3Z+j}}, \tag{43}$$

$$\frac{\partial M_{y'j}}{\partial\delta_h}-\frac{\partial M_{yij}}{\partial\delta_h}-\frac{\partial M_{yoj}}{\partial\delta_h}=0. \tag{44}$$

The derivatives of $F_{z'j}$, F_{xij}, F_{xoj}, F_{yij}, F_{yoj}, M_{sij}, M_{soj}, M_{Rij}, M_{Roj}, $M_{z'j}$, M_{yij}, M_{yoj} and $M_{y'j}$ with respect δ_h, h = 9Z+1, …, 9Z+3, to be used in (33)-(44) are given by (12)-(24). In (13)-(20), (22)-(23) γ_{ij}, γ_{oj} are given by (42) of [13]. The derivatives of γ_{ij}, γ_{oj} with respect δ_h, h = 9Z+1, …, 9Z+3, to be used in (13)-(20), (22)-(23) are given by (25), with V_{xij}/ω_{sij}, V_{yij}/ω_{sij}, V_{xoj}/ω_{soj} and V_{yoj}/ω_{soj} given by (43)-(46) of [13]. The derivatives of V_{xij}/ω_{sij}, V_{yij}/ω_{sij}, V_{xoj}/ω_{soj} and V_{yoj}/ω_{soj} with respect δ_h, h = 9Z+1, …, 9Z+3, to be used in (25) are given by (26)-(29).

For outer race to be stationary ω_{mj}/ω and ω_{Rj}/ω are given by (33)-(34) of [13] and for inner race to be stationary are given by (35)-(34) of [13], the last with opposite sign. The derivatives of (33)-(34) of [13] with respect δ_h, h = 9Z+1, …, 9Z+3, to be used in (12), (21) and (24) are given by (30)-(31) and for (32) and (31), the last with opposite sign, are given by (32)-(31), the last with opposite sign.

The (36)-(44) linear system's solutions – $\partial\delta_{kZ+j}/\partial\delta_h$, k = 0, …, 8; j = 1, …, Z – are to be used in (33)-(35) for the new estimates of δ_{9Z+1}, δ_{9Z+2} and δ_{9Z+3}.

4. References

[1] Tomović R 2012 *Mechanism and Mach. Theory* **47** 74-88

[2] Tomović R 2012 *Int. J. Mech. Sci.* **60** 23-33

[3] Tomović R 2013 *Adv. Mat. Res.* **633** 103-16

[4] Tudose C, Rusu F and Tudose L 2013 *Appl. Math. Mech.* **56** 469–74

[5] Rasolofondraibe L, Pottier B, Marconnet P and Chiementin X 2012 *IEEE Sens. J.* **12** 2186–91

[6] Rasolofondraibe L, Pottier B, Marconnet P and Perrin E 2013 *IEEE Sens. J.* **13** 3067–72

[7] Murer S, Bogard F, Rasolofondraibe L, Pottier B and Marconnet P 2015 *Mech. Syst. Signal Process.* **54-5** 306–13

[8] Starvin M S, Christopher A S and Manisekar K 2011 *CiiT Int. J. Automation and Autonomous Syst.* **3** 389–94

[9] Rajasekhar M, Srinivas J, Divekar A 2013 *Proc 1st Int. and 16th Nat. Conf. on Machines and Mechanisms (iNaCoMM2013)* 941-46

[10] Muszynska A 1986 *J. sound vib.***110** 443-62

[11] Seong S, Kim W, Bae D and Lee S 2014 *Korean Soc. Mech. Eng. Fall Annual Meet.* 776-79

[12] Zhenguo S, Huimin D, Fanhai M and Hua W 2011 *Trans. CSAE* **27** 52-6

[13] Ricci M C 2015 *Journal of Physics: Conference Series* **641** 012017

Attitude control system design using a flywheel suspended by two gimbals

R W Peres[1], **M C Ricci**[1,2]

[1] Faculdade de Tecnologia São Francisco (FATESF), Brazil.
[2] Divisão de Mecânica Espacial e Controle (DMC), Instituto Nacional de Pesquisas Espaciais (INPE), Brazil.

E-mail: raphael-peres@hotmail.com

Abstract. This work presents the attitude control system design procedures for a three axis stabilized satellite in geostationary orbit, which contains a flywheel suspended by two gimbals. The use of a flywheel with two DOFs is an interesting option because with only one device it's possible to control the torques about vehicle's three axes; through the wheel speed control and gyrotorquing phenomenon with two DOFs. If the wheel size and speed are determined properly it's possible to cancel cyclic torques using gas jets only periodically to cancel secular disturbance torques. The system, based on a flywheel, takes only one pitch/roll (earth) sensor to maintain precise attitude, unlike mass expulsion based control systems, which uses propellants continuously, beyond roll, pitch and yaw sensors. It is considered the satellite is in nominal orbit and, therefore, that the attitude's acquisition phase has already elapsed. Control laws and system parameters are determined in order to cancel the solar pressure radiation disturbance torque and the torque due to misalignment of the thrusters. Stability is analyzed and step and cyclic responses are obtained.

1. Introduction

A satellite's pictorial view is shown in figure 1 in the orbital nominal position, with the body fixed axis x, y, z - the principal axes of inertia - aligned with the orbital reference frame. The satellite parameters as well as the requirements for the torque disturbance and pointing accuracy are given in Table 1.

Table 1. Parameters and design requirements

Satellite mass	716 kg
Moments of inertia	$I_x = I_z = 2000\text{Nms}^2$, $I_y = 400\ \text{Nms}^2$
Attitude Accuracy requirements	Roll (ϕ) and pitch (θ) = 0,05°, yaw (ψ) = 0,40°
Solar pressure torques (t = 0 at 6 AM or 6 PM orbital position)	$T_x = 2 \times 10^{-5}(1 - 2\sin\omega_o t)$ Nm $T_y = 10^{-4}(\cos\omega_o t)$ Nm $T_z = -5 \times 10^{-5}(\cos\omega_o t)$ Nm
Thruster misalignment torque	$T_F = 8{,}5 \times 10^{-5}$ Nm

Where ω_o is the geostationary orbit rate with constant value of 7.28×10^{-5} rad/sec.

The nominal orientation of the momentum wheel is shown in figure 2, with the spin axis coinciding with the pitch axis. The components of the angular momentum of the wheel with respect to the principal axes of inertia are obtained with help of the figure 3.

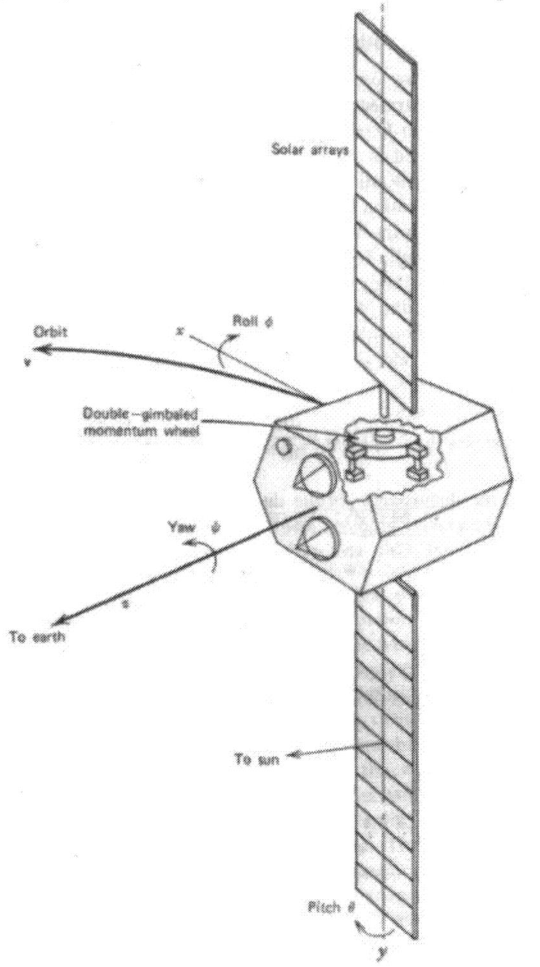

Figure 1. Artist's conception of the three axes stabilized satellite.
SOURCE: Kaplan (1976), p. 242.

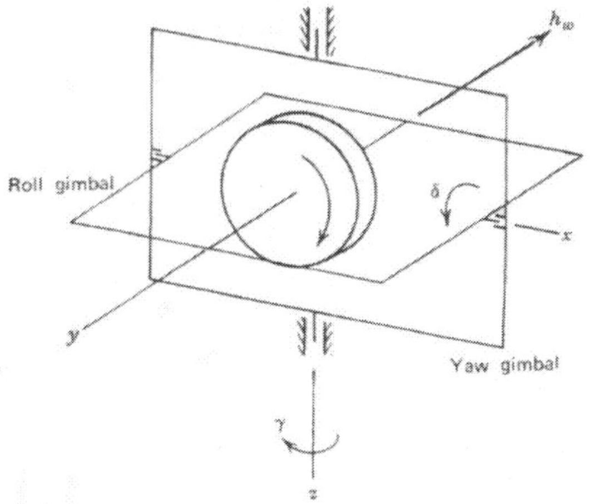

Figure 2. Nominal orientation of flywheel with two DOFs.
SOURCE: Kaplan (1976), p. 243.

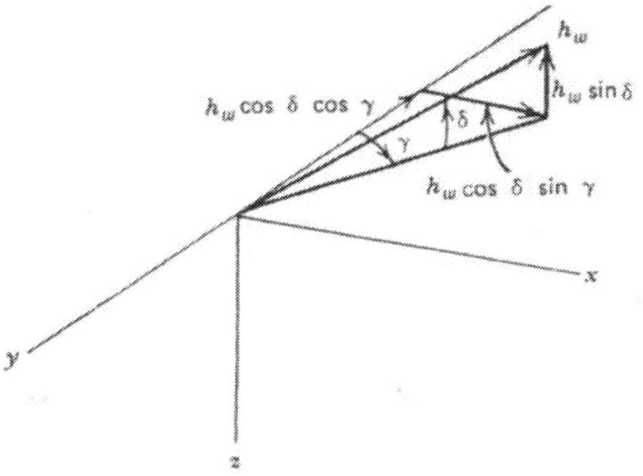

Figure 3. Angular momentum components
SOURCE: Kaplan (1976), p. 243.

2. Linearized equations of motion

Separating the environmental torques in two instalments, the Euler equations of motion are given by

$$\mathbf{T} + \mathbf{G} = \frac{d\mathbf{h}}{dt} = \left[\frac{d\mathbf{h}}{dt}\right]_B + \boldsymbol{\omega} \times \mathbf{h}, \tag{1}$$

where $\mathbf{T}$ are the solar radiation pressure and thrust's misalignment torques; $\mathbf{G}$ is the gravity gradient torque given by $\mathbf{G} = -3\omega_0^2(I_y - I_z)\phi\mathbf{i} - 3\omega_0^2(I_x - I_z)\theta\mathbf{j}$; $\boldsymbol{\omega}$ is the angular velocity vector of the satellite body. The vector $\mathbf{h}$ represents the total angular momentum of the system, $\mathbf{h} = \mathbf{h}_v + \mathbf{h}_w$, wherein $\mathbf{h}_v$ is the angular momentum of the vehicle and $\mathbf{h}_w$ is the angular momentum of the inertia wheel. The components of the vehicle's angular momentum vector are given with respect to the principal axis of inertia as $\mathbf{h}_v = I_x\omega_x\mathbf{i} + I_y\omega_y\mathbf{j} + I_z\omega_z\mathbf{k}$.. From figure 3 the components of the wheel angular momentum with respect to *x*, *y*, *z* are given by $\mathbf{h}_w = (\cos\delta\sin\gamma)h_w\mathbf{i} - (\cos\delta\cos\gamma)h_w\mathbf{j} - (\sin\delta)h_w\mathbf{k}$, where δ and γ are the gimbals deflection angles on the roll and yaw axis, respectively.

Expanding equation (1) yields the general equation of motion

$$\mathbf{T} = \left[I_x\dot{\omega}_x - \dot{\delta}(\sin\delta\sin\gamma)h_w + \dot{\gamma}(\cos\delta\cos\gamma)h_w + (\cos\delta\sin\gamma)\dot{h}_w + \omega_y(\omega_z I_z - h_w\sin\delta) - \omega_z\left(\omega_y I_y - h_w\cos\delta\cos\gamma\right) - 3\omega_0^2\left(I_y - I_z\right)\phi\right]\mathbf{i} + \left[I_y\dot{\omega}_y + \dot{\delta}\,(\sin\delta\cos\gamma)h_w + \dot{\gamma}(\cos\delta\sin\gamma)h_w - (\cos\delta\cos\gamma)\dot{h}_w + \omega_z(I_x\omega_x + h_w\cos\delta\sin\gamma) - \omega_x(I_z\omega_z - h_w\sin\delta) - 3\omega_0^2(I_x - I_z)\theta\right]\mathbf{j} + \left[I_z\dot{\omega}_z - \dot{\delta}(\cos\delta)h_w - (\sin\delta)\dot{h}_w + \omega_x\left(I_y\omega_y - h_w\cos\delta\cos\gamma\right) - \omega_y(I_x\omega_x + h_w\cos\delta\sin\gamma)\right]\mathbf{k}. \quad (2)$$

Whereas the vehicle's (ϕ, θ and ψ) and the gimbal's (δ and γ) angular deviations, with respect to the nominal frame shown in figure 1, are considerably smaller. Deviations from nominal wheel momentum h_n are also assumed small, allowing $h_w = h_n$ and applying the Euler angles transformation (sequence 3-2-1) in the $\boldsymbol{\omega}$ vector components, thus yields the linearized equations of motion

$$T_x = I_x\ddot{\phi} + \left[4\omega_0^2\left(I_y - I_z\right) + \omega_o h_n\right]\phi + \left[-\omega_o\left(I_x - I_y + I_z\right) + h_n\right]\dot{\psi} + \dot{h}_{xc} - \omega_o h_{zc}, \quad (3)$$

$$T_y = I_y\ddot{\theta} + \left[3\omega_0^2(I_x - I_z)\right]\theta + \dot{h}_{yc}, \quad (4)$$

$$T_z = I_z\ddot{\psi} + \left[\omega_0^2\left(I_y - I_x\right) + \omega_o h_n\right]\psi - \left[-\omega_o\left(I_x - I_y + I_z\right) + h_n\right]\dot{\phi} + \dot{h}_{zc} + \omega_o h_{xc}. \quad (5)$$

3. Yaw axis control system design

For satellite specifications given in Table 1, it is observed that $I_x = I_z$ and thus, the pitch axis equation of motion given by equation (4) can be simplified to

$$T_y = I_y\ddot{\theta} + \dot{h}_{yc}, \quad (6)$$

in which $\dot{h}_{yc}$ is the inertia wheel's angular momentum rate of change, whose aim is to impose direct torque control about the pitch axis. A pseudorate modulator is to be used, which serves to modulate the torque control $\dot{h}_{yc}$ and synthesizing the angular rate of change $\dot{\theta}$, ie, the control device makes torque being dependent on the angular rate of change. A satisfactory control form is given by

$$\dot{h}_{yc} = K_p(\tau_p\dot{\theta} + \theta), \quad (7)$$

wherein the term $\dot{\theta}$ introduces the necessary damping for the system, K_p is the pitch autopilot gain and τ_p is the pitch constant time. Substituting equation (7) into equation (6), applying the Laplace transform and rearranging the terms, the pitch axis control system transfer function results as being

$$\frac{\Theta(s)}{T_y(s)} = \frac{1}{I_y s^2 + K_p\tau_p s + K_p}. \quad (8)$$

The pitch loop root locus diagram for the variable K_p is shown in figure 4, and it indicates that the system is stable for all values of K_p. Adjusting the output to behave critically damped (no overshoot) and according to the attitude accuracy requirements given in Table 1, obtains the most suitable parameters pitch autopilot gain K_p = 0.275 Nm/rad and pitch constant time τ_p = 80 s. Applying the parameters found in equation (8) yields the step and cyclic disturbance torques responses, shown in figures 5 and 6, respectively. The step type input is due to thruster misalignment and the cyclical type input is characteristic of the solar radiation pressure torque.

4. Roll/yaw control system design

In order to simplify the equation of motion for roll and yaw axis, it is assumed that $\boldsymbol{h_n} \gg \mathbf{max}\,[\boldsymbol{I_x\omega_o}, \boldsymbol{I_y\omega_o}, \boldsymbol{I_z\omega_o}]$, so the equations (3) and (5) are rewritten as

$$T_x = I_x\ddot{\phi} + \omega_o h_n\phi + h_n\dot{\psi} + \dot{h}_{xc} - \omega_o h_{zc}, \quad (9)$$

Journal of Physics: Conference Series **641** (2015) 012019 doi:10.1088/1742-6596/641/1/012019

$$T_z = I_z\ddot{\psi} + \omega_o h_n \psi - h_n\dot{\phi} + \dot{h}_{zc} + \omega_o h_{xc}. \tag{10}$$

Again using a pseudorate modulator and based on the same principles used to prepare the pitch axis control system, the used roll/yaw control laws are

$$M_{xc} = \dot{h}_{xc} - \omega_o h_{zc} + \omega_o h_n \phi = K(\tau\dot{\phi} + \phi), \tag{11}$$

$$M_{zc} = \dot{h}_{zc} + \omega_o h_{xc} - h_n\dot{\phi} = -kK(\tau\dot{\phi} + \phi), \tag{12}$$

where M_{xc} and M_{zc} represent the roll and yaw control moments, respectively, K is the roll autopilot gain, τ is the roll time constant and k is the yaw-to-roll gain ratio.

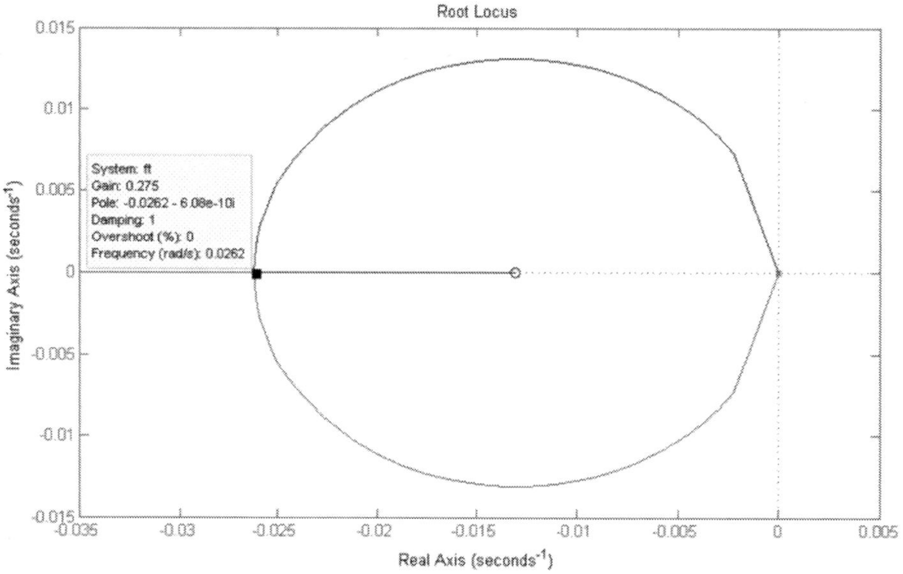

Figure 4. Pitch loop root locus diagram for variable K_p.

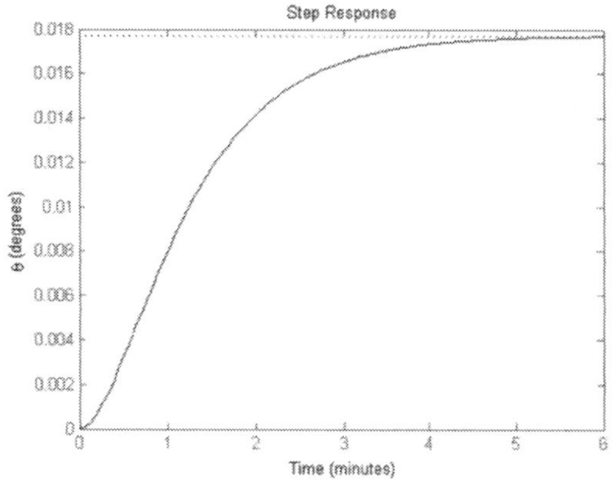

Figure 5. Pitch axis output for a step input.

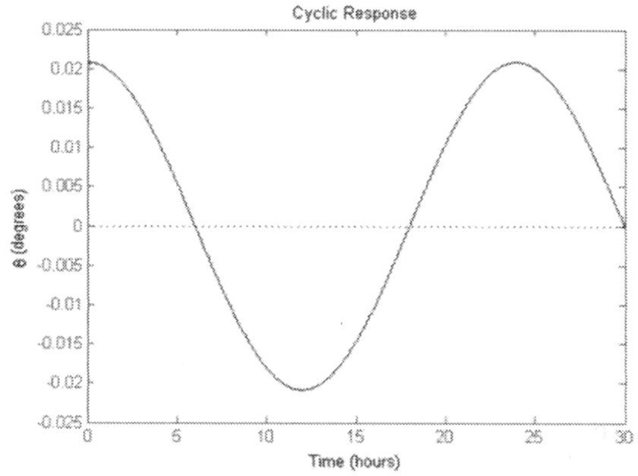

Figure 6. Pitch axis output for a cyclical input.

Replacing control law in equations (9) and (10), applying the Laplace transform and rearranging the terms, the angular responses are obtained

$$\Phi(s) = \frac{T_x(s)(I_z s^2 + \omega_o h_n) - T_z(s) h_n s}{I_x I_z s^4 + K I_z \tau s^3 + (K I_z + I_x \omega_o h_n + kK\tau h_n)s^2 + (\omega_o h_n K\tau + kK h_n)s + \omega_o h_n K}, \tag{13}$$

Journal of Physics: Conference Series **641** (2015) 012019 doi:10.1088/1742-6596/641/1/012019

$$\Psi(s) = \frac{T_z(s)\left(I_x s^2 + K\tau s + K\right) + kK(\tau s + 1)T_x(s)}{I_x I_z s^4 + K I_z \tau s^3 + (K I_z + I_x \omega_o h_n + kK\tau h_n)s^2 + (\omega_o h_n K\tau + kKh_n)s + \omega_o h_n K}. \tag{14}$$

The roll/yaw system root locus diagram for the variable K is shown in figure 7. Adjusting the output to behave critically damped (no overshoot) and according to the attitude accuracy requirements given in Table 1, obtains the most suitable parameters for roll autopilot gain K = 1.56 Nm/rad, roll constant time τ = 80 s and yaw-to-roll gain ratio k = 0.054. Due to the coupling between the motions, a disturbance on one axis (roll or yaw) directly affects the motion of both axes. The figures 8 and 9 show the output of the roll axis due to a step disturbance in the roll and yaw axes, respectively, caused by the thrust misalignment; while the figures 10 and 11 show the output of the yaw axis due to a step disturbance in the roll and yaw axes, respectively, caused by the same effect. The figures 12 and 13 show the roll and yaw axes responses due to a disturbance caused by a cyclic solar pressure torque type.

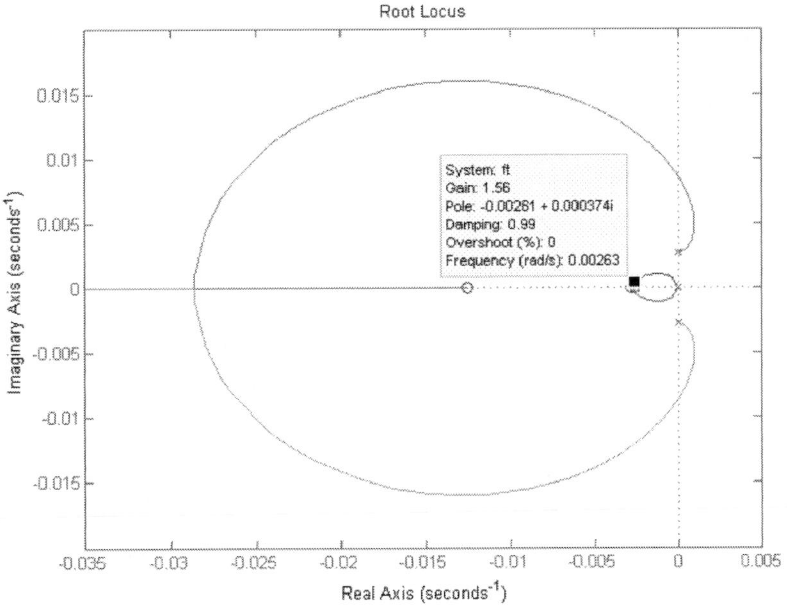

Figure 7. Roll/yaw system root locus diagram for variable K.

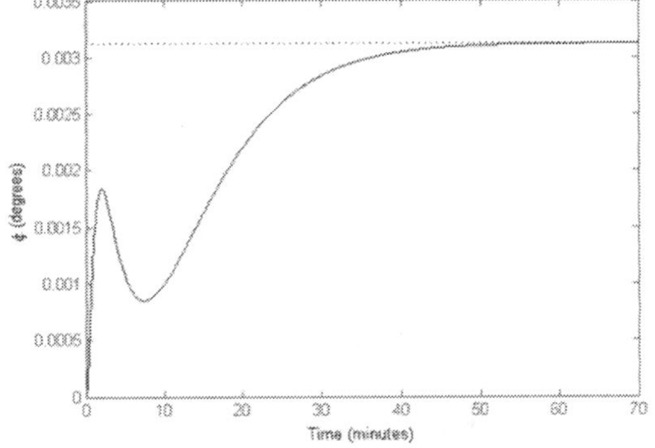

Figure 8. Roll axis output for a step input in the roll axis.

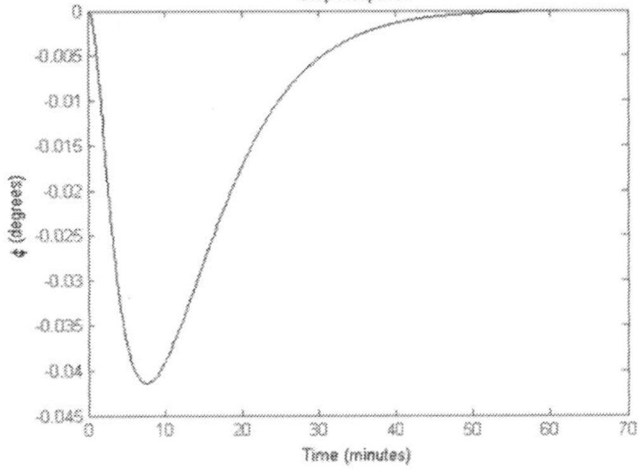

Figure 9. Roll axis output for a step input in the yaw axis.

Journal of Physics: Conference Series **641** (2015) 012019 doi:10.1088/1742-6596/641/1/012019

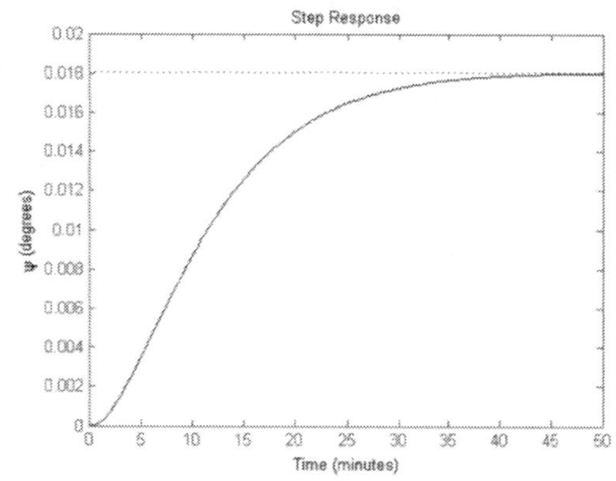

Figure 10. Yaw axis output for a step input in the roll axis.

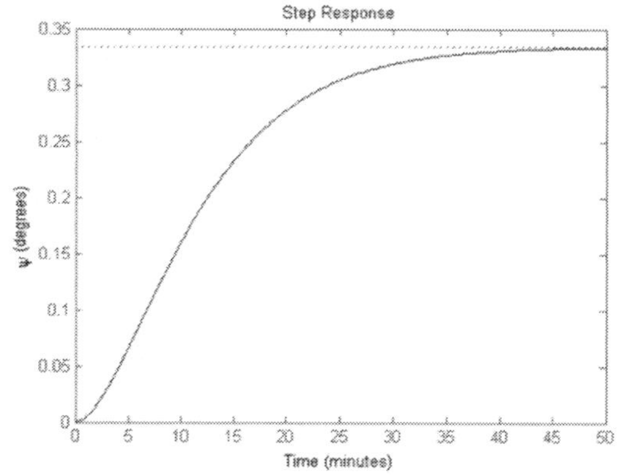

Figure 11. Yaw axis output for a step input in the yaw axis.

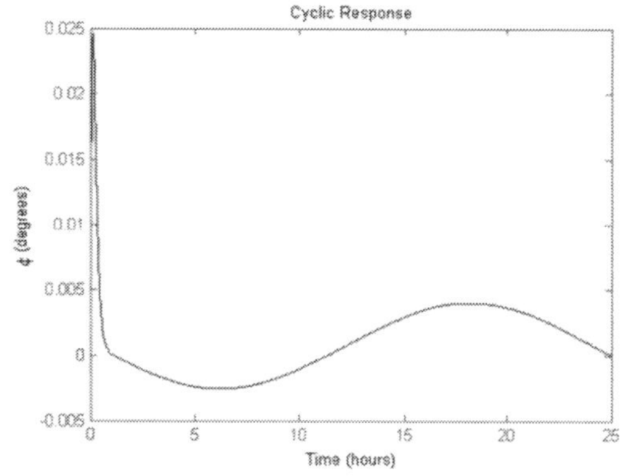

Figure 12. Roll axis output for a cyclical input.

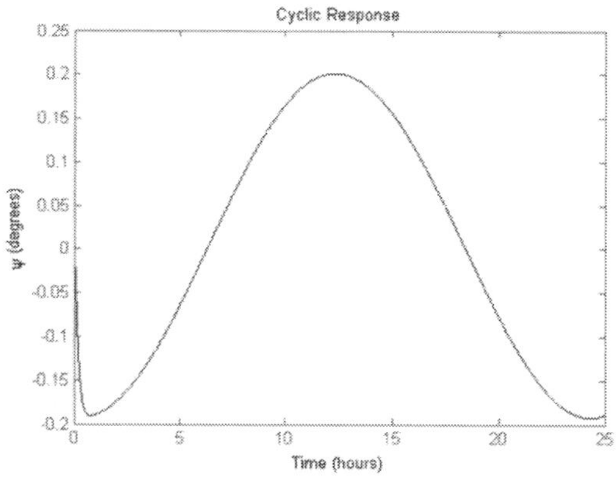

Figure 13. Yaw axis output for a cyclical input.

5. Conclusions

The linearized equation of motion for the system under consideration was obtained, which in turn separate the pitch motion of roll and yaw motions, thus simplifying the application of control system theory. Unlike most designs, which use three inertia wheels, one for each axis, this design has achieved successful results in accordance with any attitude of accuracy requirements only with one inertia wheel, minimizing costs, improving the dynamics and reducing satellite weight.

References

[1] Kaplan M H 1976 *Modern Spacecraft Dynamics & Control* (New York: John Wiley & Sons)

Journal of Physics: Conference Series **641** (2015) 012020 doi:10.1088/1742-6596/641/1/012020

Ionosphere influence on success rate of GPS ambiguity resolution in a satellite formation flying

Leandro Baroni

Federal University of ABC – Rua Arcturus, 3, São Bernardo do Campo, SP, Brazil, 09606-070

E-mail: `leandro.baroni@ufabc.edu.br`

Abstract. Satellite formation flying is one of the most promising technologies for future space missions. The distribution of sensors and payloads among different satellites provides more redundancy, flexibility, improved communication coverage, among other advantages. One of the fundamental issues in spacecraft formation flying is precise position and velocity determination between satellites. For missions in low Earth orbits, GPS system can meet the precision requirement in relative positioning, since the satellite dynamics is modeled properly. The key for high accuracy GPS relative positioning is to resolve the ambiguities to their integer values. Ambiguities resolved successfully can improve the positioning accuracy to decimetre or even millimetre-level. So, integer carrier phase ambiguity resolution is often a prerequisite for high precision GPS positioning. The determination of relative position was made using an extended Kalman filter. The filter must take into account imperfections in dynamic modeling of perturbations affecting the orbital flight, and changes in solar activity that affects the GPS signal propagation, for mitigating these effects on relative positioning accuracy. Thus, this work aims to evaluate the impact of ionosphere variation, caused by changes in solar activity, in success rate of ambiguity resolution. Using the Ambiguity Dilution of Precision (ADOP) concept, the ambiguity success rate is analyzed and the expected precision of the ambiguity-fixed solution is calculated. Evaluations were performed using actual data from GRACE mission and analyzed for their performance in real scenarios. Analyses were conducted in different configurations of relative position and during different levels of solar activity. Results bring the impact of various disturbances and modeling of solar activity level on the success rate of ambiguity resolution.

1. Introduction

Formation flying is one of the most promising technologies for future space missions. The sensor and payload distribution among different satellites provides more redundancy, exibility, improve communications and surveillance coverage and new applications which are not possible with a single satellite [1]. This distributed approach provides enhanced performance and operational advantages.

One of the fundamental issues of formation flying spacecraft is determination of the state (position and velocity) among satellites belonging to formation. For missions in low earth orbit (LEO), the Global Positioning System (GPS) can meet the accuracy requirement in relative positioning, since satellites in LEO are covered by the GPS constellation. GPS receivers have a relatively low cost compared with other navigation sensors and their performance is generally reliable and is always available. Therefore, GPS is often regarded as the main instrument for navigation in satellite formation missions. The key for high accuracy GPS relative positioning is to resolve the ambiguities to their integer values. Ambiguities resolved successfully can

Journal of Physics: Conference Series **641** (2015) 012020 doi:10.1088/1742-6596/641/1/012020

improve the positioning accuracy to decimetre or even centimetre-level. So, integer carrier phase ambiguity resolution is often a prerequisite for high precision GPS positioning [2].

Solar activity level varies with a period of 11 years on average, known as the solar cycle. A peak in solar activity affects the behavior of the ionosphere, causing disturbances in the propagation of signals from GPS (ionospheric scintillation) and reducing positioning accuracy. The next peak of solar activity (cycle 24) is foreseen for May, 2013 [4].

Thus, this work aims to evaluate the impact of ionosphere variation, caused by changes in solar activity, in success rate of ambiguity resolution. Using the Ambiguity Dilution of Precision (ADOP) factor, the ambiguity success rate is analyzed and the expected precision of the ambiguity-fixed solution is calculated.

This work uses real data delivered from GRACE mission. The GRACE mission (Gravity Recovery and Climate Experiment) is made up two identical satellites (GRACE A and GRACE B) in a leader-follower formation, orbiting Earth at the same orbital plane, with inclination of 89°. Initial altitude was about 500 km above Earth's surface, after launch. Due to atmospheric drag, the altitude decreases to about 300 km. The nominal separation between the satellites is 220 km.

Results show the influence of ionosphere activity and satellite separation on success rate of GPS ambiguity resolution, using single and double frequency GPS measurements.

2. Measurement models

Ambiguity resolution will be tested in two situations: *i*) solving L1 ambiguities only; *ii*) solving L1 and L2 ambiguities. The following models, which are given referred to the state vector (geometric distance, in double difference form, and ambiguities), are linear. The measurement of carrier phase in single difference form is shown in Fig. 1. The double difference measurements are obtained by subtracting two single difference measurement referred to two different satellites.

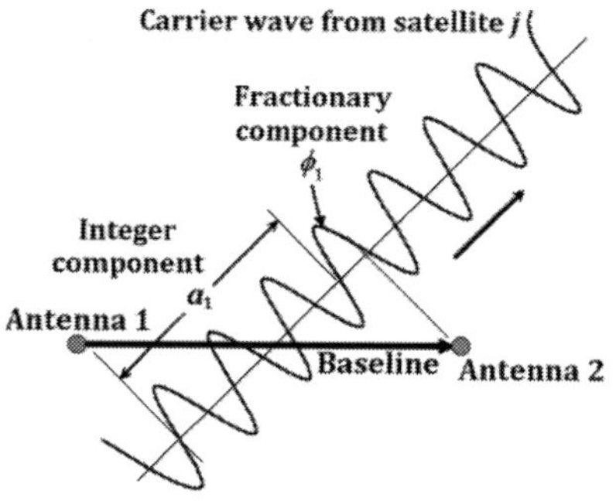

Figure 1. Carrier phase measurement in single difference form.

Single frequency model uses double differenced pseudorange and carrier phase on L1:

$$\begin{bmatrix} \rho \\ \phi_1 \end{bmatrix} = \begin{bmatrix} 1 & 0 \\ 1 & \lambda_1 \end{bmatrix} \begin{bmatrix} D \\ a_1 \end{bmatrix} + \epsilon \tag{1}$$

where ρ is double differenced pseudorange, ϕ_1 is double differenced carrier phase (in distance units) in L1, λ_1 is L1 wavelength, D is geometric distance in double difference form, a_1 is double difference integer ambiguity and ϵ represents unmodeled errors.

Using L1 and L2 carrier measurements, the observation model for a_1 and a_2 ambiguities is given by:

$$\begin{bmatrix} \rho \\ \phi_1 \\ \phi_2 \end{bmatrix} = \begin{bmatrix} 1 & 0 & 0 \\ 1 & \lambda_1 & 0 \\ 1 & 0 & \lambda_2 \end{bmatrix} \begin{bmatrix} D \\ a_1 \\ a_2 \end{bmatrix} + \epsilon \tag{2}$$

where ρ is double differenced pseudorange, ϕ_1 and ϕ_2 are double differenced carrier phase measurements in L1 and L2, whose wavelengths are λ_1 and λ_2, respectively, a_1 and a_2 are double differenced integer ambiguity, and ϵ represents unmodeled errors.

If baseline is greater than about 5 km, ionospheric effect becomes more significant. An ionospheric parameter is inserted together with geometric distance and ambiguities in state vector. In this case, it is introduced a pseudo-observation of ionospheric error I_0, zero-valued and variance σ_0^2 in each epoch [5]. This pseudo-observation helps narrow the ionospheric error to reasonable values during the initial phase of the filter, leading to distinguish between the ambiguities and ionospheric error. Therefore, it is expected the position estimate will not be influenced by the ionospheric error. The measurement model, with the introduction of this pseudo-observation, is described by:

$$\begin{bmatrix} \rho \\ I_0 \\ \phi_1 \\ \phi_2 \end{bmatrix} = \begin{bmatrix} 1 & 1 & 0 & 0 \\ 0 & 1 & 0 & 0 \\ 1 & -1 & \lambda_1 & 0 \\ 1 & -\beta & 0 & \lambda_2 \end{bmatrix} \begin{bmatrix} D \\ I \\ a_1 \\ a_2 \end{bmatrix} + \epsilon \tag{3}$$

where $\beta = (\lambda_2/\lambda_1)^2 = (77/60)^2$.

The estimation is carried out in three steps: *i*) float solution; *ii*) integer solution; *iii*) correction of parameters from resolved integer ambiguities.

In the first step, the integer constraint $a \in \mathbb{Z}^n$ is disregarded on the ambiguities and a standard estimation procedure is performed. As a result, the *float solution* is obtained, which is a real-valued estimate of a and b, together with their covariance matrix:

$$\begin{bmatrix} \hat{a} \\ \hat{b} \end{bmatrix}, \quad \begin{bmatrix} P_{\hat{a}} & P_{\hat{a}\hat{b}} \\ P_{\hat{b}\hat{a}} & P_{\hat{b}} \end{bmatrix} \tag{4}$$

where $P_{\hat{a}}$ is the partition of covariance matrix regarding to $\hat{a}$ estimate, $P_{\hat{b}}$ is the partition regarding to $\hat{b}$ estimate, and $P_{\hat{a}\hat{b}}$ and $P_{\hat{b}\hat{a}}$ are matrix of cross correlated terms.

In the second step, the float ambiguity estimate $\hat{a}$ is used to compute the corresponding integer ambiguity estimate, denoted as $\check{a}$. Once the integer ambiguities are computed, the third step consists in correcting the float estimate of real-valued parameters b. As a result, the *fixed solution* is obtained:

$$\check{b} = \hat{b} - P_{\hat{b}\hat{a}} P_{\hat{a}}^{-1} (\hat{a} - \check{a}) \tag{5}$$

If the success rate, *i.e.*, probability that estimated integer ambiguities are equal the true ambiguities, is sufficiently close to 1, the precision of the fixed solution can be described by the following covariance matrix:

$$P_{\check{b}} = P_{\hat{b}} - P_{\hat{b}\hat{a}} P_{\hat{a}}^{-1} P_{\hat{a}\hat{b}} \tag{6}$$

2.1. *Ambiguity Dilution of Precision*

The success of ambiguity resolution process depends on quality of float ambiguity estimation. This quality can be measured by covariance matrix of float ambiguities. However, instead analyzing each entry of covariance matrix (Eq. 6), [9] introduced the Ambiguity Dilution of Precision (*ADOP*) concept as a way of quality measure. It is a scalar value, easier to evaluate than taking each entry of covariance matrix. The *ADOP* concept is defined as:

$$ADOP = \left(\sqrt{|P_{\hat{a}}|}\right)^{1/n} \tag{7}$$

where $P_{\hat{a}}$ is ambiguities covariance matrix, n is its order, and $|\cdot|$ means the determinant of a matrix. *ADOP* is a scalar quantity given in cycles unit. This *ADOP* value not only depends on the variances of the ambiguities, but also on their covariances.

The $ADOP$ is equal to the geometric average of the sequential conditional standard deviations of the ambiguities. This follows from the fact that the determinant of the covariance matrix of the ambiguities is equal to the product of the n sequential conditional variances. The $ADOP$ is thus a simple measure for the average precision of the ambiguities [7].

The $ADOP$ is invariant against the choice of ambiguity parametrization. $ADOP$ does not change when the definition of the ambiguities is changed, due to all admissible ambiguity transformations can be shown to have a determinant equal to one. If the ambiguities are transformed by:

$$\hat{z} = Z^T \hat{a}, \quad P_{\hat{z}} = Z^T P_{\hat{a}} Z \tag{8}$$

The determinant of the transformed ambiguity covariance matrix remains invariant, since:

$$|P_{\hat{z}}| = |Z^T P_{\hat{a}} Z| = |Z^T||P_{\hat{a}}||Z| = |P_{\hat{a}}| \tag{9}$$

once $|Z| = 1$ for all admissible transformations. Because of this property, using the decorrelating Z-transform of the LAMBDA method [8, 3], the $ADOP$ remains the same.

Since the $ADOP$ gives a good approximation to the average precision of the ambiguities, it also provides a good approximation to the integer ambiguity success rate:

$$P(\check{a} = a) \leq \left[2\Phi\left(\frac{1}{2ADOP}\right) - 1\right]^n \tag{10}$$

where $\Phi(x)$ is the standard normal cumulative distribution function:

$$\Phi(x) = \int_{-\infty}^{x} \frac{1}{\sqrt{2\pi}} \exp\left(-\frac{1}{2}v^2\right) dv \tag{11}$$

In general, if $ADOP$ is smaller than about 0.12 cycles, the success rate becomes larger than 0.999, while for $ADOP$ smaller than 0.14 cycles, the success rate is always better than 0.99 [7].

3. Results

All tests were carried out using real data from GRACE mission. GRACE satellites performed a maneuver of switching positions on December 10th, 2005. At this event, the relative distance between satellites was less than 10 km for several hours. So, the first data set was chosen at this period of closest approximation. A second approaching maneuver was carried out on July 14th, 2014, and the second data set was chosen at this date. The period of first data set was close to a solar minimum, while second data set was close to a solar maximum [6]. It was assumed that ambiguities can be considered resolved if success rate is greater than 0.97.

Observation models in Section 2 were used to compute, using a Kalman filter, float ambiguities and their covariance matrix for the two data sets. For all tests, pseudorange standard deviation was set to 1.0 m and standard deviation of L1 and L2 carrier phase measurement was set to 0.005 m.

For the first test, graphs in Figs. 2(a) and 2(b) show $ADOP$ value along time of observation for L1 only and L1 and L2 frequencies, respectively.

As seen on Fig. 2(a), $ADOP$ reaches values as low as 1 cycle, giving a success rate of only 0.003. This means ambiguity cannot be resolved successfully in this time span. Otherwise, using double frequency measurements, $ADOP$ reaches 0.1 cycle, so ambiguities can be resolved after 180 seconds, with success rate of 0.97. Figures 3(a) and 3(b) show success rates for single and double frequency measurements. In this test, ionosphere pseudo-observation standard deviation was set to 0.04 m.

The second test shows that using only L1 frequency, it was not possible to resolve ambiguities within observation time, as in the first test. However, using L1 and L2 frequencies, ambiguities

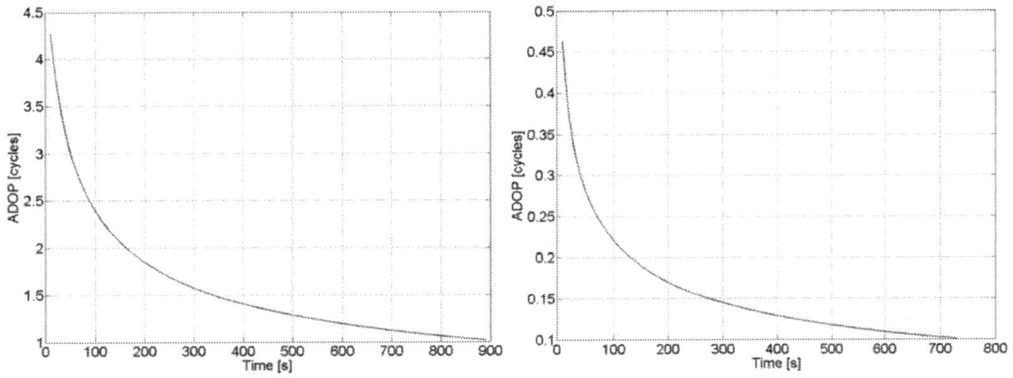

(a) (b)

Figure 2. First data set: *ADOP* values for (a) L1 only; (b) L1 and L2 frequencies.

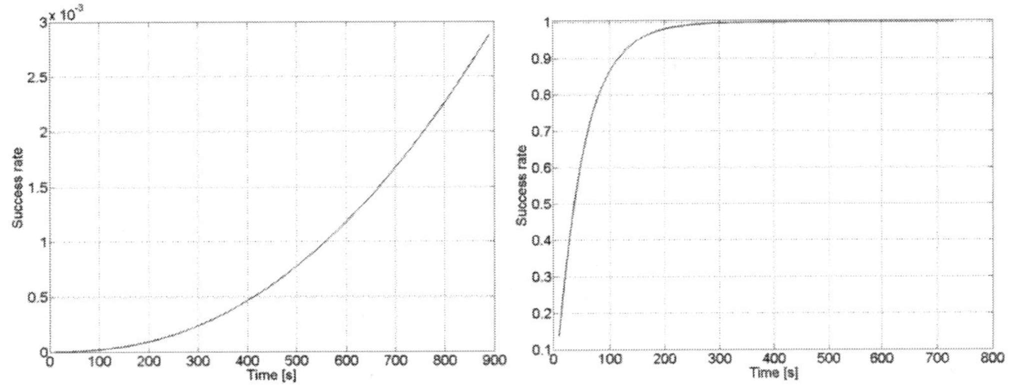

(a) (b)

Figure 3. First data set: Success rate values for (a) L1 only; (b) L1 and L2 frequencies.

could be resolved after 280 seconds (success rate of 0.97). *ADOP* values are shown on graphs of Figs. 4(a) and 4(b). Ionosphere pseudo-observation standard deviation was set to 0.25 m, due to the higher solar activity. One factor which can explain the slower ambiguity resolution is the ionosphere was more disturbed, once solar activity was increasing. So ionosphere pseudo-observation standard deviation was set to 0.25 m in this case. Thus, the success rate took a longer time to reach 0.97. Figures 5(a) and 5(b) show the success rate for second data set.

4. Conclusion

Two tests were conducted using real GPS data from GRACE mission. First data set was chosen in a period of minimum solar activity. Second data set was chosen in a period of higher solar activity, and when the satellites were close each other.

Both tests showed that L1 only measurements are not sufficient to resolve ambiguities in neither case. However, using L1 and L2 measurements, ambiguities could be resolved in both cases. The tests showed that in first test, ambiguities could be resolved faster than in second test. Lower solar activity causing less disturbances in ionosphere can explain this fact.

These results are important in developing filters for future navigation applications that use GPS receivers to determine relative position of satellites in formation flying configurations, supporting filter parameters tuning, in particular on ionospheric error components.

Journal of Physics: Conference Series **641** (2015) 012020 doi:10.1088/1742-6596/641/1/012020

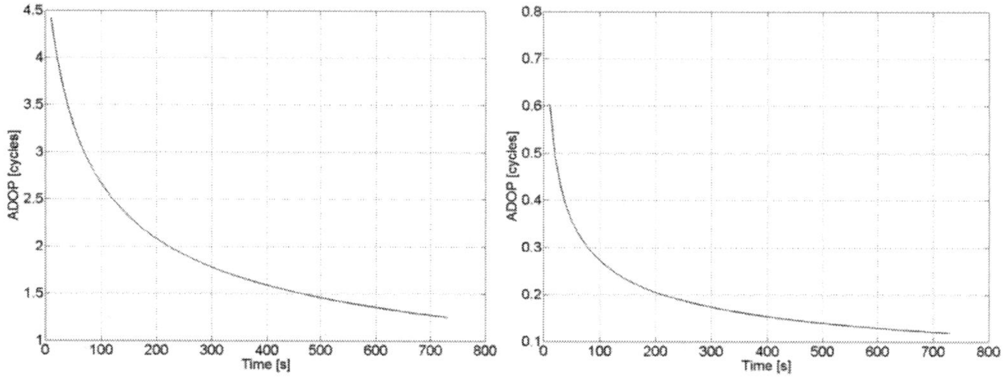

(a) (b)

Figure 4. Second data set: *ADOP* values for (a) L1 only; (b) L1 and L2 frequencies.

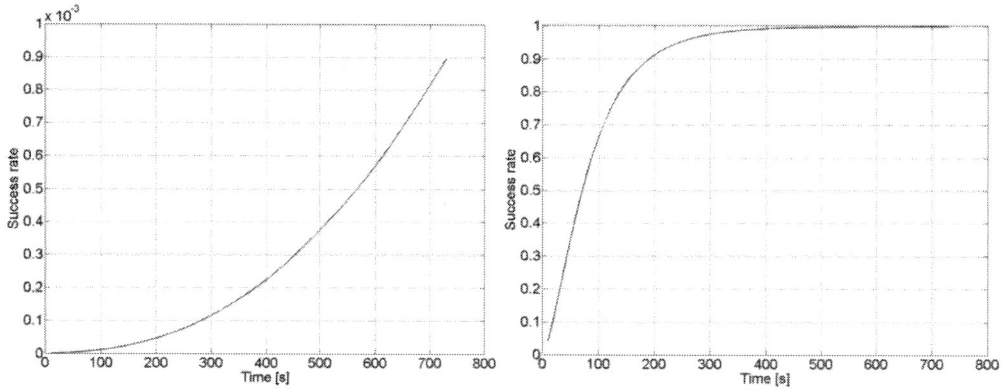

(a) (b)

Figure 5. Second data set: Success rate values for (a) L1 only; (b) L1 and L2 frequencies.

References

[1] Ardaens J-S and D'Amico S 2009 Spaceborne autonomous relative control system for dual satellite formations. *J. Guid. Ctrl. Dynam.* **32**(6).

[2] Baroni L 2009 *Algoritmos de Navegação em Tempo Real Para um Sistema GPS de Posicionamento Relativo de Precisão* INPE-16598-TDI/1584 (São José dos Campos: Instituto Nacional de Pesquisas Espaciais) p 125

[3] de Jonge P J, Tiberius C C J M and Teunissen P J G 1996 Computational aspects of the LAMBDA method for GPS ambiguity resolution *Proc. 9. Int. Tech. Meeting of the Satellite Division of the Institute of Navigation* (Kansas City: ION) pp 935–944

[4] Kintner Jr. P M, Humphreys T E and Hinks J C 2009 GNSS and ionospheric scintillation - how to survive the next solar maximum *InsideGNSS* (July/August) p 22–30

[5] Liu J, Cannon M E, Alves P, Petovello M G, Lachapelle G, Macgougan G and Degroot L 2003 A performance comparison of single and dual frequency GPS ambiguity resolution strategies *GPS Solutions* **7**(2) pp 87–100

[6] Marshall Space Flight Center 2015 *Solar Cycle Prediction* `http://solarscience.msfc.nasa.gov/predict.shtml`

[7] Odijk D and Teunissen P J G 2008 ADOP in closed form for a hierarchy of multi-frequency single-baseline GNSS models *J. Geod.* **82** pp 473–492

[8] Teunissen P J G 1993 Least-squares estimation of the integer GPS ambiguities *Proc. IAG General Meeting* (Beijing, China: IAG)

[9] Teunissen P J G 1997 A canonical theory for short GPS baselines. Part IV: precision versus reliability *J. Geod* **71** pp 513–525

XVII Colóquio Brasileiro de Dinâmica Orbital – CBDO IOP Publishing
Journal of Physics: Conference Series **641** (2015) 012021 doi:10.1088/1742-6596/641/1/012021

Evasive Maneuvers in Route Collision With Space Debris Cloud

A D C Jesus[1], **R R Sousa**[2], **E V Neto**[3]

[1] Department of Physics, University of State of Feira de Santana, Brazil
[2] Department of Physics, University of State of São Paulo, Brazil
[3] Department of Mathematics, University of State of São Paulo, Brazil

E-mail: a1d1j1@gmail.com

Abstract. Collisions between operational vehicles and space debris can completely derail the continuity of space missions, especially if there is chain collisions between debris, which generate even smaller fragments. In this paper, we investigate the dynamics on between an operational vehicle and space debris that form a cloud, considering the possibility of collisions between debris during an evasive maneuver the vehicle. For a radius of 3 km celestial sphere, we find possibilities of collision between debris up to 10 m, while the vehicle performs an evasive maneuver in time 3,000 s range. These results depend on the time collision, the angular positions of the collisional objects and the amount of debris that form the cloud.

1. Introduction

Debris clouds potentially represent the biggest nightmare in space missions in the regions where they occur. In many regions these clouds are generated mainly due to secondary collisions (with debris) and also fragmentation of the rockets and satellites bodies. The satellite fragmentation total is 37.7 % of cataloged objects and due to collisions are only 0.6 % (NASA - JSC, 2004). With the space activities, the distribution of these objects in the operating regions increased, even with mitigation measures and the natural reduction in the atmospheric drag of objects in LEO. Collisions with debris larger than 10 cm are considered catastrophic and still trigger a collision cascade process. This type of collision produces a long term critical density, without perspective of reduction, unless the quantity of large objects is sufficiently reduced. Control techniques should be adopted, in view of the growing demands of space campaign perpetuating along with the growth of debris in the operating regions (Kessler et al, 1978 Kessler, 1991). Simulations indicate that in a few decades the debris from fragments of the collisions will be dominant, at least in altitudes 800-1400 km (ESA, 2013). Operate in LEO regions, MEO and GEO necessarily imposes concerns about the safety of space activities due to the distribution of space debris in these altitudes. In MEO focus navigation constellations (GPS, GLONASS, etc.) and about 16,000 debris with diameters larger than 1 cm are predicted to cross the orbits in the region. Most of them have non-zero eccentricities, leading them to achieve GPS altitude at apogee. The energy produced in a collision (10^4 J) with such objects would cause severe damage to the spacecraft (Klinkrad 2006; Rossi 2005; Rossi and Valsecchi, 2006; Smirnov, 2002). Disturbances occurring in GEO, unlike what happens in LEO due to atmospheric drag, does not reduce the amount of debris (Milani et al., 1987). The growth of GEO debris without removal

Journal of Physics: Conference Series **641** (2015) 012021	doi:10.1088/1742-6596/641/1/012021

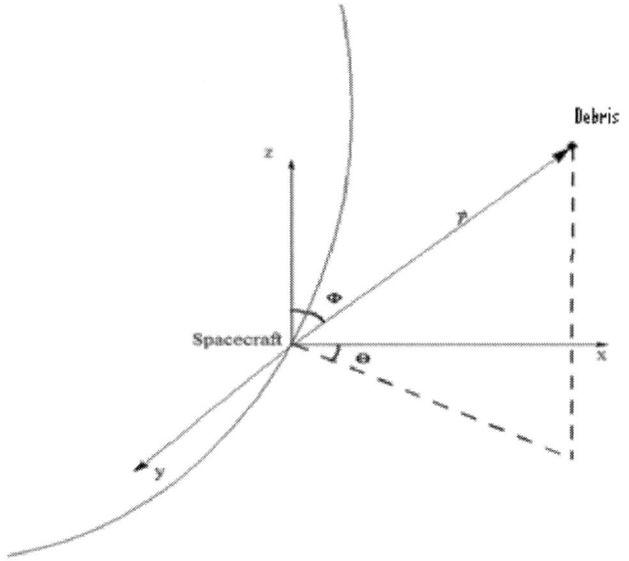

Figure 1. Orbital Configuration and Reference System Vehicle and Space Debris around the Earth

increases in 30 debris 30 to year (Valk et al., 2009). This paper studies evasive maneuvers in debris cloud environment with the possibility of collision between them.

2. Mathematical model

The relative dynamic equations between the collisional bodies in Cartesian coordinates are:

$$\ddot{x} - 2\omega\dot{y} - 3\omega^2 x = -v_{ex}\frac{d}{dt}\ln(M(t)), \quad (1)$$

$$\ddot{y} + 2\omega\dot{x} = -v_{ey}\frac{d}{dt}\ln(M(t)), \quad (2)$$

$$\ddot{z} + \omega^2 z = -v_{ez}\frac{d}{dt}\ln(M(t)). \quad (3)$$

These questions are Hills Equations (Hill,1878) adapted in this paper for the relative dynamics between the operational vehicle and space debris, subject to two forces: the Earth's gravitational and the vehicle propellant (non-gravitational force). The vehicle orbits the Earth with circular $\vec{\omega} = \omega\hat{k}$, and the y-axis is taken in the direction of its radius, as shown in Figure 1. The coordinates ϕ and θ are spherical angles to locate the debris in a system centered on the satellite with a radius r. The propulsion force is required to implement the evasive maneuvers in the orbital regions (LEO , MEO, GEO) in which the angular speed assumes specific values.

This system describes the motion and is fixed on the vehicle such that all technical control evasive maneuver is performed from it. The dynamic equations are valid in that the relative distance between objects ($r = |\vec{r}|$) is very small compared to the distance from the vehicle to the center of the Earth ($R = |\vec{R}|$), that is, ($r/R << 1$). M(t) is the total mass of space vehicle at time t. For this dynamic, we adopted exponential variation of the vehicle mass in time, such

Journal of Physics: Conference Series **641** (2015) 012021 doi:10.1088/1742-6596/641/1/012021

that,

$$M(t) = M_o + m(t), \tag{4}$$

$$\chi = \frac{M_o}{m_o}, \tag{5}$$

$$M(t) = m_o(\chi + \exp(-\gamma t)), \gamma > 0. \tag{6}$$

The exhaust velocity (v_{ex}, v_{ey}, v_{ez}), the motor power factor of the engine $\gamma > 0$ and the mass factor $(\chi > 1)$,ie, the ratio of the mass of the spacecraft mass (M_o) and the initial mass of the propellant (m_o) , call the technological parameters. For this model, the components of the propulsion force are:

$$F_x = \gamma v_{ex} m_o e^{-\gamma t}, \tag{7}$$

$$F_y = \gamma v_{ey} m_o e^{-\gamma t}, \tag{8}$$

$$F_z = \gamma v_{ez} m_o e^{-\gamma t}. \tag{9}$$

These equations show that the propulsion force on the exponential model mass is proportional to the power factor. This factor can be controlled such that the evasive maneuver can be implemented in a given time interval, maintaining the circular orbit vehicle or operating nearly circular.

2.1. Strategy for evasive maneuvers

The evasive maneuvers are performed by acting propellants under appropriate technological strategies for each mission. The internal computer data requires the implementation of software that will use the equations of dynamics on between the vehicle and the debris. Initially, the homogeneous solution of equations (1) to (3), provide the relatives initial velocity and position on favorable to the collision. The coordinates of the relative position of these objects are obtained by scanning the spherical angles in space that contains them. The solution r(t) of the non-homogeneous equations (1) to (3) is a function of technological parameters that characterize the evasive maneuvers for our exponential mass model, Equation (4). The relative position vector Cartesian components are, therefore,

$$x(t) = 2A\sin(nt) - 2B\cos(nt) + Et + \sum_{n=1}^{\infty} F_n e^{-n\gamma t} + G, \tag{10}$$

$$y(t) = A\cos(nt) + B\sin(nt) - \sum_{n=1}^{\infty} C_n e^{-n\gamma t} + D, \tag{11}$$

$$z(t) = H\cos(nt) + I\sin(nt) - \sum_{n=1}^{\infty} J_n e^{-n\gamma t}. \tag{12}$$

This solution was obtained before by Jesus et al (2012). All these equations coefficients depend on the initial conditions and the technological parameters which in general can be written as $L = L(\vec{r}_o, \vec{v}_o, \vec{v}_e, \gamma, \chi, n)$. Thus, the relative final position between the collisional objects can be controlled from the adjustment of these coefficients through the suitable technological parameters to implement of evasive maneuvers. These maneuvers are possible whenever the relative final distance (on time to collision) is compared with the size of collisional objects.

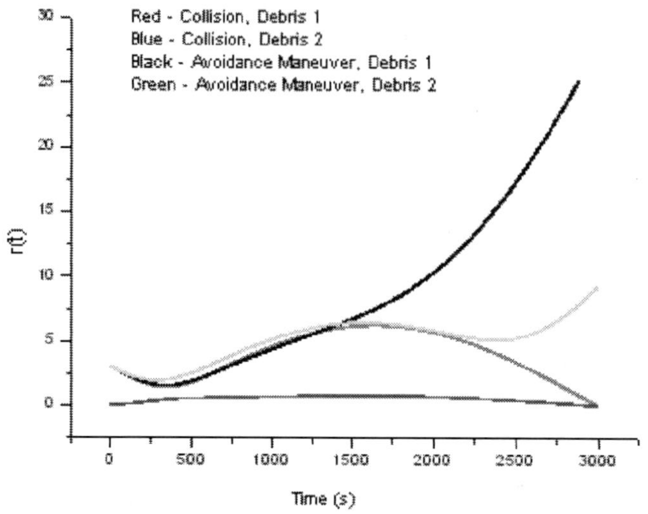

Figure 2. Collision and Avoidance: Two space debris with a spacecraft [r(t) in km].

So, initially without the propulsion, find the favorable initial conditions to and collision time. This can be done with a radar on board the vehicle operational. During the collision time the propulsion is burned to perform an evasive maneuver. The technological parameters must be such that the evasive maneuver can be implemented in the collision interval. We chose a "nominal solution" in the set of possible collision and the collision time. This solution is characterized by initial velocities of interest for technological applications and for a time allowing the computer implementation of the evasive maneuver. Specifically, we have chosen the initial relative velocity equal to (7.76 km/s, 4.25 km/s, 1.01 km/s) to the regions LEO, MEO and GEO, respectively, in non-planar maneuvers. The time interval is of the order 45 minutes.

3. Study of collision with debris cloud

The study of evasive maneuvers in debris cloud environment is not trivial, because if we suppose that a spacecraft escaped from a cloud of debris, not imply that it will escape from others. In addition, the debris can themselves collide during the time of the evasive maneuver, generating smaller debris, increasing the cloud and the difficulty to complete the space mission. Depending on the sizes of objects, the first collision may occur by more than one at a time. Our results show situations in which the vehicle escapes from debris colliding with each other and further showing the angular regions that are favorable to collisions between them. We found the distribution of the approximations of positions between the debris during the collision time or evasive maneuvers time. The initial relative position between the debris and the spacecraft is 3 km and they are in different angular positions, all on a collision course with the vehicle. The initial relative velocities (extracted from histogram initial conditions) are such that spatial objects are on a collision course. Figure 2 shows the approach and avoidance trajectories between two debris and an operational vehicle.

In the Figure 2, the curves in red and blue represent the course of a collision of the debris with the space vehicle, that is, the collision of the debris 1 with the satellite and the debris 2 with the satellite, respectively. The curves in black and green are escape routes of the vehicle in front of the collision debris possibility. A cross between black and green paths are approximations

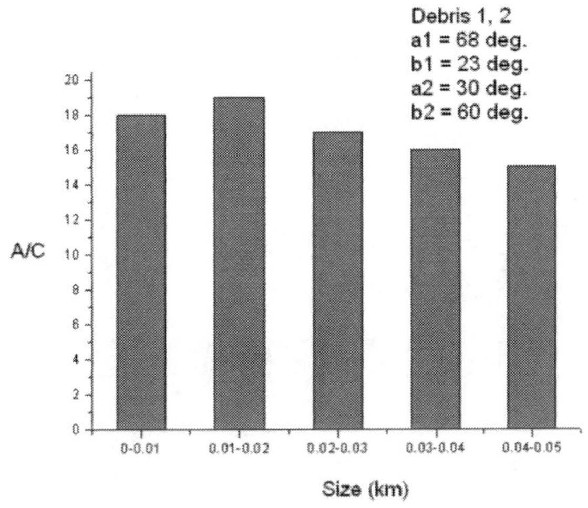

Figure 3. Approaches/Collisions (A/C) between Debris and the Spacecraft during Evasive Maneuvers - Fixed Angular Positions.

between the space debris and, depending on their sizes and angular positions, represent collisions between them, while spacecraft escapes of both. In the histogram of Figure 3, show the amount 90 approaches between two debris with sizes up to 50 meters. These approaches would be collisions if the debris sizes within this range. For example, in the range between 0 and 10 meters, exist 18 approaches that would be collisions if the debris is smaller than or equal to 10 meters. These results were obtained for both debris located on two pairs of fixed angle ($\phi = a$ and $\theta = b$) and a relative initial distance of 3 km with the spacecraft.

It is important to be said also that these collisions are dangerous because debris can fragment and generate a cloud of debris that further damage the evasive maneuvers of the spacecraft. These results were obtained for two pairs of fixed angles. In the Figure 4, shows a histogram in which there is a distribution of debris sizes (between 0 and 50 cm) in an angular region where the angles between 0 and 90 degrees (I Octant Figure 1).

In this graph, we see that many collisions and/or approaches (more than 104) occur often and the highest concentration of debris of them occurs for up to 1 cm in size.

3.1. Dynamic with debris cloud
In this case, we simulated collision dynamics for 5 debris. The Figure 5 shows results for the distribution of debris collisions between 5 and 10 cm in size. The angular coordinates of the debris are: $a = 0°$ for all the debris and $b = 1, ..., 5°$ to 5 debris, respectively. The graph shows collisions between these 5 debris without discriminating which of them collide. The statistics were performed to register any collisions between two debris and even between 5 debris. For this configuration of the angular distribution were more than 160 collisions of debris sizes between 1 and 10 cm, while an evasive maneuver is implemented by the spacecraft.

4. Conclusions
Our results show that occur approaches debris, while the operational vehicle performs an evasive maneuver. These approaches may be collisions between debris, depending on their sizes. We found 90 approaches between two debris up to 50 meters for fixed spherical angles for both and initial relative distance between them and the vehicle equal to 3 km. Setting the spherical angles

Journal of Physics: Conference Series **641** (2015) 012021 doi:10.1088/1742-6596/641/1/012021

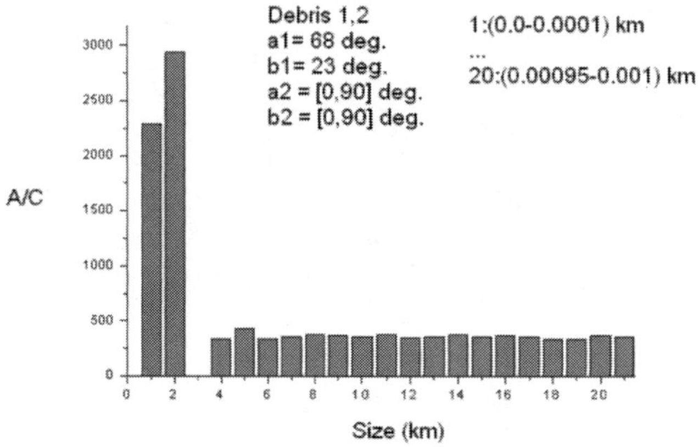

Figure 4. Approaches/Collisions (A/C) depending on the Size Distribution of the Debris - I Octant.

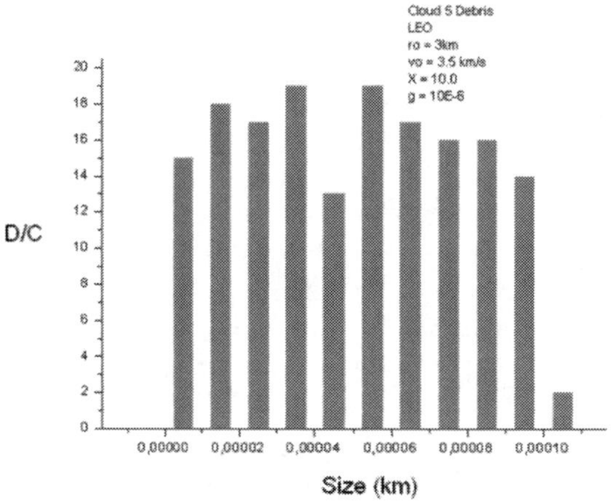

Figure 5. Collisions between 5 Debris (D/C): Sizes up to 10 cm, b = [10,50].

of the debris and varying angles in the first octant of the second debris encountered over 104 approaches for size distribution up to 50 cm and an increased incidence of collision of debris to 1 cm. For a cloud of debris with 5 fixed angles in the plane and out of plane ranging between 1 and 5 degrees, found 160 approaches and/or collisions between debris of the 5 and 10 cm sizes. The debris cloud environment is confirmed to be highly dangerous for space operations, even if they are evasive maneuvers of spacecraft, especially if they happen in octant where debris are localized initially.

5. References

[1] ESA European Space Agency, 20, April, 2013- http://www.esa.int
[2] 13 Ed. History of On-Orbit Satellite Fragmentations, NASA Johnson Space Center (JSC 62530) Orbital Debris Program Office (May, 2004).
[3] Hill, G. W. Research in Lunar Theory. American Journal of Mathematics, vol. 1, no. 1, pp. 5-26, 1878.

Journal of Physics: Conference Series **641** (2015) 012021 doi:10.1088/1742-6596/641/1/012021

[4] Jesus, D. C. A., Ribeiro, R.S., Rossi, A. and Neto, V. E. Evasive maneuvers in space debris environment and technological parameters. Mathematical Problems in Engineering, vol. 2012, pp. 1-13.

[5] Kessler, D. J. Collisional Cascading: The limits of Population Growth in Low Earth Orbit. Adv. Space Res.Volume 11, No. 12, pp. 12(63) 12(66), 1991.

[6] Kessler, D.J. and B.G. Cour-Palais. Collision Frequency of Artificial Satellites: The Creation of a Debris Belt. Journal of Geophysical of Research, Vol.83,No.A6, June 1, 1978, pp 2637-2646.

[7] Klinkrad, H.: Space Debris: Models and Risk Analysis. Springer Praxis, Berlin-Heidelberg (2006).

[8] Rossi, A.: The earth orbiting space debris. Serbian Astron. J. 170, 112 (2005)

[9] Rossi, A., Valsecchi, G.B.: Collision risk against space debris in Earth orbits. Cel. Mech. Dyn. Astron. 95, 345356 (2006)

[10] Smirnov, N.N. (ed.): Space Debris Hazard Evaluation and Mitigation. Taylor Francis, London-New York, (2002).

[11] Milani, A., Nobili, A. and Farinella, P.: 1987, Non gravitational perturbations and satellite geodesy, Adam Hilger Ltd., Bristol and Boston.

[12] Valk, S., Lematre, A., Deleflie, F. Semi-analytical theory of mean orbital motion for geosynchronous space debris under gravitational influence. Advances in Space Research, V.3. Issue 7, pp. 1070-1082, 1 april 2009.

Influence of the external torques in the angle between the spin axis and the Sun direction for spin stabilized satellite

G. B. MOTTA[1], M. C. ZANARDI [2]

[1]São Paulo State University - UNESP, Av. Dr. Ariberto Pereira da Cunha, 333, Guaratinguetá, Sp, Brazil
[2]Federal University of ABC - UFABC, Av. dos Estados, 5001, Santo André - Brazil

E-mail: mceciliazanardi@gmail.com

Abstract: The goal of this paper is the study of the influence of the environmental torques in the angle between the spin axis and the Sun direction (solar aspect angle) for spin stabilized satellite. The theory uses a cylindrical satellite in an illumined orbit, considering the gravity gradient, aerodynamic, solar radiation, residual magnetic and eddy current torques. The mathematic model for each torque is shown. The dynamic equations are represented in a reference system fixed in the satellite and described by spin velocity and the right ascension and declination angles of the spin axis. An analytical solution for the spin velocity and the attitude angles is used to study the behavior of the solar aspect angle. The theory is applied for the real data of the Brazilian Satellite of Data Collection - SCD1 and SCD2. Two approaches are presented. The results agree with the real satellite behavior for specific time simulation. Then the theory has consistency and can be applied to predict the behavior of the solar aspect angle.

1 Introduction

The goal of this paper is to analyze the influence of the some external torques in the solar aspect angle, considering a cylindrical spin stabilized satellite in an illumined orbit. The Earth's shadow is not considered. The adopted mathematical model for the gravity gradient, aerodynamic, solar radiation, residual magnetic and eddy current torques can be found in [3,4,6].

The gravity gradient torque (GGT) is created by the difference of the Earth gravity force direction and intensity acting on each satellite mass element and is inversely proportional to the cube of the satellite geocentric distance. Therefore it decreases when the altitude increases.

The aerodynamic torque (AT) is created by the interactions of rarefied air particles with the satellite surface and is predominant in satellites with low altitude, because it depends on the quantity of air molecules in the Earth atmosphere. Calculation of aerodynamic torques for realistic spacecraft is not very accurate because of existing uncertainties in the atmospheric density and in drag coefficient. In this paper TD-88 model is used to describe the atmospheric density.

The solar radiation pressure (SRT) is created by the continuous photons collisions with the satellite surface, which can be able to absorb or reflect on this flow. The total change of the momentum of all the incident photons on the satellite surface originates from the solar radiation force and a torque can be produce.

Magnetic disturbance torques result from the interaction between the spacecraft's residual magnetic field and the Earth´s magnetic field. The residual magnetic torque (RMT) results from the interaction between the spacecraft's residual magnetic moment and the Earth magnetic field and its main effect are to produce a spin axis orientation drift. The torque induced by eddy currents (ECT) is caused by the spacecraft spinning motion and produces a reduction in the satellite spin rate with time.

A spherical coordinate system fixed in the satellite (*body system*) is used to locate the spin axis of the satellite in relation to the terrestrial equatorial system, with the axis z being along the direction of the spin velocity vector. The directions of the spin axis are specified by the right ascension (α) and the declination (δ) as represented in the Fig. 1.

The dynamic equations are represented in a body system and described by spin velocity and the right ascension and declination of the spin axis. These equations depend on the torques components in the body system. The averages of each torque components are determined over on orbital period and are substituted in the equations of motion. Due to the cylindrical form of the satellite and others simplifications, only the gravity gradient and eddy currents torques have a non-null z-components.

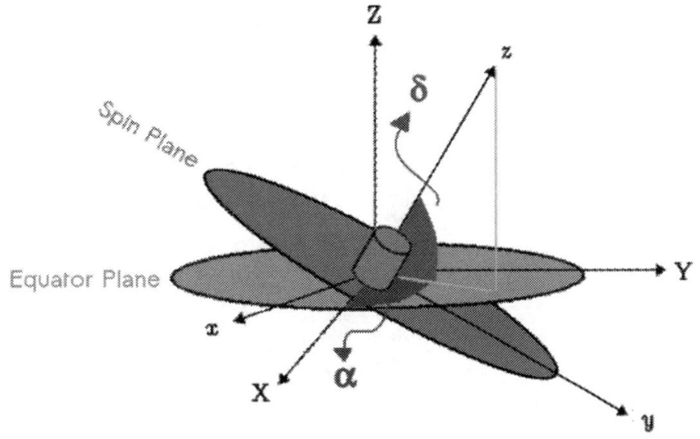

Figure 1. Body System (x,y,z) and Equatorial System (X,Y,Z): right ascension (α) and declination (δ) of the spin axis (z).

An analytical solution was gotten for the spin velocity and the attitude angles for one orbit period [3]. These analytical solutions are used to study the behavior of the solar aspect angle, which is gotten by the dot product of the Sun direction and the spin axis, and depend on the declination and the right ascension of the spin axis and declination and the right ascension of the Sun direction [4].

The theory is applied for the real data of the Brazilian Satellite of Data Collection - SCD1 and SCD2, which are quite appropriated for verification and comparison of the data generated and processed by the Satellite Control Center (SCC) of the Brazil National Research Institute (INPE). Two approaches were presented. In the first one the attitude and orbital data are daily updated with real attitude data supplied by INPE. In the second approach the attitude and orbital data are not updated.

Others analysis of the solar aspect angles [1, 4] have been developed but they haven't considered the all torques acting together. Then the present theory could be more realistic than the other ones.

2. Analytical Solution for the Equations of Rotational Motion

The variations of the spin velocity, the declination and the right ascension of the spin axis for spin stabilized artificial satellites are given by the Euler equations in spherical coordinates [10], when the mean components of the external torque are included.

$$\frac{dW}{dt} = \frac{N_{gzm}}{I_z} + \frac{N_{izm} W}{I_z}, \tag{1}$$

Journal of Physics: Conference Series **641** (2015) 012022 doi:10.1088/1742-6596/641/1/012022

$$\frac{d\delta}{dt} = \frac{N_{tym}}{I_z W} + \frac{N_{iym}}{I_z}, \quad (2)$$

$$\frac{d\alpha}{dt} = \frac{N_{txm}}{I_z W \cos\delta} + \frac{N_{ixm}}{I_z \cos\delta}, \quad (3)$$

where I_z is the moment of inertia along the spin axis, N_{ixm}, N_{iym}, N_{izm} are the mean components of eddy currents torques and $N_{tym} = N_{Ay} + N_{sy} + N_{gym} + N_{rym}$ and $N_{txm} = N_{Ax} + N_{sxm} + N_{gxm} + N_{rxm}$, being N_{Axm}, N_{Aym}, N_{Azm}, N_{sxm}, N_{sym}, N_{szm}, N_{gxm}, N_{gym}, N_{gzm}, N_{rxm}, N_{rym}, N_{rzm} the mean components of the aerodynamic, solar radiation, gravity gradient and residual magnetic torques, respectively. The details of each mean external torque can be observed in [1,2,3,9,10].

The differential equations of Eqs. (1) - (3) can be integrated assuming that mean components of each torque is constant over one orbital period. Then for one orbit period the solution of the differential equation of spin velocity is given by Eq. (4).

$$W = \left(W_0 + \frac{N_{gzm}}{N_{izm}}\right) e^{\frac{N_{izm}}{I_z}t} - \frac{N_{gzm}}{N_{izn}}. \quad (4)$$

In order to get the solution of the others equations, it is necessary to substitute the solution given by Eq. (4) in the Eqs. (2) and (3) for declination and right ascension of spin axis, respectively, and after some algebraic manipulations the analytical solution can simply be expressed by Eqs. (5) and (6).

$$\delta = \frac{t}{I_z}\left(N_{iym} - \frac{N_{tym} N_{izm}}{N_{gzm}}\right) + \frac{N_{tym}}{N_{gzm}} \ln\left(\frac{W}{W_0}\right) + \delta_0 , \quad (5)$$

$$\alpha = \frac{t}{I_z \cos\bar{\delta}}\left(N_{ixm} - \frac{N_{txm} N_{izm}}{N_{gzm}}\right) + \frac{N_{txm}}{N_{gzm} \cos\bar{\delta}} \ln\left(\frac{W}{W_0}\right) + \alpha_0 , \quad (6)$$

with: $\bar{\delta} = \frac{\delta - \delta_0}{2}$, δ is the computed declination, W_0, δ_0 and α_0 are the initial values for spin velocity, declination and right ascension of the spin axis.

These solutions are valid for one orbital period. Thus, for every orbital period, the orbital data must be updated, taking into account at least the main influences of the Earth oblateness. With this approach, the analytical theory will be close to the real attitude behavior of the satellite.

The goal of this paper is to analyze the solar aspect angle η, which is the angle between the spin axis $\hat{k}$ and the Sun direction $\hat{u}$, which can be computed by the dot product of the Sun direction $\hat{u}$ and the spin axis $\hat{k}$, and depend on the declination (δ) and the right ascension (α) angles of the satellite spin direction and the declination (δ_S) and the right ascension (α_S) angles of the Sun direction. They are represented in the Fig.2.

Then solar aspect angle can be computed by Eq.(7).

$$\cos\eta = \hat{u} \bullet \hat{k} = M \quad \text{and} \quad \eta = acos\, M \quad \text{with } 0^\circ < \eta < 180^\circ . \quad (7)$$

For the mission of the satellite SCD1 and SCD2 there is a restriction for the solar aspect angle: $60^\circ < \eta < 90^\circ$ for SCD1 and $80^\circ < \eta < 100^\circ$ for SCD2 [4,6].

3. Applications

The theory developed has been applied to the spin stabilized Brazilian Satellite SCD1 and SCD2 for verification and comparison of the theory against data generated by the SCC of INPE. Operationally, SCC attitude determination comprises: sensors data pre-processing, preliminary attitude determination and fine attitude determination. The pre-processing is applied to each set of data of the attitude sensors collected from every satellite that pass over the ground station. Afterwards, from the whole pre-processed data, the preliminary attitude determination produces estimative to the spin velocity vector from every satellite that pass over a given ground station. The fine attitude determination takes (one week) a set of spin velocity vector and estimates dynamical parameters (spin velocity vector, residual magnetic moment and Foucault parameter). Those parameters are further used in the attitude propagation to predict the need of attitude corrections.

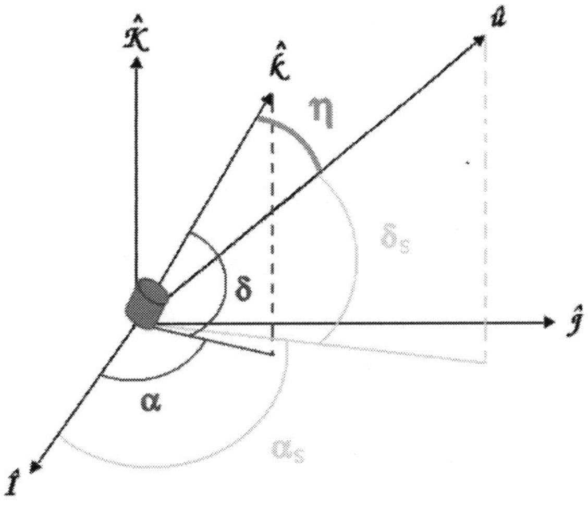

Figure 2. Solar angle η.

Two approaches are assumed to examine the influence of the considered external torques actuating during the evolution of rotational motion of the satellite. In the first approach the attitude and orbit data are updated every 24 hours with the data generated by SCC of INPE. In the second approach the computed attitude and orbit data aren't updated in order to determine the validate period of the analytical solution. In all numerical simulation the orbital elements are updated taking into account the main influence of the Earth oblateness.

3.1. Results for SCD1 satellite

The initial conditions of attitude had been taken for date of July, 24th, 1993 at 00:00:00 GMT, supplied by the INPE's Satellite Control Center (SCC) for 40 days and can been found in [3].

The results for the first approach are shown in Fig. 3. Figure 3a presents the results for temporal behavior of the solar aspect angle. Figure 3b represents the deviation between the computed values and real values of this angle and it is important to observe that they are smaller than 0.5° for the time simulation, which is within the dispersion range of the attitude determination system performance by SCC (0.5°). Then the mean error deviation for the solar aspect angle was 0.23127° for the time simulation.

Table 1 presents the results for this approach for 2 days of two different periods. The simulations were interrupted in the 3rd day because the mean deviations errors between the computed values and real values for the spin velocity, right ascension and declination were bigger than INPE's required precision. Other simulations were done for different initial data but in all of them the results were similar, which means that the computed values have a good agreement with the real data only for the 2 days simulations.

Table 1. Deviation between computed and real values without the daily updated data for SCD1.

Day	$\Delta\eta$ (°)	Day	$\Delta\eta$ (°)
07/27/1993	0	08/25/1993	0
07/ 28/1993	0.13293	08/26/1993	0.18526
07/29/1993	0.13945	08/27/1993	0.48049

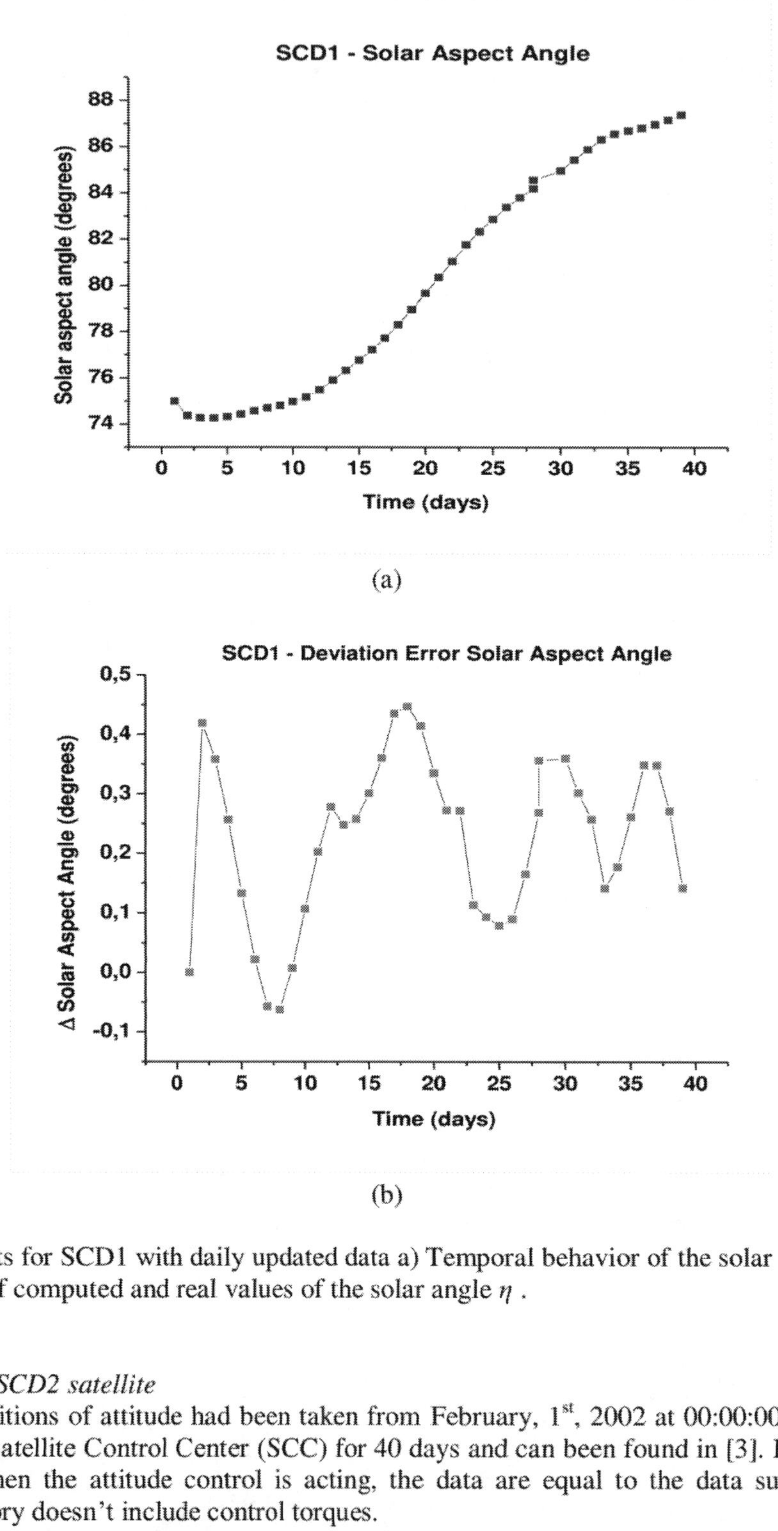

Figure 3. Results for SCD1 with daily updated data a) Temporal behavior of the solar aspect angle η, (b) Difference of computed and real values of the solar angle η .

3.2. Results for SCD2 satellite

The initial conditions of attitude had been taken from February, 1st, 2002 at 00:00:00 GMT, supplied by the INPE's Satellite Control Center (SCC) for 40 days and can been found in [3]. It is important to observe that when the attitude control is acting, the data are equal to the data supplied by SCC, because the theory doesn't include control torques.

Figure 4 present the results for temporal behavior of the solar aspect angle and the deviation between the computed values and real values of this angle, respectively. It can also be observed that they are smaller than 0.25°, with the mean error deviation for the solar aspect angle is 0.10614° for the

Journal of Physics: Conference Series **641** (2015) 012022 doi:10.1088/1742-6596/641/1/012022

time simulation. Note that the differences are zero when the attitude control is acting. Then they are within the dispersion range of the attitude determination system performance by SCC.

Table 2 presents the results for this approach for 2 days of two different periods. The same way of the SCD1, the simulations were interrupted in the 3rd days because the mean deviations errors between the computed values and real values for the others parameters were bigger than INPE's required precision.

Table 2. Deviation between computed and real values without the daily updated data for SCD2.

Day	$\Delta\eta$ (°)	Day	$\Delta\eta$ (°)
02/09/2002	0	03/05/2002	0
02/ 10/2002	-0.24611	03/ 06/2002	0.01108
02/11/2002	-0.24029	03/07/2002	-0.04099

4. Conclusions

In this paper an analytical approach for the spin-stabilized satellite rotational motion was presented taking into account the influence of torques: aerodynamic, gravity gradient, solar radiation, residual magnetic and eddy currents. The goal was to analyze the influence of these torques in the solar aspect angle.

The analytical solution shows that the spin velocity has an exponential variation due to the eddy current torque and a linear variation due to the gravity gradient torque. The combinations of the others torques contribute for a precession and drift on the spin axis, due to the temporal variation on the right ascension and declination angles respectively. The theory was applied to the spin stabilized Brazilian's satellites SCD1 and SCD2. Two approaches were developed.

In the first one the attitude and orbital data are daily updated with real attitude data supplied by INPE. The results showed a good agreement between the computed and real data during 40 days. The mean error deviation for the solar aspect angle was 0.23127° for SCD1 and 0.10614° for SCD2, which are within the dispersion range of the attitude determination system used for this satellite.

In the second approach the attitude and orbital data are not updated. The results presented a good agreement between the analytical solution and the actual satellite behavior only for two days simulation. For more than 2 days the mean deviation of the right ascension, declination and pointing deviation were higher than the precision required for SCC (0.5°).

For both approaches it is possible to note the influence of the declination of the spin axis in the calculation of the solar aspect angles. Results have shown the agreement between the analytical solution and the real satellite behavior for specific time simulation and two approaches were presented. Then the theory has consistency and can be applied to predict the behavior of the solar aspect angle.

Others analysis of the solar aspect angles [1, 4] have been developed but they haven't considered the all torques acting together. Then the present theory could be more realistic than the other ones.

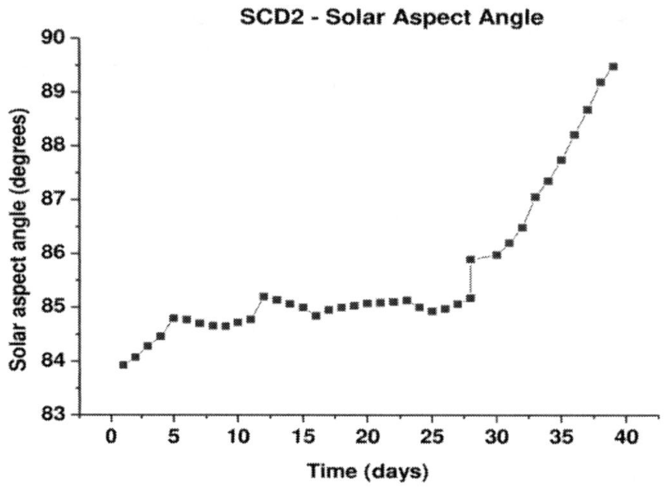

(a)

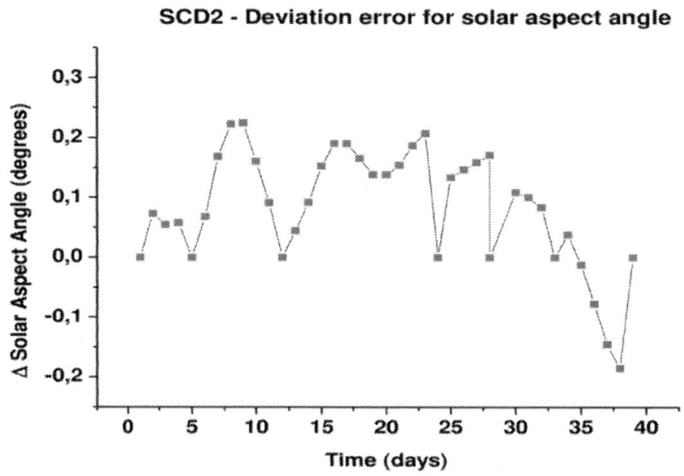

(b)

Figure 4. Results for SCD2 with daily updated data: (a) Temporal behavior of the solar aspect angle η, (b) Difference of computed and real values of the solar angle η .

ACKOWLEDGMENTS
This present work was supported by FAPESP (Process. No. 2012/21023-6) and by CAPES/PVNS.

References

[1] Garcia R, V Zanardi M C and Kuga H K 2009 Spin-stabilized spacecrafts: analytical attitude propagation using magnetic torques *Math. Prob. Eng.*
[2] Kuga H K, Silva W C C, Guedes U T V 1987 Attitude dynamic of the spin stabilized satellite *Technical Report, INPE-4403-NTE/275* (in Portuguese)

[3] Motta G B and Zanardi M C 2014 Environmental torques and the solar radiation *Proceed. 24th ISSFD*

[4] Motta G B, Carvalho M V and Zanardi M C 2013 Analytical prediction of the spin stabilized satellite's attitude using the solar radiation torque *J. Phys: Conf. Serie.* **465** 012009

[5] Pisacane V L and Moore R C 1994 *Fundamentals of Space System* (Oxford University Press)

[6] Vilhena de Moraes R 1994 Non-gravitational disturbing forces *Adv. Space Res.* **14**

[7] Wertz J R 1978 *Spacecraft Attitude Determination and Control* (Dordrecht, DD. Reidel)

[8] Zanardi M C, Vilhena de Moraes R 1999 Analytical and semi-analytical propagation of artificial a satellite's rotational motion *Celes. Mech. and Dyn. Astr.* **75**

[9] Zanardi M C, Quirelli I M P and Kuga H K 2003 Analytical attitude propagation of spin stabilized Earth artificial satellites *Proceed. 17th ISSFD* **2**

[10] Zanardi M C, Quirelli I M P and Kuga H K 2005 Analytical attitude prediction of spin stabilized spacecrafts perturbed by magnetic residual torques *Adv. Space Res.* **36**

Journal of Physics: Conference Series **641** (2015) 012023 doi:10.1088/1742-6596/641/1/012023

Magnitude of Solar Radiation Torque in the Transition Region from the Umbra to the Dark Shadow of the Earth

R E S Cabette[1], M C Zanardi[2] and I Kolesnikov[1]

[1]São Paulo Salesian University, UNISAL, Dom Bosco Street, 284, Lorena, SP, Brazil
[2]Federal University of ABC, UFABC, Avenue of the States, 5001, Santo André, SP Brazil

E-mail: recabette@uol.com.br

Abstract. The analysis of solar radiation pressure force and its influence on the motion of artificial satellites has been developed by researchers. Accurate models to describe the influence of the Earth's shadow on the torque and force due to solar radiation pressure have been presented. In this work the solar radiation torque (SRT) and its influence on the attitude of an artificial satellite are taken into account by the introduction of the Earth's shadow function in the equations of motion. This function assumes a unitary value when the satellite is in the fully illuminated region of its orbit, and the value zero for the full shade region. The main objective of this study is to analyze the magnitude of SRT using the equations described by quaternions during a 35 day period and to compare the results with the satellite transition through the shadow region and the time interval in this region. The duration and transition through the shadow region were obtained using the software "Shadow Conditions of Earth Satellites". The formulation is applied to the Brazilian Data Collection Satellites SCD1 and SCD2, and the torque model is presented in terms of the satellite attitude quaternion, distance of the satellite to the Sun, orbital elements, right ascension and declination of the Sun.

1. Introduction

The study of the rotational motion of artificial satellites leads to a more precise results when the torques caused by external forces that influence their motion are included. In this paper, the force caused by the solar radiation pressure is one of the external forces that has been modeled, taking into account the effects of Earth´s shadow, and inserted in the equations in order to propagate the rotational motion of the artificial satellites.

In the literature several papers considered the influence of the solar radiation torque (SRT) on the attitude of the artificial satellite [3,6,7,9,11,12]. Most of them modeled the torque acting on the artificial satellite and determine analytical and numerical solutions for the dynamic system of equations describing the rotational motion of the satellite, including the shadow region of the Earth.

The goal of this paper is the analysis of the magnitude of the SRT during a 35 day period using the Brazilian Satellites SCD1 and SCD2 data. Here it is assumed that the orbital motion is known and given by the problem of two bodies and quaternions are used to represent the space orientation (attitude) of the satellite. The equations of motion are given by Euler's equation, relating the rate of change of the components of the satellite angular velocity with the external torques, and by the attitude

Journal of Physics: Conference Series **641** (2015) 012023 doi:10.1088/1742-6596/641/1/012023

kinematic equations, relating the rate of change of the components of the quaternion with the components of the angular velocity. To determine the components of the solar radiation torque a cylindrical satellite is admitted, so that the component of the torque on the Z axis of the system fixed in the satellite is zero. Therefore, the magnitude of the angular velocity is not affected by this torque. Also, the applications are made for a spin stabilized satellite.

In spite of the applications were performed for the spin stabilized satellites SCD1 and SCD2, this formulation can be apply for any kind of satellite since it uses the attitude quaternions and the components of the angular velocity.

2. Attitude Motion

The motion of a satellite is specified by its position, velocity, space orientation and angular velocity. The last two quantities describing the rotational motion around its center of mass, and the space orientation is called attitude and they are the focus of this paper.

The attitude expresses the relation between two coordinate systems, one system fixed in the satellite (herein given by principal axis of inertia of the satellite and named principal system) and another system having the same origin, but with axes parallel to a inertial reference system (here called equatorial system, with axes parallel to the Earth's equatorial system, with Z-axis along the north pole, XY plane being the Earth equatorial plane and the X-axis along the spring equinox). The Equator system, the orbital system and the orbital elements are represented in the Figure (1).

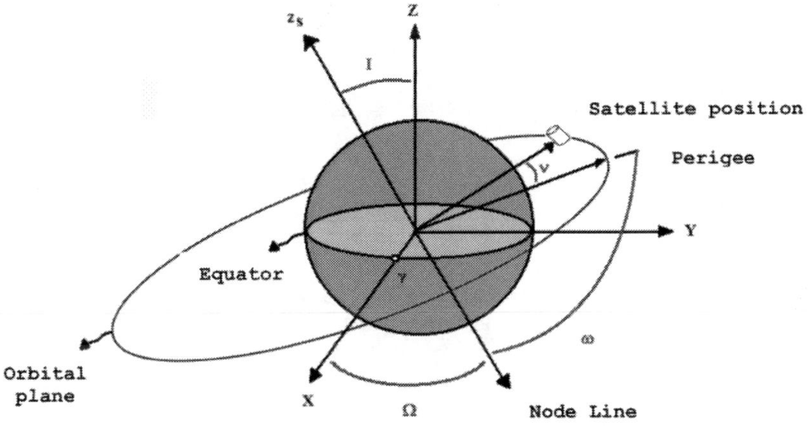

Figure 1: Equatorial and Orbital systems: Ω - longitude of the ascending node, ω - argument of perigee, I - orbital inclination, ν - true anomaly.

If the satellite has a rotation around some axis, then its orientation or attitude changes every instant. Thus, the angular velocity and the attitude time derivatives are related through the kinetics equations.

There are many ways to represent the attitude, but in this study, the quaternions are used. The attitude quaternion q is a 4x1 vector given by [8,12]:

$$q=[q_1 \quad q_2 \quad q_3 \quad q_4]^t=[\vec{q} \quad q_4]^t, \tag{1}$$

where t represents the transposed of the matrix and it can be expressed in function of the rotation angle (ϕ) and of the axis of rotation $\hat{w}$:

$$\vec{q}=[q_1 \quad q_2 \quad q_3]^t=sin(\phi/2)\,\hat{w} \quad ; \quad q_4=cos(\phi/2). \tag{2}$$

Journal of Physics: Conference Series **641** (2015) 012023 doi:10.1088/1742-6596/641/1/012023

There are two reasons to use the quaternion formulation: first one is to save computational time by avoiding integrating trigonometric functions and the second is to avoid the singularity problem that is characteristic of the Euler's angles formulations [8,12].

2.1. Equation of the rotational motion

2.1.1. Dynamic Equations

The dynamics equations of the artificial satellite rotational motion express in the principal system (Oxyz), are given by [8,12]:

$$\dot{p} = \frac{N_x}{I_x} + \left(\frac{I_y - I_z}{I_x}\right) qr \qquad \dot{q} = \frac{N_y}{I_y} + \left(\frac{I_z - I_x}{I_y}\right) pr \qquad \dot{r} = \frac{N_z}{I_z} + \left(\frac{I_x - I_y}{I_z}\right) pq \tag{3}$$

where: I_x, I_y and I_z are the Principal Moments of Inertia of the satellite; , p,q,r and N_x, N_y, N_z are the components of the angular velocity and the external torques in the principal system, respectively. These equations will be used in the attitude propagation with the inclusion of the components SRT in these equations. The mathematic model for SRT is discussed in the topic 2.2.

2.1.2. Kinematic Equation for the quaternion

The kinematic equations, describing the time derivatives of the attitude quaternion components, due to the rotation of the satellite, are given by [8,10]:

$$\dot{q}_1 = \frac{1}{2}[pq_4 - qq_3 + rq_2] \qquad \dot{q}_2 = \frac{1}{2}[qq_4 - rq_1 + pq_3] \tag{4}$$

$$\dot{q}_3 = \frac{1}{2}[rq_4 - pq_2 + qq_1] \qquad \dot{q}_4 = -\frac{1}{2}[pq_1 + qq_2 + rq_3] \tag{5}$$

The great advantage of the use of quaternions in kinematic equations is the no existence of a zero denominator, which means there isn't singularity, such as those occur in Euler angles [4].

2.2. Direct Solar Radiation Torque

Solar radiation pressure is generated by the continuous stream of photons that collide with the surface of the satellite. The rate of momentum of all incident photons in the satellite's surface originates the Solar Radiation Force, which can cause disturbances in the satellite orbit and may result in a torque depending on the physical and geometric characteristics of the satellite. However, the solar radiation torque exists only when the satellite is partially or fully illuminated [11,12]. Thus, there are three transition phases, represented in Figure 2:

- Transition through dark region (shadow);
- Transition through partially illuminated region (penumbra);
- Transition through fully illuminated region.

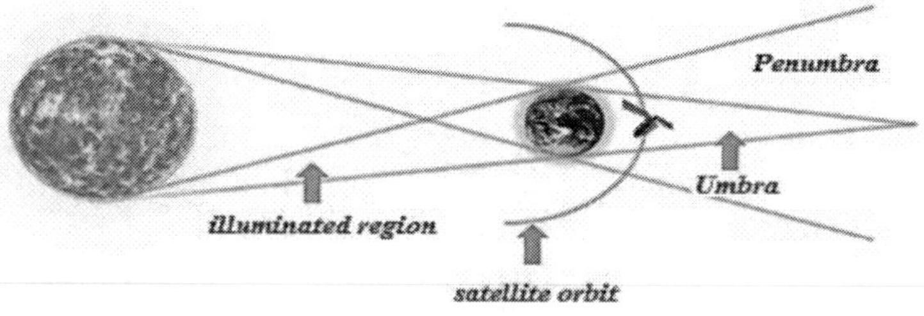

Figure 2: Transition Phases: umbra, penumbra and illuminated regions.

Journal of Physics: Conference Series **641** (2015) 012023 doi:10.1088/1742-6596/641/1/012023

Solar radiation pressure force on the satellite is given by the surface integral over the entire surface of the satellite in which there is the incidence of sunlight [11]:

$$\vec{F} = -\int_S \frac{\overline{K}}{R^2}\left\{\left[\frac{2\gamma}{3}(1-\beta)\cos\theta + 4\beta\gamma\cos^2\theta\right]\hat{n} + [(1-\beta\gamma)\cos\theta]\hat{u}\right\}dS \quad (6)$$

where $\overline{K}$ is the solar parameter and assume the value $1{,}01\mathrm{x}10^{17}$ kgm/s [6], γ is the total reflection coefficient, β is the specular reflection coefficient, $\hat{n}$ is the normal unit vector of the satellite surface, $\hat{u}$ is the direction of the incident solar flux over the element of surface dS, θ is the angle between $\hat{n}$ and $\hat{u}$ (see Figure 3), R is the distance of de element dS and the Sun (here is adopted equal to the distance of the satellite center to the Sun).

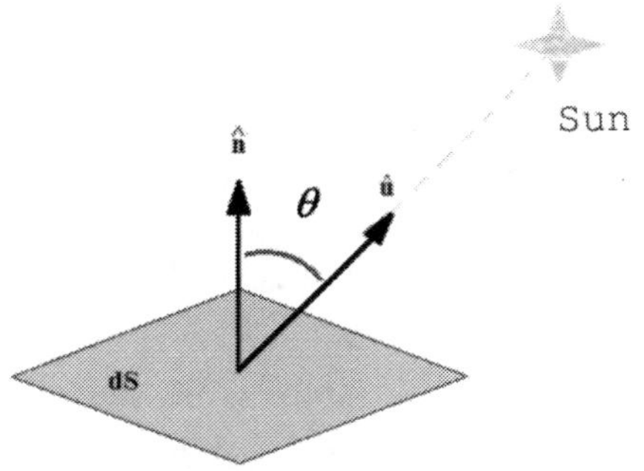

Figure 3: Geometry of the solar incident light over satellite surface dS.

The element of Radiation Solar Torque ($d\vec{N}$) around the satellite center of mass (CM) due to the solar radiation force ($\vec{F}$) is given by:

$$d\vec{N} = \vec{r} \times d\vec{F} \quad (7)$$

where $\vec{r}$ is the position vector of the surface element dS of the satellite in relation to the CM. After performing the cross product, the solar radiation torque (SRT) over all the satellite surface can be given by [11]:

$$\vec{N} = -\int_S \frac{\overline{K}}{R^2}\left\{\left[\frac{2\gamma}{3}(1-\beta)\cos\theta + 4\beta\gamma\cos^2\theta\right]\vec{r}\times\hat{n} + [(1-\beta\gamma)\cos\theta]\vec{r}\times\hat{u}\right\}dS \quad (8)$$

In the development of the integrals of the Eq. (8) it is necessary to consider the shape of the satellite. Depending on the satellite shape, the SRT can be zero if there is symmetry in the satellite or if the center of mass is coincident with the application point of the radiation pressure force. In this paper the considered satellite is cylindrical, with symmetry along the z-axis.

In order to get the components of the SRT in the body system it is assumed that the position vector $\vec{R}$, is equal to the position vector $\vec{R}*$ of the satellite CM in relation to the Sun (see Figure 4) since the satellite dimensions are negligeble in relation to the satellite distance to the Sun.

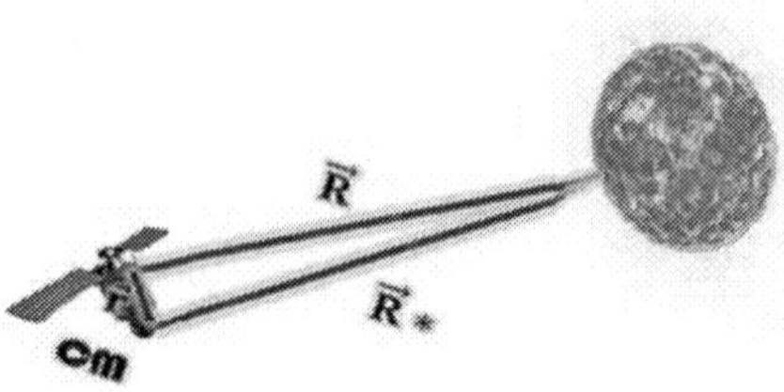

Figure 4: CM satellite positioning and surface element relative to the Sun

After some algebraic manipulations, which are associated with the developments of the integral in Eq. (8) for a cylindrical satellite, the z-axis component of the SRT is zero (because of the satellite symmetry) and the others SRT components in the principal system [3,5,6] are given by:

$$\vec{N} = N_x \vec{i} + N_y \vec{j} \tag{9}$$

$$N_x = -\frac{\overline{K}}{R^4}(\beta_1\gamma_1 - \beta_2\gamma_2)\frac{h}{2}\pi\sigma^2\{a_S{}^2 R_y R_z + a_S r'(R_y a_{31} + R_z a_{21}) + r'^2\, a_{21}a_{31}\} \tag{10}$$

$$N_y = \frac{\overline{K}}{R^4}(\beta_1\gamma_1 - \beta_2\gamma_2)\frac{h}{2}\pi\sigma^2\{a_S{}^2 R_x R_z + a_S r'(R_x a_{31} + R_z a_{11}) + r'^2\, a_{11}a_{31}\} \tag{11}$$

where h and σ are the height and radius of the circular cylindrical satellite; $\beta_i, \gamma_i, i = 1,2$ are specular and total reflection coefficients respectively, for each satellite surface (which assume constant values); a_{11}, a_{12}, a_{13}, $a_{21}, a_{22}, a_{23}, a_{31}, a_{32}$ and a_{33} are the elements of the rotation matrix between the principal system and orbital system [3,5,8] and depend on the attitude quaternion, (q_1, q_2, q_3, q_4) and the orbital elements (orbital inclination, perigee argument, longitude of the ascendant node and true anomaly). R_x, R_y, R_z give the Sun direction in the principal system and are expressed by:

$$R_x = q_A \cos\delta_s \cos\alpha_s + q_D \cos\delta_s sen\alpha_s + q_G sen\delta_s \tag{12}$$

$$R_y = q_B \cos\delta_s \cos\alpha_s + q_E \cos\delta_s sen\alpha_s + q_H sen\delta_s \tag{13}$$

$$R_z = q_C \cos\delta_s \cos\alpha_s + q_F \cos\delta_s sen\alpha_s + q_I sen\delta_s \tag{14}$$

where α_S, δ_S are the Sun's right ascension and declination, respectively; q_A, q_B, q_C, q_D, q_E, q_F, q_G, q_H, q_I are the elements of the rotation matrix between the principal system and Equatorial system, and depends on the attitude quaternion [8,10].

3. Numerical Simulations

In this paper the magnitude of the SRT is analyzed, with the numerical simulation developed in a FORTRAN code for a 35 day period, using the data of Brazilian Collecting Data Satellites (SCD1 and SCD2), supplied by INPE's Satellite Control Center [1,4]. In order to get the satellite transition through umbra region and the permanence time in this region, it was used a MATLAB program available at the *Orbital Mechanics with MATLAB* website. This site can be used for many practical problems associated with orbital mechanic, applied astrodynamics and astrometry [2].

3.1. Data Satellites

a) SCD 1 – Initial Date: 1993 – July - 24 00:00:00 GMT

Semi-major axis= 7139615.83m Eccentricity = 0.00453 Inclination = 25.00°
Right ascension of the ascending node= 260.43° Argument of perigee= 260.23 °
Mean anomaly = 102.89 ° Moment of inertia in the x-axis=11.06kg.m²
Moment of inertia in the y axis=10.67kg.m² Moment of inertia in the z axis=13.00kg.m²

Journal of Physics: Conference Series **641** (2015) 012023 doi:10.1088/1742-6596/641/1/012023

b) SCD 2 – Initial Date: 2002 – February - 01 00:00:00 GMT

Semi-major axis = 7133679.70m
Eccentricity = 0.00175 Inclination = 25.006 °
Right ascension of the ascending node = 88.303 °
Argument of perigee = 288.207 °
Mean anomaly = 300.030 °
Moment of inertia in the x-axis =12.33kg.m^2
Moment of inertia in the y axis =12.35kg.m^2
Moment of inertia in the z axis =14.50kg.m^2

Structural data (similar to SCD1 and SCD2):
Total reflection coefficient of the octagonal base = 0.7
Specular reflection coefficient of the octagonal base = 0.1
Total reflection coefficient of the shell of the prism = 0.5
Specular reflectance of the dark of the prism = 0.1

First we use the script available at the Orbital Mechanics with MATLAB website called "Shadow Conditions of Earth Satellites" that generates the results for the satellite passes through the shadow region and the permanence time in this region. This information is included in the FORTRAN code and the simulations are realized for each day.

3.2. Results for SCD1 and SCD2 Satellites

This section presents the results obtained in the simulations by MATLAB programs (satellite passage through the shadow region) and FORTRAN (calculus of the magnitude of solar radiation torque). These results are compared to verify the variation of the magnitude of the torque during the permanence in the shadow region.

The data for SCD1 satellite passage through the shadow region covers the period from 07/25/1993 to 26/08/1993 (35 days) and for SCD2 satellite from 01/02/2002 to 08/03/2002 (35 days). The complete data can be accessed in the references [4, 5, 6].

Simulations in MATLAB show the passage of SCD1 and SCD2 satellites by the umbra region and its permanence in this region. The results of the simulations for the considered period are presented in the Figure 5 for each satellite.

Using the Software FORTRAN, the SRT magnitude was computed, including the computed values of the Earth umbra. Figure 6 shows the SRT magnitude for the SCD1 and SCD2.

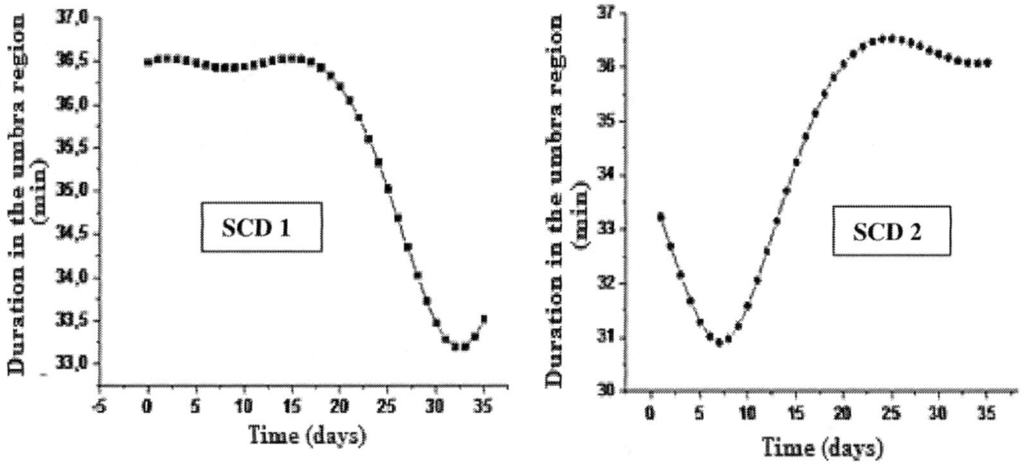

Figure 5: Time duration of the passage of the satellite umbra region.

By the comparison of the umbra time of duration (Figure 5) with the SRT magnitude it is possible to observe that this magnitude increase (decrease) when the time of duration of the satellite transition

Journal of Physics: Conference Series **641** (2015) 012023 doi:10.1088/1742-6596/641/1/012023

decrease (increase). It is important to observe that the simulation time interval is different for each satellite, and as the position of the Sun is different for each interval, then the SRT magnitude has a little different value for each satellite.

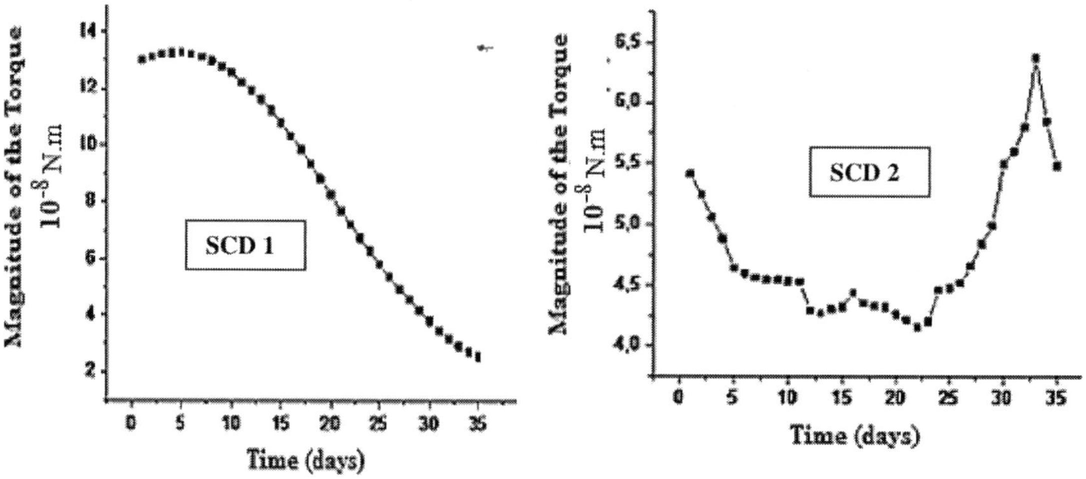

Figure 6: Magnitude of Solar Radiation Torque

The SCD1 and SCD2 are LEO (Low Earth Satellite) and small in size. Because of that the SRT magnitude is also small and negligible. It is well known that the magnetic torques are more important for this kind of satellite [1,4]. Despite this fact, as the attitude data for real satellites are hard to be obtained, and as the SCD1 and SCD2 data sets were kindly provided by the CRC/INPE, they were used here to validate the developed model. It is important to be pointed out that torque can be significant for other satellite (larger and in high altitude orbit) and this model can be applied for them.

4. Conclusion

The objective of this work was to verify the magnitude variation of the SRT considering the period that the satellite stays inside the umbra. In order to determine the satellite entrance and exit times in this region, a MATLAB program [2] was used. The simulations data were compared and the results showed that the magnitude of the SRT decreases when umbra duration increase while the magnitude increase with the decreasing of umbra duration.

The result analysis confirms the importance of the introduction of the Earth shadow influence in the equation of the rotational motion. It can be done by using the Euler equation and the kinetic equation in terms of quaternions, in order to get a propagated attitude closer to the real attitude of the satellite.

The mathematic expression of the components of the SRT is useful for the numerical and analytical propagation of the satellite attitude. However, depending on the size and altitude of the considered satellite, better results for the attitude propagation must include the others external torques.

Also, the obtained results permitted to verify that the analytical equations used to direct solar radiation torque are valid and the simulations demonstrated the expected behavior. As can be observed in the figures, the torque magnitude calculated using the analytical equations (presented in details in [3]) and the simulations during its passage through the Earth's shadow (period in which the satellite does not experience the strength of solar radiation) are consistent .

This fact confirms the validity of the numerical integration of the equations of rotational motion carried out using the FORTRAN code.

It is also important to observe that, in spite of the applications were performed for the spin stabilized satellites SCD1 and SCD2, this theory can also be applied for any kind of satellite since it uses the attitude quaternions and the components of the angular velocity.

ACKOWLEDGMENTS

This present work was supported by FAPESP (Process No. 2012/21023-6) and by CNPq/PIBITI.

5. Referências

[1] KUGA, H. K.; ORLANDO, V.; LOPES, R. V. F. Flight dynamics operations during leop for the inpe'ssecundenviromental data collecting satellite SCD2. RBCM _ Journal of the Brazilian Society of Mechanical Sciences, v.21-SP ISS, p.339-344, 1999

[2] Orbital Mechanics with MATLAB. Disponível em: https://cdeaglejr.wordpress.com/?ref=spelling. Acesso em: 07/07/2015

[3] MOTTA, G B ; CARVALHO, M V ; ZANARDI, M C . Analytical Prediction of the Spin Stabilized Satellite's Attitude Using The Solar Radiation Torque. Journal of Physics. Conference Series (Print), v. 465, p. 012009, 2013

[4] ORLANDO, V.; LOPES, R. V. F.; KUGA, H. K. INPE'S flight dynamics team experience thought four years of scd1 in orbit operations: main issues, improvements and tends, ESA International Symposium on Spaceflight Dynamics. Darmstadt, Alemanha, p.433-437, 1997

[5] SANTOS, J. C. (2011) Análise da estabilidade do movimento rotacional com quaternions e sob a influência de torques externos. Trabalho de Graduação.UNESP _ Campus de Guaratinguetá

[6] SANTOS, J. C. ;ZANARDI, M. C. F. P. S.; MATOS, N. F. O. Semi-analytical study of the rotational motion stability of artificial satellites using quaternions.Journal of Physics.Conference Series (Online), v. 465, p. 012012, 2013

[7] SHIVASTAVA, S. K.; MODI, V. J. Satellite Attitude Dynamics and Control in the Presence Of Enviromental Torques - A Brief Survey _ AIAA Journal, v. 6, no6,1993.

[8] SHUSTER, M. D. A survey of attitude representation. Journal of Astron. Sciences. v. 41. 4, 1993

[9] TAKEICHI, N. Parametric Excitation Induced by Solar Pressure torque on the Roll-Yaw attitude Motion of a Gravity-Gradient Stabilized Spacecraft Celestial Mechanics and DynamicalAstronomy, 2010

[10] WERTZ, J.R.; Spacecraft attitude determination and control. London: Reidel, 1978,v. 73

[11] ZANARDI, M. C. F. P. S. ; VILHENA de MORAES, R. Analytical and Semi-Analytical Analysis an Artificial Satellite's Rotational Motion. Celestial Mechanics & Dynamical Astronomy, Klhumer Academic Publishers, v. 75, n.4, p. 227-250, 1999

[12] ZANARDI, M. C. F. P. S. ; VILHENA de MORAES, R.; CABETTE, R. E. S.;GARCIA, R. V. Spacecraf's Attitude Prediction: Solar Radiation Torque and the Earth's Shadow. Advances in Space Research, Elsevier Science, v. 36, p. 466-4471, 2005

Journal of Physics: Conference Series **641** (2015) 012024 doi:10.1088/1742-6596/641/1/012024

Development of a helicon double layer thruster

Jose Leonardo Ferreira[1], Felipe Nathan de Oliveira[1], Ernesto Gonçalves Costa[1], Helbert de Oliveira Coelho Junior[2] and Artur Castelo Branco[1]

[1] University of Brasilia, Brasilia, DF, 70910-900, Brazil
[2] University of Brasilia, Gama, DF, 72444-240, Brazil

E-mail: `jleonardoferreira@uol.com.br`

Abstract. This work describes the development of a Helicon Double Layer Thruster and its acquisition system at the University of Brasilia Plasma Physics Laboratory. Together with the data analysis process, we are able to measure important plasma characteristics in order to compound an Integrated Plasma Diagnostics System. Experimental results can be compared with computational simulations in order to improve parameters and optimize the performance of the thruster.

1. Introduction

The Plasma Physics Laboratory at the University of Brasilia (PPLUnB) has been working with Plasma Thrusters designs since 2001, when the first PHALL thruster was developed and, subsequently, the models PHALL II and III were simulated, designed and tested. A Hall Effect Thruster (HET) is a type of plasma thruster in which the propellant gas is ionized and accelerated by a magneto hydrodynamic effect combined with electrostatic ion acceleration. The essential operating principle of the HET is that it uses a $\mathbf{J} \times \mathbf{B}$ force and an electrostatic potential to accelerate ions up to high speeds [1]. The Laboratory has been, also, in the after years, carrying researches in the field of characterization of Double Layer for Thrust Engines, the scope of this paper.

Helicon waves belong to the category of whistler waves, which are right-handed circularly polarized electromagnetic waves in free space [5]. They have two main properties, such as: Low frequency, leading to the electrons gyrations slightness and therefore only their guiding center motions being kept, and they are modes of bounded systems, in which their purely electromagnetic character cannot be maintained. In [5], Francis F. Chen gives the theoretical basis to show that the higher then expected gain in plasma acceleration with Helicon waves in Argon gas is possible if the ionized electrons are directly accelerated by the wave interaction with the particles rather than by a random heating process.

Double Layers are the spatial separation of charge existent in nearly colisionless plasmas created due to perturbation of the plasma environment. This perturbation may be generated by different sources, including disruption of a neutral current sheet [2], current-driven instabilities [3] or constraint by a magnetic mirror [4]. At the PPLUnB, the Double Layer is studied by the magnetic mirror constraint principle. More recently, the electric field embedded in the Double Layer has been proposed as a thrust producing mechanism for space propulsion, especially in light of the relative simplicity and anticipated reliability of the underlying technology [7].

Journal of Physics: Conference Series **641** (2015) 012024 doi:10.1088/1742-6596/641/1/012024

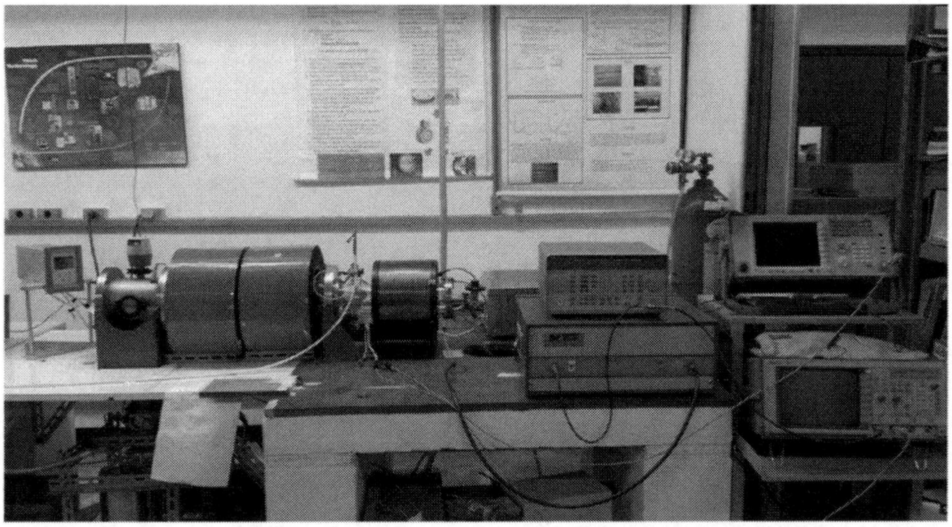

Figure 1. Helicon Double Layer Thruster at the PPLUnB.

2. Description of the Helicon Double Layer device

Based on most of the Helicon Double Layer Devices on the literature [4][5][6][11], the HDLT (Helicon Double Layer Thruster) of UnB has three types of magnetic coils containing 730, 912 and 710 loops of 3mm diameter (10 AWG) copper insulated wire. The coils are connected in a DC voltage supply, model Tectrol TCA 15-50D-1A, which releases a maximum current of 30 Amperes per a maximum of 40 Volts to the coil system. Two vacuum chambers Pyrex made (Borosilicate) with approximately 1 meter and 10 centimeters long with several flanges ports for optical and instrumentation access constitute the device. The First and Large one is about 70 cm long and 16 cm diameter, the small one is 11 cm diameter per 40 cm long. All the connections are stainless non-magnetic made. Horizontally, from the right to the left of the Pyrex Chamber, Argon gas cylinder is connected using a Teflon hose and a leak valve, feeding the system.

The vacuum system is generated in two steps, using two distinct vacuum pumps: one mechanical pump (Edwards RV5 E2M5), being able to pump at a rate of 5.1 m^3/h, used for the initial low vacuum; and a diffusion pump for high vacuum (Edwards Diffstak 63/150M) pumping at a rate of 135 L/s. With this system, a vacuum of the order of 10^{-6} Torr can be reached in approximately two or three hours. To track how the vacuum evolve, it is used a system of pressure gauges two Edwards APG-M NW16 for the low vacuum, and a Edwards AIM-S-NW25 for the high vacuum and a controller where it is shown the information collected by the gauges Edwards TIC Instrument Controller 3 Head RS232/RS485. When the system reaches the necessary pressure, approximately $5{\cdot}10^{-6}$ Torr, it is started the process of Argon release into the chamber the pressure increases to approximately $5{\cdot}10^{-3}$ Torr.

After the right pressure conditions are achieved, the RF system is turned on. A signal generator, model hp 8648A, is used to generate RF wave with a frequency of 29,959 MHz and an amplitude of 7,5 dBm. This signal is measured with the help of an oscilloscope, model hp 54540A, and amplified by a 25 WATTS amplifier, model Amplifier Research 25W1000M7. The coupling with the antenna is made using a capacitive impedance matching circuit for maximum power transfer.

The RF antenna is made using a ring shaped copper bar. This inductive-capacitive ring is used to excite the plasma by electromagnetic radiation. All the connections in the RF assembly are made with coaxial cables.

Journal of Physics: Conference Series **641** (2015) 012024 | doi:10.1088/1742-6596/641/1/012024

Once the RF is turned on, the electromagnetic waves heat the Argon gas, ionizing it and creating the plasma environment. The ionization by electromagnetic radiation is a process in which electrons are pulled out of the atoms by interaction with the field. By absorbing the photons, electrons gain energy enough to disrupt the atomic bound and by collision increase the rate of ionization. The difference between Normal RF waves and helicon waves, described in [6], rests upon the fact that the latter has a higher penetration power, and when coupled with the magnetic field, generates a efficient production of a controllable fast electrons population.

After the Plasma is created, the Coils are turned on and the Magnetostatic field of an maximum of 500 Gauss on the axis is generated, squeezing the plasma and then, if adjusted with the right conditions, generating the ambipolar field.

Interesting to notice that, the property of the expanding magnetic field gives strong conditions for the ambipolar field appearance and consequently the drop of the potential inside of the plasma, creating the Double Layer region.

The expansion of the Magnetic field lines and its action upon the ions and electrons creates a charge separation on the plasma. This charge separation generates a Electric field that accelerate ions in the direction of the axis, giving the impulse necessary for a propulsion principle. As demonstrated by [5], there may be found a direct connection between the use of Helicon waves with magnetic fields and the increase of power deposited on the plasma from electromagnetic fields, justifying the use of such system configuration.

3. Finite Element Magnetics Simulations (FEMM)

FEMM is a suite of programs for solving low frequency electromagnetic problems on two-dimensional planar and axisymmetric domains. The program currently addresses linear/nonlinear magnetostatic problems, linear/nonlinear time harmonic magnetic problems, linear electrostatic problems and steady-state heat flow problems [8].

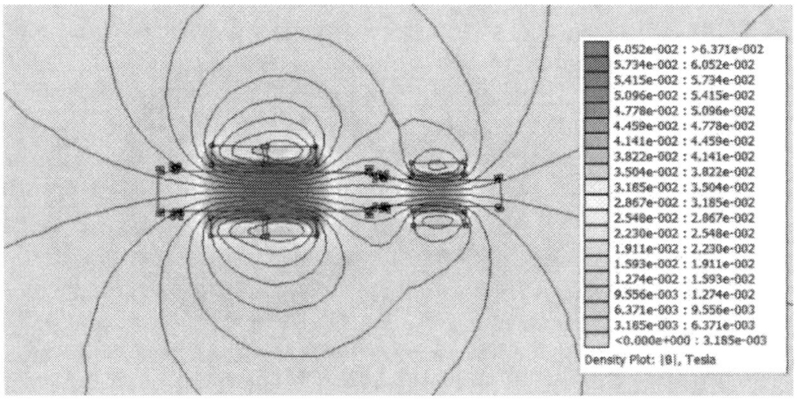

Figure 2. Magnetic Field Distribution simulated in FEMM.

Magnetostatic simulations of the profile of the devices surroundings were made with the help of the program. For plasma constrained by a magnetic mirror, the position of the double Layer is linearly tied to the magnetic field at a position close to the maximum in the gradient and half the maximum of the field [6], stressing so the importance in modeling the magnetic field profile of the device.

With the help of the program, it is also possible to plot a graph showing the intensity of the Magnetic field in the axial direction. This graphic is relevant for helping to find the position established in [6]. In the case of the UnBPPL, the Double Layer must be found approximately at the position 0,60 meter, where the magnetic field lines expansion occurs.

Journal of Physics: Conference Series **641** (2015) 012024 doi:10.1088/1742-6596/641/1/012024

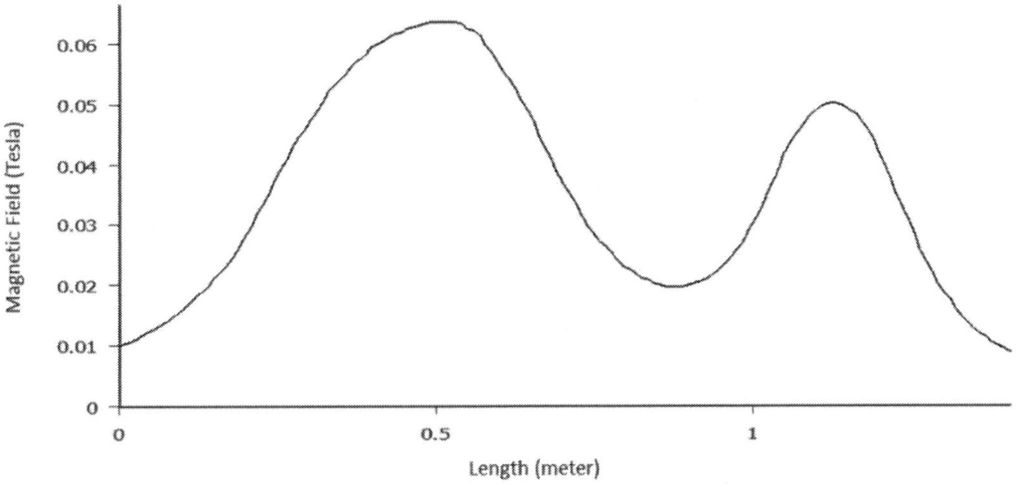

Figure 3. Axial Magnetic Intensity.

Although is known the relation of the position of the Double Layer and the magnetic field distribution, in the future, other variables may be considered in the process of Double Layer generation.

4. Acquisition system

4.1. Probe

The probes used in the Laboratory are basically separated in two groups: Langmuir (electric probes), and the Energy Analyzer, to be explained in next section.

An electric probe is a cylindrical metallic device (generally tungsten or molybdenum) insert in the plasma and posteriorly biased with a potential, in order to discriminate ions and get the curve from which the Temperature, floating potential and Plasma potential can be inferred.

The edge of the probe is firstly negatively biased, in order to generate an ionic saturation section. Slowly, the potential is increased. At the Plasma Space Potential, the electron saturation occurs and the Plasma temperature can be measured[10]. The Langmuir method is well known and highly effective in Plasma Diagnostics.

4.2. Energy analyzer

The energy analyzer is composed of three parts: 2 grids and a plate. The first grid, the one in contact with the plasma, is left at the plasma floating potential, so that it let the ions pass through it but do not let the electrons pass, *i. e.*, the first grid select the ions. The second grid, which is between the first grid and the plate, is charged positively, letting only the high-energy ions pass through it, *i. e.*, the second grid selects the high-energy ions. The plate is called collector, it is charged very negatively, collecting the ions that were able to pass through the second grid and eliminating the few electrons that were able to pass through the first grid. So, it is possible to develop experiments in order to measure the Ion Temperature, to observe the ion beams and measure the decay of the charged exchange cross-section of the ion beam, and measure, also, the ion distribution function.

The PPLUnB is designing one energy analyzer for the Helicon Double Layer Thruster. A cup of Teflon holds the analyzer altogether and then it is put the collector plate on the background, but in between the plate and the second grid, and in between the second and first grid, there is

Journal of Physics: Conference Series **641** (2015) 012024 doi:10.1088/1742-6596/641/1/012024

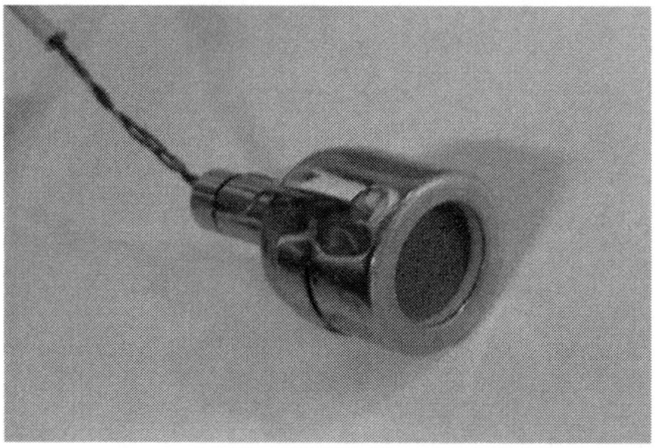

Figure 4. Energy Analyzer designed and produced at the UnBPPL.

a Teflon ring that will isolate the grids and the plate from each other. The Teflon is dielectric but it is not enough resistant for the plasma environment, so, it was necessary to make a case to protect it. A stainless steel case was construct in order to protect the Teflon.

4.3. Power supplies and circuit design

For the Langmuir probe, a DC voltage power supply, model Agilent 6035A, is used to create the desired potential on the probes edge. The terminals are connected to a relay circuit, responsible to invert the polarity from the negative to positive voltages. The probe is linked using a coaxial cable to the relay circuit and the bench multimeter, model hp 34401A, is connected to the ground and to the other exit of the relay circuit. The relay is controlled by a signal from the control program. The figure bellow illustrates the circuit.

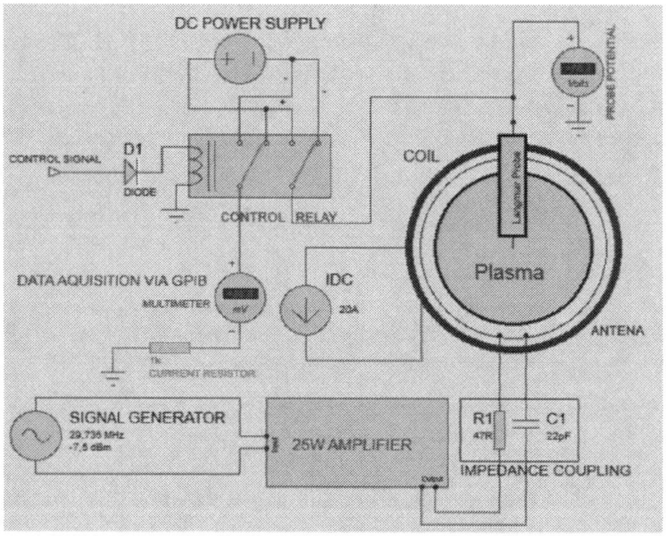

Figure 5. Acquisition system circuit design.

Journal of Physics: Conference Series **641** (2015) 012024 doi:10.1088/1742-6596/641/1/012024

4.4. LabView acquisition and aata treatment

The control program is created using the LabView platform by National Instruments and it is responsible for controlling the voltage of the power supply and recording the voltage data from the multimeter. The test usually goes from -20 to 80 volts. When the voltage of the power supply reaches zero, the relay circuit is triggered and the polarity of the power supply is inverted. This next figure illustrates the block diagram of the control program.

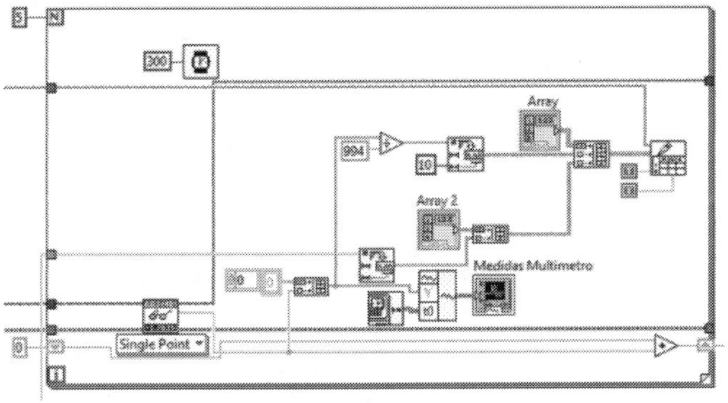

Figure 6. Section of the Block diagram of the control program.

First the control program requests a filename were the data will be saved in a txt format. The control values for the power supply (minimum and maximum voltage), the VISA address for the power supply and the multimeter are settled. The figure bellow illustrates the front panel of the control program.

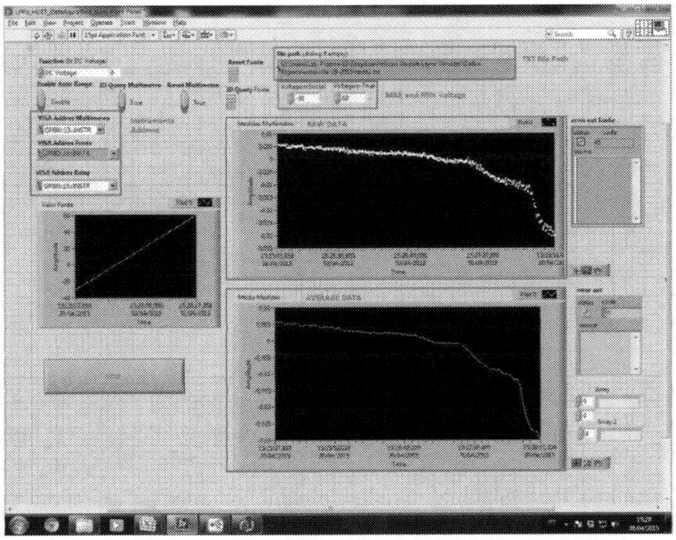

Figure 7. LabView front painel.

The routine starts by pressing RUN. All the communications are made using GPIB protocols. The power supply starts at the minimum voltage and steps towards zero, one volt at a time. The program then reads five values from the multimeter and stores them in a txt file for each voltage level. Two graphs are plotted using these points.

Journal of Physics: Conference Series **641** (2015) 012024 doi:10.1088/1742-6596/641/1/012024

When the power supply reaches zero, the relay circuit is triggered and the connection polarity is inverted. This is necessary to cover a range from negative to positive voltages. With the polarity inverted, the power supply goes from zero to the maximum value.

At each voltage level, the program reads five values and calculates the average. Simultaneously two graphs are plotted, one with every value from the multimeter and another with the average of those five points per voltage level. The user can choose which collection of data (the raw data, the average values or both) to be saved on the txt file.

When the routine reaches its stopping condition, or the user stops the execution of the routine, the control program for safety reasons turns off the power supply automatically. With the previously generated txt file Fig. 8 can be plotted using a suitable program.

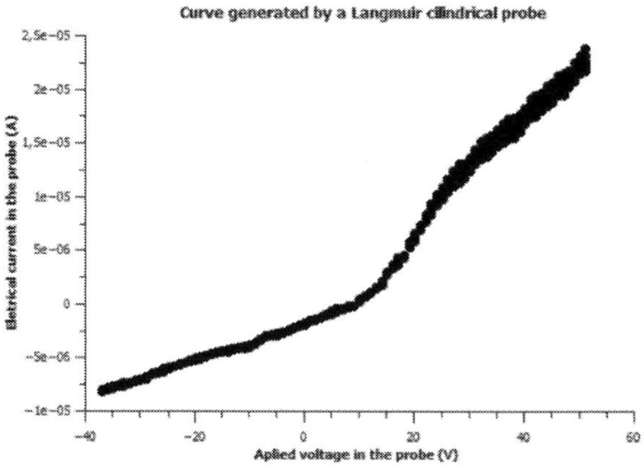

Figure 8. Langmuir curve.

The data treatment is done with the help of the program QtiPlot in a Linux environment. A data treatment and diagnostic is being conceived on LabView environment so the plasma parameters can be more easily and faster calculated.

5. Preliminary results

Using method [10], the data generated can be properly derived and analyzed.

The data used to generate Fig. 8 can be used, with the help of QtiPlot, for data treatment, and the derivative of the linearized Langmuir curve may be obtained:

$$\frac{d \ln i}{dV} = \frac{e}{K_b T_e} \tag{1}$$

Based in [10], the electronic temperature and the electronic density can be measured.

In Fig. 9 and Fig. 10, a change in the plasma potential and temperature in function of the density is observed, being so inferred a linear relation between the plasma electronic density and potential, Fig.9, and the electronic density versus temperature, Fig. 10.

6. Conclusions

The aim of the Helicon Double Layer project at the UnBPPL is to find evidences of the Double Layer in a expanding magnetized plasma and show its feasibility when compared with other thruster models.

Improvements need to be done in the measurement apparatus in order to achieve an axial analyses of the plasma potential. A mechanism, aimed to cross the plasma axially, using a movable probe, is being constructed in order to measure such a potential profile.An improvement

Journal of Physics: Conference Series **641** (2015) 012024 doi:10.1088/1742-6596/641/1/012024

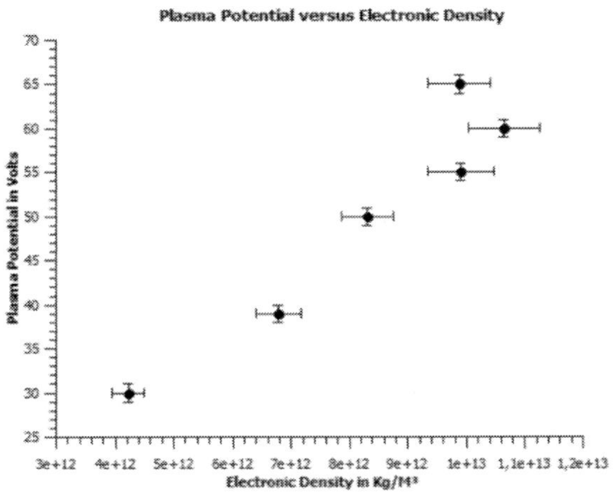

Figure 9. Plasma Potential vs. Electronic Density.

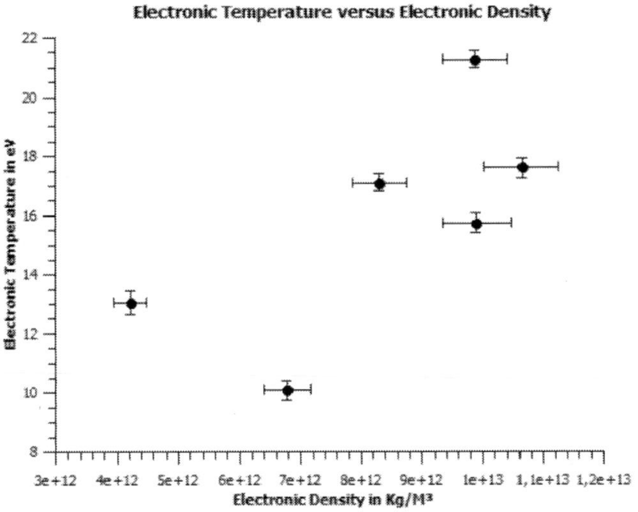

Figure 10. Electronic Temperature vs.Electronic Density.

in the RF system may also be created. Chokes for blocking higher-frequency alternating current in the electrical circuit are in construction. The magnetic coils are being reconfigured in order to increase the magnetic line expansion. A faster algorithm in LabView is being implemented in the system for more precise data analyses and treatment.

Further analysis ought to be done when the complete apparatus is settled.

7. References

[1] Jose Leonardo Ferreira, Ivan Soares Ferreira, Jean Carlo Santos, Rodrigo Miranda, Paolo Gessini, Gabriela Possa and Lui T. C. Habl, Plasma Diagnostics System Development for Permanent Magnet Hall Thrusters, IEPC-2013-428.

[2] L. R. Stenzel. "Double layer formation during current sheet disruptions in a recnnection experiment", Geophysical Research Letters, vol. 9, June 1982, pp. 680-683.

[3] Michael A. Raardu, J. Juul Rasmussen. "Dynamical aspects of electrostatic double layers", Atrophysucs and Space Science (ISSN 0004-640X), vol. 144, n. 1-2, May 1988, pp. 43-71.

Journal of Physics: Conference Series **641** (2015) 012024 doi:10.1088/1742-6596/641/1/012024

[4] W. Lennartsson. "Some aspects of double layer formation in plasma constrained by a magnetic mirror", NASA Conference Publication, no.2468. Washington, D.C.: NASA Scientific and Technical Information Branch, p. 275, 1987.
[5] Francis F. Chen. "Plasma Ionisation by Helicon Waves", Plasma Physics and Controlled Fusion, vol. 33. No 4. pp. 339 to 354, 1991.
[6] O. Sutherland, C. Charles, N. Plihon, R. W. Boswell. "Experimental evidance of a Double Layer in a large volume Helicon Reactor", Phy.Rev.Let.95, 205002 (2005).
[7] C. Chan, N. Hershkowitz. Phys. Fluids 25, 2135 (1982).
[8] David Meeker. "Finite Element Method Magnetics, Version 4.2" User's Manual. 2010.
[9] Alfred Y. Wong. "Introduction to Experimental Plasma Physics, Volume 1". Physics Department, University of California at Los Angeles. 1977.
[10] Lochte-Holtgreven, W. (editor) - Plasma Diagnostic, AIP, 1995.
[11] Charles C. and Boswell R. "Current-free double-layer formation in a high-density helicon discharge", Applied Physics Letter, Vol 82, no 9. March, 2003.

The State-of-the-art in Space Robotics

Ijar M da Fonseca[1,2] **and Maurício N Pontuschka**[3]

[1]Aeronautical Mechanics, Department. Mechatronics, ITA/DCTA, São José dos Campos, SP, Brazil.
[2] Space Mechanics and Control Division, INPE/DMC, P.O. box 515, São José dos Campos, SP, Brasil
[3] Computer Science Department, PUC-SP, São Paulo, SP, Brazil

E-mail: ijar@uol.com.br

Abstract. This paper deals with the space robotics and associate space applications. An overview of the space era and the robotic space probes is presented to contextualize the space robotics in the space exploration scenario. Concepts, classification and key-questions associated with robotics for space applications are presented and discussed. Safety-critical aspects of the space robotics are discussed as well the human limitation to operate in the hostile space environment and long time duration missions. The paper also focuses on the state-of-the-art of robotics for the International Space Station EVA operations, for the planetary exploration such as the ongoing Mars exploration, Hayabusa rendezvous and landing in asteroids and the robotic probe Rosetta landed in a comet recently. The paper also includes a discussion of the applications of new concepts like the robonauts, the space tugs applications and robots for future planetary exploration.

1. Introduction

The space robotics [1] deeply marked the beginning and the evolution of the space age with the spaceships and probes placed around the Earth and in an amazing evolution to the Moon, other planets and their moons, the Sun and finally asteroids and comets. Remarkably, the space journey reached the interstellar region with Voyager 1 and the most distant planets with Voyager 2. The robots are appropriate machines to substitute the astronauts in risky activities in the hostile space environment. The integration of artificial intelligence [2] to the robot design and development is getting more and more importance and will be a key issue for the future of the space exploration. The orbital robotics requires special attention to problems of radiation, strong temperature variation, micro-gravity environment, and magnetic fields associated to some celestial bodies. Specifically, for planetary exploration one special aspect of the robotics applications include surface exploration where the local gravity, soil characteristics, local pressure, and atmosphere must be taken into account.

Some fields of space applications require robot and astronaut to work in a micro-gravity environment where man experiences the apparent absence of weight and fluctuates. This means that any work platform is a non fixed base sustaining the man and machines while in activities in orbit. The apparent absence of weight represents a key issue to be analyzed in order to assure good and safe performance of on orbit operations, be it done by astronauts or mobile robots. Presently the International Space Station labs are playing a very important role for tests in micro-gravity environment [3-4].

The space conquer has been evolved in the scenario of the cold war having the US and the former Soviet Union as the main parts of a strong competition characterizing the space exploration as a space race. The parties did their best to prove that their political and economic regimes were the best for the world (Democracy and Capitalism & Communism and Socialism). In such scenario the Earth orbital missions, followed by missions to the Moon, Venus and Mars were developed in parallel by both the US and the former URSS, with all sort of secrets, misinformation, and spying technique. For years both countries persuaded the firsts aiming to beat each other. The first satellite, the first man in space, the first woman in space and even the first animal in orbit, the first walk in space, and so on. The results of the competition are the space exploration as we have today, characterized by the landing of man in the moon, landing o robotic spacecraft, probes, and rovers on the Moon, Venus, Mars, asteroids and comet. Scientific missions have been sent to outer space covering all distant planets and their moons. It is worth to point out the incredible missions of the Voyager 1 and 2, with the first reaching the interstellar region of space, resulting in the discovery of teens of moons in Jupiter and Saturn, new knowledge about the atmosphere and soil composition of the planets. It is worth to point the discovery of the exoplanets and so many galaxies thanks to the space telescopes (Kepler and Hubble). The Mars exploration is another extraordinary accomplishment of the man toward the planetary exploration [6-9].

The very first space robots were spacecraft and probes and the main goal of those machines was the study of the Moon surface aiming manned missions to the Earth natural satellite. The capability of those robotic spacecraft included photographing and recording the surface and sending television images to the Earth. The first spacecraft sent to the Moon had not even the capability to land in the surface. They were lunar impact spacecraft. Television images were obtained and send to the Earth while falling down on the moon. Other robotic space achievements by using probes were directed to Venus, Mars, and Mercury, Jupiter and Saturn.

2. An overview on Robotic Functionality for Space Applications

The main space robot functionalities can be classified into two categories, planetary surface exploration and in-space exploration [10]. The planetary surface operations involve surface mobility, science planning and perception [11], instrument, deployment and sample manipulation. Exploration of asteroids and comets also belongs to this category.

In-space operations involve operations functionalities like in-space assembly, in-space inspection in-space maintenance, EVA (Extra-Vehicular Activities), and in-orbit scientific experiment manipulation.

2.1. Planetary Exploration

Planetary exploration robotics applications are remarkably operating on Mars surface. Robotic space probes also have reached a comet (Rosetta, mission, ESA) and asteroids Hayabusa mission (JAXA) and Dawn spacecraft in mission at the asteroid belt (NASA) to study protoplanets Vesta and Ceres.

The state-of-the-art for deployed planetary surface exploration robots remounts the Sojourner [5], the Pathfinder Mars rover that landed on o Mars on July 4, 1997 and explored Mars for around three months. It has been followed by the Mars exploration rovers (MER), and the Martian Scientific Laboratory (MSL) Curiosity rover. The MER launched in 2003 (Opportunity and Spirit) [10] and the MSL Curiosity rover (2012) [12], are the best and most complex space robots (or a space robot system) ever constructed and in operation. Opportunity and Spirit are the NASA's twin robot geologists. Table 1 shows the scientific Spirit rover instruments and functionalities carried by both rovers [4].

Journal of Physics: Conference Series **641** (2015) 012025 doi:10.1088/1742-6596/641/1/012025

Table 1. Opportunity and Spirit rover Instruments/Functionalities [10]

Scientific Instrument	Functionality
Pancam (Panoramic camera)	For determining the mineralogy, texture and structure of the local terrain
Mini-TES (Miniature Thermal Emission Spectrometer)	For identifying promising rocks and soils for closer examination and for determining the processes that formed Martian rocks.
MB (Mössbauer Spectrometer)	For close-up investigations of the mineralogy of iron-bearing rocks and soils
APXS (Alpha Particle X-ray Spectrometer)	For close-up analysis of the abundances of elements that make up rocks and soils.
Magnet	For collecting magnetic dust particles to be analyzed by MB and APXS
MI (Microscopic Imager)	For obtaining close-up, high resolution images of rocks and soils
RAT (Rock Abrasion Tool)	For removing dust and weathered rock surfaces and exposing fresh material for examination by instruments onboard

Figure 1. Mars rovers and spacecraft landed on Mars

The Curiosity is the Mars Science Laboratory (MSL) rover, the most technologically challenging interplanetary rover ever designed. To accomplish the science objectives of the Mars Scientific Laboratory mission, the rover carries a set of payloads with instruments that are shown in table 1. The set of instruments and associate functionalities pictures the complexity of the space robot and also the high level of technology ever used before in planetary exploration. Curiosity really represents the stat-of-the-art of space robotic applications for surface exploration. Figure 2 shows the MSL and associated robotic subsystems. Table 2 shows the Curiosity rover instruments by categories and functionalities.

Journal of Physics: Conference Series **641** (2015) 012025 doi:10.1088/1742-6596/641/1/012025

Table 2 Curiosity Instruments/categories and functionalities [12]

Instrument	categories	functionalities
Mast Camera (MastCam)	Remote Sensing	The MastCam takes color images and color videos of the Martian terrain. The objectives are to study of the Martian landscape, rocks and soil, to view frost and weather phenomena and to support the driving and sampling operations of the rover.
Chemistry and Camera (ChemCam)	Remote Sensing	ChemCam fires a laser and analyzes the elemental composition of vaporized materials from areas smaller than 1 millimeter on the surface of Martian rocks and soil.
Mars Hand Lens Imager (MAHLI)	In-Situ	This system is equivalent to a geologist's hand lens. It provides the earthbound scientists with close-up views of the minerals, textures, and structure in Martian rocks and surface layers of rocky debris and dust.
Alpha-Particle X-ray Spectrometer (APXS)	In-Situ	APXS measure the chemical elements in rocks and soils.
Chemistry and Mineralogy Instrument (CheMin)	Analytical	CheMin identifies and measures various mineral on Mars. The search for minerals may lead scientists to get indication of the past Martian environment that might have supported life.
Sample Analysis at Mars (SAM)	Analytical	SAM searches for compounds of elements carbon, including methane. This element is associated with life and explores ways in which they are generated and destroyed in the Martian ecosphere.
Radiation Assessment Detector (RAD)	Environmental	RAD is one of the first instruments sent to Mars specifically to prepare for future human exploration. RAD measures and identifies all high-energy radiation on the Martian surface, such as protons, energetic ions of various elements, neutrons, and gamma rays. Also RAD includes the capability to investigate how radiation affects the chemical and isotopic composition of the Martian rocks and soil.
Mars Decent Imager (MARDI)	Environmental	The system was used to take color video during the rover descent toward the Martian surface, providing a view of the local environment as a view of an on-board astronaut. Also it includes the capability to help Earthbound planners select an optimum path of exploration and capability to provide information about the larger ecologic context surrounding the landing site.

Dynamics Albedo of Neutrons, DAN	Environmental	Cosmic rays constantly bombard the Mars surface. In this scenario if liquid or frozen water happens to be present, hydrogen atoms slow the neutrons down. The rover pulsating neutron generator called DAN that is sensitive enough to detect water content as low as one-tenth of 1 percent resolves layers of water and ice beneath the surface.
Rover Environmental Monitoring Station (REMS)	Environmental	REMS measures and provides daily and seasonal reports on atmospheric pressure, humidity, ultraviolet radiation at the Mars surface. Also it provides the wind speed and direction, air temperature and ground temperature around the rover. In addition its infrared sensors measures the infrared radiation emitted by ground.

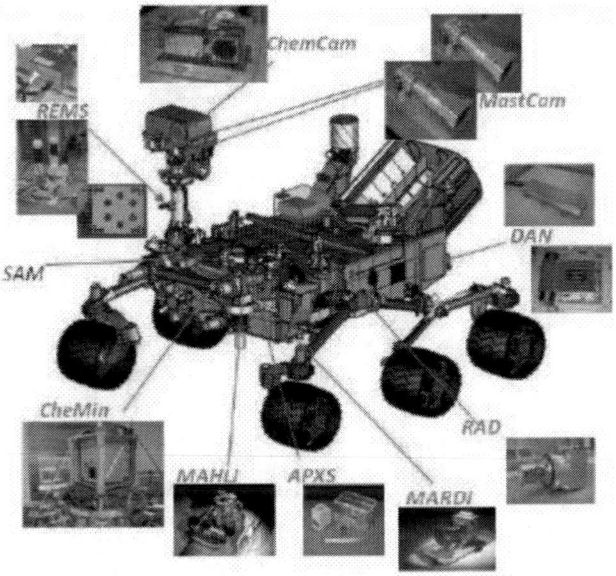

Figure 2. Curiosity Scientific Instruments

In the next 10 years it is expected that the space robots for surface exploration be not constrained anymore by navigation and mobility. Long traverses and access to most locations on a planetary surface will be possible [13]. On the other side it is expected that ground based planning and visualization tools enable scientists from ground station to interact straightforward with the robots in the surface of celestial bodies. However, robotic performance at the level of a space suited human scientist is and will continue to be a major challenge.

2.2. *In-Space Operations*
The automation by using robots in space operation got new impulse with advent of space stations and the space tugs concepts for orbital servicing. One of the most important accomplishment with an extraordinary advance in robotic was the space shuttle series of spacecraft. Space shuttle can be considered the first space tug in that it really implemented orbital servicing of grasping satellite for maintenance, maintenance and assembly for the International Space Station (ISS), and inserting new satellites into Low Earth Orbit (LEO). In 1993 astronauts based on the space shuttle grasped and

Journal of Physics: Conference Series **641** (2015) 012025 doi:10.1088/1742-6596/641/1/012025

repaired the mirrors of the Hubble Telescope. In-space operations and orbital servicing require rendezvous & docking and berthing operations [14,15].

The Space Shuttle was replaced by the ATV, the automated transfer vehicle (a robotic space tug) that executed several rendezvous & docking operation with the ISS for maintenance and scientific purposes. Presently the SpaceX Dragon spacecraft is replacing the ATV. Dragon is a partially reusable spacecraft developed by SpaceX, an American private space transportation company. Figure 3 (a) and (b) shows the ATV and Dragon spacecraft, respectively. The space shuttle cannot be considered as a robotic spacecraft as his successors. However, it carried a long robotic manipulator capable to grasp satellites to provide OOS. ATV and Dragon were/are spacecraft with large autonomy. Again, most operations include some level of teleoperation and straightforward commands from ground based control station. The teleoperation is still part of the Space Station Remote Manipulator System from the ISS, currently in orbit.

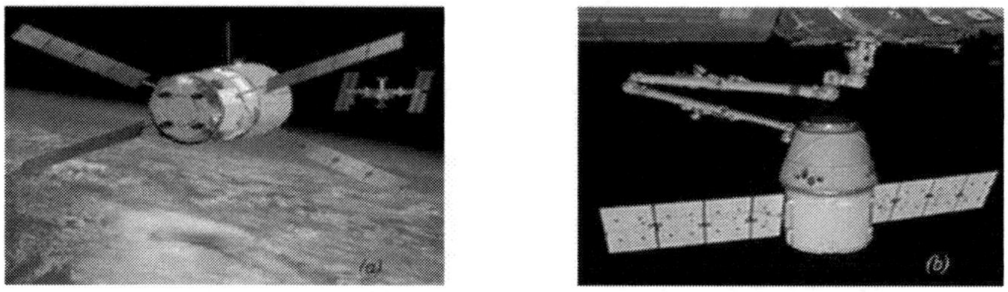

Figure 3. (a) and (b) showing the ATV and Dragon, respectively

The robonauts [16, 17] represent another state-of-the-art in the space robotics area. They resemble a human in shape and size. The hands are like human hands capable to operation the same tools human operates. Some robonauts versions include the capability to climb on the external surfaces of ISS like a spider. Some other activities are being planned to the robonauts so that it can help astronauts by manipulating task such as turn on/off systems and helping with scientific experiment manipulation. The robonauts also is being trained to execute medical task for future support to astronauts. Leap motion technology is being used to train the robonauts to grasp and hold tools and instruments.

The in-orbit inspection aims to use robots for external inspection of the space station. However, so far this kind of robotic operation has not been implemented. A test of a free-flying camera, AERCam Sprint was conducted during the STS-87 in 1997. The robot was purely teleoperated. In same line the Germans designed a robot called Inspector to implement inspection operation at MIR space station but it failed in flight. Presently tests are being conducted in the ISS with free-flying smart spheres. In a near future the in-orbit operations will be part of the space operations. The microgravity environment at ISS is appropriate to conduct the test of this kind of robots.

In-space maintenance refers to the so called OOS and the services include refueling, repairing satellites, replacing pieces, providing food and water to astronauts, correcting orbits, and in a near future, towing a spacecraft to the a space station for repairing and fixing problems and then, take it back to its orbit.

3. Conclusion

This paper presented the state-of-the art of space robots with focus on spacecraft and probes like robots, in-orbit robotic applications, and planetary surface robotic applications. It was placed more emphasis on the robotic applications for planetary exploration for it represents currently the most risky type of exploration for humans. The risky is not associated only with the hostile environment of planets like Mars, Venus, and other but also with the long duration of the space trips that involves aspects that may impact the human health to deaf and not ready investigated and clarified by the scientists. The robots represent the only alternative for those long trips and risky activities in the space environment. The space robots are designed by using ultimate technology aiming performance and autonomy. However, in terms of intelligence much is still to be done. Activities where decision taking is required, mainly those in dangerous and risky situations require still the presence of humans. The progress reached in space area is remarkable considering the beginning to space era by 1957 with the launching of the Sputnik.

References

[1] Qureshi, F 2006 *PhD thesis*, University of Toronto Intelligent perception in virtual sensor networks and space robotics
[2] *Machine Intelligence and Robotics: Report of the NASA Study Gruoup, NASA/Jet Propulsion laboratory, Final Report 715-32, March 1980*
[3] Enright, J, Saenz-Otero A, et. al., 2004 The SPHERES Guest Scientist Program: Collaborative Science On the ISS", *IEEE Aerospace Conference 2004, Big Sky, Montana*, Paper #1296, March 2004
[4] https://directory.eoportal.org/web/eoportal/satellite-missions/i/iss-spheres acces on April 30 2015
[5] Wilcox, B and Nguyen T 1997, *Sojourner on Mars and Lessons Learned for Future Planetary Rovers*, in *ICES*, 1997
[6] Mirza W. I. B 2014, Mars Planetary Rover Control Protocol, *International Journal of Multidisciplinary Sciences and Engineering*, Vol. 5, No. 2, February 2014
[7] Ari K. J, Paul H M, Kanna R John L. B, Activity Planning for the Mars Exploration Rovers, in *15th International Conference on Automated Planning and Scheduling* (ICAPS 2005), 2005.
[8] Forest F, Daniel G, Caroline C, Steve S, and Issa N T E 2002 Continuous Planning and Execution for an Autonomous Rover, in 3rd *International NASA Workshop on Planning and Scheduling for Space*, 2002
[9] Alexis O, Kostas B R L 2010 Network-Guided Multi-Robot Path Planning, in *10th International Symposium on Artificial Intelligence, Robotics and Automation in Space* (i-SAIRAS) 2010
[10] http://mars.nasa.gov/mer10/, access April 30 2015
[11] Fasisal Q and Demetri T *Intelligent perception and control for space robotics Autonomous Satellite and Rendezvous and Docking, Machine Vision and Applications*, Spring Verlag, DOI 10.1007/s00138-007-0085-z, 2007
[12] *Nasa's Management of the Mars Science Laboratory Project 2011, Office of Audits, Audit Report*, june 8 2011
[13] Liam P David K, David W, Illah. N, *Trey S NASA Exploration Team (NEXT) Space Robotics Technology Assesssment Report, NASA/TM-2004,* 2004
[14] Wertz, J and Bell, R Autonomous rendezvous and docking technologies - status and prospects. In: *SPIE's 17th Annual International Symposium on Aerospace/Defense Sensing, Simulation, and Controls*, Orlando, FL (21–25 April 2003)
[15] Polites M An assessment of the technology of automated rendezvous and capture in space. *Technical Report NASA/TP-1998-208528*, Marshall Space Flight Center, Alabama, USA (1998)
[16] Diftler M A and Robert O. A Robonaut: A Robotic Astronaut Assistant," *in Proceedings of the 6th International Symposium on Artificial Intelligence, Robotics and Automation in Space* (i-SAIRAS), 2001
[17] Staritz P, Skaff C U and Whittaker W.L., Skyworker: a robot for assembly, inspection and maintenance of large scale orbital facilities, *Proceedings of the 2001 IEEE International Conference on Robotics and Automation (ICRA '01)*, Vol. 4, May, 2001

Journal of Physics: Conference Series **641** (2015) 012026 doi:10.1088/1742-6596/641/1/012026

Study of the decay time of a CubeSat type satellite considering perturbations due to the Earth's oblateness and atmospheric drag

T P Brito[1], C C Celestino[1], R V Moraes[2]

[1] Universidade Federal do ABC - UFABC, Av. dos Estados, 5001, 09210-588, Santo André, SP, Brasil
[2] Instituto de Ciência e Tecnologia -Universidade Federal de São Paulo - UNIFESP, Rua Talim, 330, 12231-280, São José dos Campos, SP, Brasil

E-mail: claudia.celeste@ufabc.edu.br

Abstract: In this work the time required for the occurrence re-entry of a CubeSat satellite type is studied. Perturbations due to the Earth's oblateness and atmospheric drag are considered. The main objective is to study the initial conditions to determine the maximum length of stay in orbit of the CubeSat satellite type given to its mass and its altitude using numerical simulations of the motion equations and an analytical solution was proposed.

1. Introduction

The first CubeSat satellites were developed at the end of the nineties of the 20th century at the California Polytechnic State University in partnership with the Stanford University's Space Systems Development Laboratory. The CubeSat satellites are based on cubic modules with 100 mm edges, known as unit, or simply U. These satellites have been widely used to sending and receiving data.

In this work, it is performed an numerical study about the effects of atmospheric drag and of the Earth's oblateness on the orbit of CubeSat satellites type aiming predict the time required for the occurrence of its re-entry.

The protection of the Earth's space environment is producing a revival of the interest in solutions to the artificial satellite problem and new analytical and semi-analytical tools for the analysis and design of end-of-life disposal strategies have been proposed [1], [2], [3], [4]. One of the main problems of these theories is that the improvements of the analytical modelling for the atmospheric drag endanger analytical solutions for the equations involved in.

The equations of motion used here are written in Cartesian coordinates, taking into account perturbations due to harmonics factored by J_2 and to the atmospheric drag [5]. The numerical integrations in the present paper were made using a Legendre-Gauss-Radau – LGR [6] integrator of fourth order and step size control, with a numerical accuracy of 10^{-8} and were considered two models for the atmospheric density: the TD-88 model [7] and NASA model [8]. The TD-88 analytical termospheric density model is easy to be implemented and agree quite well with some spread used atmospheric model [9]. For the perturbations due to the oblateness, Brouwer theory was used [10]. Initial conditions were considered depending on some usual physical and orbital characteristics (weight of the satellite, semi-major axis, eccentricity and inclination of the orbit), for a CubeSat remain in orbit for a period not longer than 25 years.

2. The TD-88 and NASA atmospheric density models

In what follows the atmospheric density models TD -88 [7] and NASA [8] are briefly described below:

2.1. TD-88

Following [7], the density can be expressed as

$$\rho = f_x f_0 k_0 + \sum_{n=1}^{7} h_n \, g_n \tag{1}$$

where,

$$f_x = 1 + a_1(F_x - F_b),$$
$$f_0 = a_2 + F_n,$$
$$k_0 = 1 + a_3(k_p - 3),$$
$$f_n = \frac{(F_b - 60)}{160},$$
$$h_n = h_{n,0} + \sum_{j=0}^{3} k_{n,j} \, exp^{\frac{(120-h)}{29j}},$$

$$g_1 = 1,$$
$$g_2 = \frac{F_n}{2+a_4},$$
$$g_3 = sen(d - p_3) sen\varphi,$$
$$g_4 = (a_5 f_n + 1) sen(d - p_4),$$
$$g_5 = (a_6 f_n + 1) sen2(d - p_5),$$
$$g_6 = (a_7 f_n + 1) sen(t - p_6) cos\varphi,$$
$$g_7 = (a_8 f_n + 1) sen2(t - p_7) cos^2\varphi$$

where: F_x is the solar flux measured on 10.7 cm/day; F_b is mean solar flux; h is the altitude; a_i (i=1,2,…8) and p_i (i=3,…7) are numerical constants; k_p is the magnetic index; d is the day count of the year; t is the local time and φ is the latitude.

2.2. NASA model

The NASA atmospheric density is given [8],

$$\rho = \frac{P}{(0.2869*(T+27301))} \tag{2}$$

Where the temperature and pressure are given by,

$$T = -131.21 + 0.00299h \qquad P = 2.488\left(\frac{T+273.1}{216.6}\right)^{-11.388}$$

2.3. Comparison of TD-88 model and the NASA model for given initial conditions

Figure 1 shows values for the atmospheric density varying with the altitudes from 200 km to 700 km considering: the TD-88 and NASA models. The conditions used in the TD-88 model were d = 80 current day, $f_x = f_0$ = 150 W, k_0 = 4 (dimensionless), t = 3 h and φ = 0 °. Note that for h = 200 km the TD-88 model has the maximum density value $\rho = 4.50 \times 10^{-10}$ kg/m^3 while for the NASA model $\rho = 9.85 \times 10^{-9}$ kg/m^3. The minimum density occurs at h = 700 km: $\rho = 6.94 \times 10^{-14}$ kg/m^3 for the TD-88 model and $\rho = 1.11 \times 10^{-14}$ kg/m^3 for the NASA model.

3. Decay of a satellite type CubeSat

The equations of motion were implemented in a FORTRAN code program using the numerical integrator Radau [6] to calculate the orbital decay. The decay time was computed using values of position and velocity obtained from the numerical integration and substituting these values in the equation for the semi-major axis updated at the time step of the numerical integration which was 10^{-8}. The used program stopping criterion was when the orbital radius of the CubeSat is less than or equal to the orbital radius of the Earth.

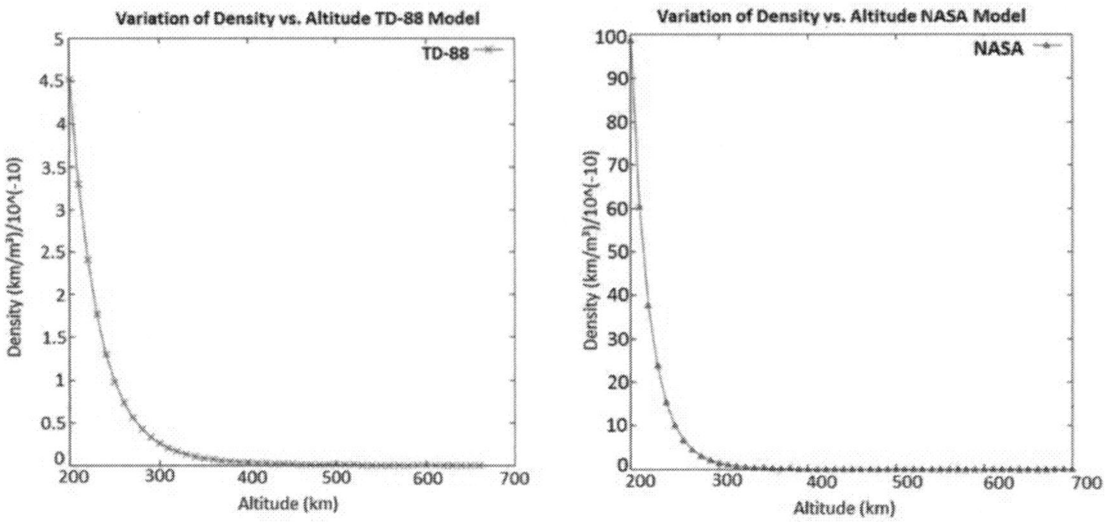

Figure 1. Atmospheric density with altitude considering: the model TD-88 and the NASA model.

3.1 Decay of a CubeSat satellite considering the atmospheric drag and Earth' oblateness perturbations

Considering some scenarios taking into account different values for the initial conditions - day of the year, local time, solar flux, magnetic index, and latitudes - some information should be placed highlighted, as follows:

- Worst case: the longer decay time. The conditions for the worst case were: day of the year, d = 172, local time, t = 3h, solar flux measured in 10.7 cm/day, f_x= 150 W, solar flux obtained after three rotations, f_0 = 150 W, geomagnetic index location, k_0 = 4 (dimensionless) and latitude, $\varphi = 0°$.
- Best case: the lowest decay time. The original terms of the best case were: day of the year, d = 80, local time, t = 3 h, solar flux measured in 10.7 cm/day, f_x = 150 W, solar flux obtained after three rotations, f_0 = 150 W, geomagnetic index location, k_0 = 4 (dimensionless) and latitude, $\varphi = 0$ °.

3.2. Atmospheric drag with density model TG-88, and Earth's oblateness

Simulations, considering the worst-case scenario for the decay period of a CubeSat type satellite considering the Earth's oblateness and atmospheric drag perturbation with atmospheric density model TD-88, are show in table 1. The results are presented according to the type of orbit, the mass of satellite and the altitude. The orbital decay period is given in years. For a satellite with a mass of 5 kg, the only altitudes that respect the European Code of mitigation are 500 km and 550 km for circular orbits, and having 7.44 and 15.81 years to decay, respectively. For a satellite with a mass of 10 kg, the only altitude respecting the European Code of mitigation is 500 km, where the decay period is 14.87 years. For all other initial conditions addressed, the decay period is longer than the 25 years stipulated in the European Code of mitigation. The following were the initial conditions used for this situation: day of the year, d = 172, local time, t =3 h, solar flux measured in 10.7 cm/day, f_x = 150 W, solar flux obtained after three rotations, f_0 = 150 W, geomagnetic index site, k_0 = 4 (dimensionless) and latitude $\varphi = 0$ °.

In table 2 contains the results considering the best setting with density model TD-88. For satellites of 5 kg, all altitudes in the range from 500 to 650 km will comply with the European Code of spatial mitigation, because none of these conditions exceeds the time limit in orbit around the Earth that is 25 years. For satellites of 10 kg, the only altitudes respecting the European Code of mitigation are between 500 and 550 km, where the decay periods are 6.13 and 12.70 years, respectively.

Table 1. Worst-case scenario for the decay period considering density model TD-88.

Orbit	Mass (kg)	Altitude (km)	Time to decay (years)
Circular and inclination of the 30 °	5	500	7.44
		550	15.81
		600	31.75
		650	61.38
	10	500	14.87
		550	31.63
		600	63.49
		650	122.76

Table 2. Best initial conditions setting for the orbital decay period considering density model TD-88.

Orbit	Mass (kg)	Altitude (km)	Time to decay (years)
Circular and inclination of the 30 °	5	500	3.07
		550	6.35
		600	12.58
		650	24.15
	10	500	6.13
		550	12.70
		600	25.15
		650	48.29

The initial conditions used in this situation were: day of the year, $d = 80$, local time, $t = 3$ h, solar flux measured in 10.7cm/day, $f_x = 150$ W, solar flux obtained after three rotations, $f_0 = 150$ W, geomagnetic index site, $k_0 = 4$ (dimensionless) and latitude $\varphi = 0$ °.

After several numerical simulations it could be observed the effects of the inclusion of the perturbation due to the Earth's oblateness in the equations of motion are negligible for the decay time.

3.3. Atmospheric drag and Earth' oblateness perturbations and atmospheric density model NASA

In table 3 is showed of the orbital decay time of a satellite type CubeSat considering the Earth's oblateness and atmospheric drag perturbations and atmospheric density model NASA. The results are presented according to the type of orbit, the satellite mass, the altitude at which it is. The orbital decay period is given in years. For a mass of 5 kg satellite, only the first two initial conditions satisfy the European Code of spatial mitigation in this case the altitudes of 500 and 550 km altitude, where the time for orbital decay is 3.06 and 9.83 years, respectively. For a 10 kg mass satellite, the only altitudes that respect the European Code of mitigation are 500 and 550 km, where the decay period is of 6.12 and 19.66 years, respectively.

4. Analytical solution

Using the Delaunay variables L_i and l_i (i= 1, 2, 3), setting P_i and Q_i as the components of the atmospheric drag, the secular perturbations in the orbital elements of an artificial Earth satellite taking into account the influence of the atmospheric drag, can be obtained solving the system [11]:

$$\dot{L}_i = \langle P_i \rangle \,;\; \dot{l}_i = -\langle {\partial H}/{\partial L_i} \rangle - \langle Q_i \rangle, \quad (i = 1,2,3) \tag{3}$$

where H is the Hamiltonian of the drag free problem. The symbol $\langle f \rangle$ means the non-periodic part of f and the ballistic coefficient is considered constant.

Table 3. The orbital decay time considering the Earth's oblateness and atmospheric drag perturbations with atmospheric density model NASA.

Orbit	Mass (kg)	Altitude (km)	Time to decay (year)
		500	3.06
		550	9.83
Circular	5	600	28.74
with		650	77.55
30 °		500	6.12
	10	550	19.66
		600	57.48
		650	155.09

After some algebraic manipulation, the TD-88 atmospheric model can be expressed in terms of the mean eccentric anomaly and we get [11]:

$$\rho = f_\chi f_0 k_0 \sum_{n=1}^{7} [k_{n,0} + \sum_{j=1}^{3} A_j \, exp^{[c_j \cos(\omega + v)]} exp^{[\beta_j \cos E]}] g_n \tag{4}$$

where

$$c_j = \frac{R\, \varepsilon sin^2 i}{2.29/_j}, \qquad \beta_j = \frac{ae}{2.29/_j}, \qquad \text{and} \quad A_j = exp\left[\frac{\left(120 + R\left(1 - \frac{\varepsilon}{2 sin^2 i}\right) - a\right) 29}{j}\right]$$

Here, R is the mean Earth's equatorial radius, ε is flattening of the Earth, a, e, i, ω, v, and E are the semi major axis, eccentricity, inclination, argument of perigee, true anomaly and eccentric anomaly, respectively.

Using the Cayley tables to express the eccentric anomaly in terms of the mean anomaly, the equations of motion can be easily solved and the decay of the semi major axis per revolution is

$$\Delta a = -2\pi a^2 \left(\frac{SC_D}{m}\right) f_\chi f_0 k_0 \sum_{n=1}^{7} [G_{n,0} \, \{k_{n,0} + \sum_{j=1}^{7} k_{n,j} A_j \, [\left(1 + \frac{c_j^2}{4}\right) + \frac{c_j^2}{4} A_{0,j}]\} \tag{5}$$

where: SC_D is the ballistic coefficient. Explicit analytical expressions for $G_{n,0}$, $k_{n,0}$ and $A_{0,j}$ as given in [11] were replaced in the above equation.

5. Conclusion

Two models namely, TD-88 and NASA, were used here to compute the atmospheric density. An analytical expression was exhibited and here adapted to be used for the first time to compute the decay of a CubeSat type satellite. According to the limitation of 25 years imposed by the European Code of spatial mitigation, considering the TD-88 model and a satellite in a circular orbit with inclination of 30° and mass of 5 kg, the maximum altitude at which this satellite can be allocated is around 550 km, where the decay time is about 15.81 years. For the same satellite considering the model of NASA, the maximum altitude is also around 550 km, but the decay period is even higher, 22.95 years. For higher altitudes the length of stay in orbit around the Earth goes beyond the stipulated by the European Code of spatial mitigation. For satellites in circular orbit with an inclination of 30 ° and mass of 10 kg for both models the maximum altitude to a minimum time for decay is about 500 km, but there is great variation in the orbital decay period in this case: 14.87 years for TD-88 model and 6.12 years for the NASA model. It is worthwhile to recognize that since the TD-88 model was inserted in the proposed analytical solution, this solution agrees better with the numerical integration of the full equations using this density model.

Acknowledgements
This work was partially supported by FAPESP under contracts 2013/24772-2 and 2012/21023-6.

References

[1] Colombo C 2014 Semi-analytical propagation for highly elliptical orbits to build maps for end-of-life disposal *Proceedings of Key Topics in Orbit Propagation Applied to SSA KePASSA*

[2] Fraysse H C, Fevre Le, Morand V, Deleflie F, Mercier P, and Dental C 2012 STELA a tool for long term orbit propagation *Proceedings of the 5th International Conference on Astrodynamics Tools and Techniques* (ESTEC/ESA Netherlands)

[3] Di Mauro G, Rasotto M, Armellin R, Di Lizia P, Madakashira H K, Lara M, and San-Juan J F 2013 End-of-Life Disposal Concepts for Lagrange-Points and HEO Missions *Final Report General Studies Programme* (ESA ITT AO/1-7210/12/F/MOS, ESA)

[4] Golikov A and *Theona* R *2012* A numerical-analytical theory of motion of artificial satellites of celestial bodies *Cosmic Research*, **50** 6 pp 449–458 10.1134/S0010952512060020

[5] Roy A E 1988 Orbital motion ed 3 (Bristol, England UK)

[6] Everhart E 1985 An efficient integrator that uses Gauss-Radau spacing In dynamics of comets: their origin and evolution *Proceedings of IAU Colloquium 83* **115** p 185

[7] Sehnal L and Pospísilová L 1988 Termospheric "Model TD 88" *Astronon- Inst. Czechosl Asad. Sci. Ondrojov* Preprint **67**

[8] NASA 2014 Atmosphere model found in: <http://www.grc.nasa.gov/WWW/k-12/airplane/atmosmet.html>. Accessed in: 01 Apr. 2014

[9] Vilhena de Moraes R and Nicoletti G W 2001 Uberlândia MG Brazil An analytical approach or evaluation of orbital perturbations due to atmospheric drag *XVI Congresso Brasileiro de Engenharia Mecânica-COBEM* pp 191-196

[10] Brouwer D 1959 Solution of the problem of artificial satellite theory without drag *The Astronomical Journal* **64** pp 378–397

[11] Vilhena de Moraes R 1989 Analytical computation of drag perturbations using thermospheric model TD 88 in orbital dynamics of natural and artificial objects eds R Vieira Martins, D Lazzaro and W Sessin (O. N. Rio de Janeiro) pp 135-144

Gravitational disturbances generated by the Sun, Phobos and Deimos in orbital maneuvers around Mars with automatic correction of the semi-major axis

E M Rocco

National Institute for Space Research – INPE, Space Mechanics and Control Division, São José dos Campos, SP-Brazil

E-mail: evandro.rocco@inpe.br

Abstract. The objective of this work is to analyze orbital maneuvers of a spacecraft orbiting Mars, considering disturbance effects due to the gravitational attraction of the Sun, Phobos and Deimos, beyond the disturbances due to the gravitational potential of Mars. To simulate the trajectory, constructive aspects of the propulsion system were considered. Initially ideal thrusters, capable of applying infinite magnitude of the thrust, were used. Thus, impulsive optimal maneuvers were obtained by scanning the solutions of the Lambert's problem in order to select the maneuver of minimum fuel consumption. Due to the impossibility of applying an impulse, the orbital maneuver must be distributed in a propulsive arc around the position of the impulse given by the solution of the Lambert's problem. However the effect of the propulsive arc is not exactly equivalent to the application of an impulse due to the errors in magnitude and direction of applied thrust. Therefore, the influence of the thrusters' capacity in the trajectory was evaluated for a more realistic model instead of the ideal case represented by the impulsive approach. Beyond the evaluation of the deviation in the orbital path, was considered an automatic correction of the semi-major axis using continuous low thrust controlled in closed loop to minimize the error in the trajectory after the application of the main thrust.

1. Introduction

This study aims to evaluate the influence of the gravitational attraction of the Sun, Phobos and Deimos, beyond the disturbances due to the gravitational potential of Mars, during orbital maneuvers of a spacecraft orbiting Mars. Some constructive aspects of the propulsion system were considered. Initially ideal thrusters, capable of applying infinite magnitude of the thrust, were used. Thus, impulsive optimal maneuvers were obtained by scanning the solutions of the Lambert's problem (Two Point Boundary Value Problem) for various values of transfer time in order to select the maneuver of minimum fuel consumption [1]. Then, the selected maneuver was simulated considering a more realistic model of the propulsion system. In fact is not possible to accomplish an impulsive maneuver because to perform this kind of maneuver it would be necessary an infinite capacity for the thrusters, because the entire velocity change of the spacecraft should occur instantly. Thus, the orbital maneuver must be distributed in a propulsive arc around the position of the impulse given by the solution of the Lambert's problem. In this arc was used a continuous thrust, limited to the capacity of the thrusters. However the effect of the propulsive arc is not exactly equivalent to the application of an impulse due to the errors in magnitude and direction of applied thrust. The difference between these approaches produces a deviation in the trajectory. The evaluation of deviation is extremely relevant in the mission

Journal of Physics: Conference Series **641** (2015) 012027 doi:10.1088/1742-6596/641/1/012027

analysis and spacecraft design of the trajectory control system. Therefore, the influence of the capacity of thrusters in the trajectory was evaluated for a more realistic model instead of the ideal case represented by the impulsive approach. So initially the bi-impulsive maneuver, which consists of finding the transfer orbit that connects a point on the initial orbit to another point in the final orbit spending a certain amount of time, is accomplished. An algorithm for solving this problem by universal variables was used [1] [2]. Then, the optimum maneuver is selected and simulated using the Spacecraft Trajectory Simulator (STRS) [3] [4] [5]. In the STRS simulation environment, the orbital movement is obtained by the solution of Kepler's equation for each simulation step. Thus, given an initial state, the Keplerian elements are obtained and propagated to the next step, to be converted into the new state. In the STRS simulator, the reference state is defined by guidance subsystem providing the ideal trajectory to be followed, according to the solution of the Lambert's problem. This reference is continuously compared with the current position of the vehicle generating an error signal, which is inserted into porporcional-integral-derivative controller, generating a signal capable of reducing errors in transition and stationary regimes. This signal is sent to the actuators to generate a signal to be applied in the dynamics model of the movement, added to the disturbing signal due to the gravitational forces of the Sun, Phobos and Deimos. Therefore, the evolution of the vehicle can be analyzed.

2. Orbital perturbation

The evaluation of the errors in the trajectory becomes even more relevant if the effects of orbital perturbations acting on the vehicle were considered. Thus, deviations in the trajectory due to disturbances caused by the Mars gravitational potential and the gravitational attractions of the Sun, Phobos and Deimos were included in the simulations, since all these perturbations models are available in the orbital simulator STRS. The mathematical models of these perturbations in the motion of a spacecraft can be found in [6] and [7]. A short discussion of the environmental perturbations considered in this work is presented below.

2.1. Mars gravitational potential

Considering two particles of mass m and M ($M >>> m$) separated by a distance r, the acceleration $\vec{a}_m$ of the mass m in relation to the center of mass of the particles, is given by:

$$\vec{a}_m = -\frac{GM}{r^3}\vec{r} \tag{1}$$

where G is the universal gravitational constant (G = $6{,}6726 \pm 0{,}0005 \times 10^{-11}$ m^3/kg s^2), $\vec{r}$ is the vector connecting the two particles, $\vec{a}_m$ is the acceleration vector in the direction of $\vec{r}$. The vector $\vec{a}_m$ can be obtained by the gradient of the gravitational potential U.

$$\vec{a}_m = \vec{\nabla} U\ ; \quad U = \frac{GM}{r} = \frac{\mu}{r}; \quad F = \frac{\partial U}{\partial r} = -\frac{\mu}{r^2} \tag{2}$$

From the second derivative of U with respect to the coordinates x, y and z the Laplace equation can be obtained.

$$\nabla^2 U = \frac{\partial^2 U}{\partial x^2} + \frac{\partial^2 U}{\partial y^2} + \frac{\partial^2 U}{\partial z^2} \quad \Rightarrow \quad \nabla^2 U = \mu\left[-\frac{3}{r^3} + \frac{3\left(x^2 + y^2 + z^2\right)}{r^5}\right] = 0 \tag{3}$$

Using spherical coordinates, the solution of the Laplace's equation is given by Eq. (4):

$$U = \sum_{n=0}^{\infty}\sum_{m=0}^{n} \frac{1}{r^{n+1}} P_{nm}(\operatorname{sen}\phi)\begin{bmatrix} C_{nm}\cos(m\lambda) + \\ + S_{nm}\operatorname{sen}(m\lambda)\end{bmatrix}; \quad P_{nm}(s) = \left(1 - s^2\right)^{m/2} \frac{d^m}{d s^m} P_n(s); \quad P_{nm}(s) = 0 \text{ for } m > n \tag{4}$$

$$P_n(s) = \frac{1}{2^n}\sum_{j=0}^{N} \frac{(-1)^j (2n-2j)!\, s^{n-2j}}{j!(n-j)!(n-2j)!}; \quad N = \frac{n}{2} \text{ for even } n; \quad N = \frac{n-1}{2} \text{ for odd } n; \quad n = 0, 1, 2, \ldots \tag{5}$$

where C_{nm}, S_{nm} are constant coefficients that represent the spherical harmonics; ϕ is the Mars-centric latitude; λ is the Mars-centric longitude; $P_{nm}(s)$ are the Legendre associated functions for degree n and order m; $P_n(s)$ are the Legendre polynomials and $s = \text{sen}\,\phi$. In this work was used the Goddard Mars model 2B. This gravitational model has been developed from data of NASA's missions [8] [9]. The maximum degree and order capable to be considered by this model is 80.

2.2. Gravitational attractions of the Sun, Phobos and Deimos

The gravitational potential due to the presence of a third celestial body is given by Eq. (6) [6]:

$$F' = \left(\frac{\mu'}{r'}\right)\left[1 + \sum_{n=2}^{\infty}\left(\frac{r}{r'}\right)^n P_n \cos\psi\right] \tag{6}$$

where: μ' is obtained by the product of the gravitational constant and the mass m' of the third body; r' is the absolute position vector of the third body related to the center of Mars; ψ is the angle between the position vector of the satellite related to the Mars $\vec{r}$ and the position vector of the satellite related to the third body; r is the absolute position vector of the satellite related to the Mars. Basically the gravitational perturbations produce variations in the right ascension of the ascending node and variations in the argument of perigee. According to [10] the general problem of three bodies provides a simple way to calculate the disturbing accelerations due to the gravitational attraction of the bodies, obtained from the Newton's law of gravitation:

$$\ddot{\vec{r}}_1 = -G\,m_2 \frac{\vec{r}_1 - \vec{r}_2}{|\vec{r}_1 - \vec{r}_2|^3} + G\,m_3 \frac{\vec{r}_3 - \vec{r}_1}{|\vec{r}_3 - \vec{r}_1|^3};\quad \ddot{\vec{r}}_2 = -G\,m_3 \frac{\vec{r}_2 - \vec{r}_3}{|\vec{r}_2 - \vec{r}_3|^3} + G\,m_1 \frac{\vec{r}_1 - \vec{r}_2}{|\vec{r}_1 - \vec{r}_2|^3};\quad \ddot{\vec{r}}_3 = -G\,m_1 \frac{\vec{r}_3 - \vec{r}_1}{|\vec{r}_3 - \vec{r}_1|^3} + G\,m_2 \frac{\vec{r}_2 - \vec{r}_3}{|\vec{r}_2 - \vec{r}_3|^3} \tag{7}$$

where $\vec{r}_1$, $\vec{r}_2$ and $\vec{r}_3$ are the positions of the bodies, m_1, and m_3 the masses of the bodies.

3. Results

Due to the impossibility of application of an infinite thrust the maneuver must be distributed in a propulsive arc around the position of the impulse. In this propulsive arc continuous thrust is applied, limited to the capacity of the thrusters. However, the effect is not equivalent to the application of an impulse. According to [6] the velocity increment applied to the spacecraft is given by Eq. (8):

$$\Delta V = g_0 I_{sp} \ln\frac{m_i}{m_f} - \int_{t_1}^{t_2} g \sin\gamma \, dt \tag{8}$$

where g_0 is the gravitational constant at planet's surface and g at the spacecraft altitude, I_{sp} is the specific impulse, γ is the flight angle formed between the direction of the velocity and the thruster pointing direction, m_i is the mass of the vehicle and m_f the final mass. From Eq. (8) it is verified that the distribution of the maneuver in an arc produces a reduction in the applied velocity increment if the direction of application of the thrust remain constant, since in this case the angle γ is different than zero. To minimize this effect the flight angle could be maintained near to zero by controlling the pointing direction of the thruster, ensuring that the thrust is always applied in the direction tangential to the path. However, this solution is more complex because it requires the use of the attitude control system, and in fact, this approach is not capable to reduce the error in the final orbit reached by the vehicle. To optimize the propulsive maneuvers distributed in arcs, it must be considered the optimization procedure for orbital maneuvers with continuous thrust: [11]. However the optimization of maneuvers with continuous thrust is a difficult task, since in most of the cases it requires numerical methods and the definition of initial values for obtaining the solution. Thus, a simple possibility to minimize the effect caused by the error in the thrust direction would be executing the maneuver in several stages, each one applying just a fraction of the total velocity increment, to reduce each propulsive arc and thereby reduce the angle γ. In this case the error in the trajectory is minimized but the total time spent to reach the final orbit is maximized, characterizing a problem of multi-objective optimization with conflicting objectives. Some studies of multi-objective problems applied to orbital trajectories can be found in [12]. Another possibility to minimize the error in the trajectory after the application of the main thrust could be the use of an automatic correction of the orbital elements using continuous low thrust controlled in closed loop, as presented by [7]. In this approach, the final orbital elements are defined, then control loops for each of these elements determine, at each step of the simulation, the magnitudes and directions of application of thrust necessary to reach the desired orbital elements. The variations of elements occur gradually until the difference between the current and reference signals do not generate errors. Then, the propulsion system is turned off automatically. To illustrate the necessity of split the maneuver in several propulsive arcs, or use the automatic correction approach, six cases are presented: case 1 for impulsive approach, with the thrust applied during 1s; cases 2 to 6 for maneuvers using thruster with capacity of 150 N to 350 N, with and without the automatic correction. The initial orbit is defined by the following keplerian elements: a = 4000 km, e = 0.001; i = 10°; Ω = 55°; ω = 25°. The final orbit is defined by: a = 5000 km, e = 0.001; i = 10°; Ω = 55°; ω = 25°. The time interval necessary to perform the transfer maneuver and the necessary velocity increment, were determined by the Lambert's problem (case 1): t = 4583 s; Δv_1 = 176.988 m/s ($\Delta\vec{v}_1 = -175.174\hat{i} - 0.226\hat{j} + 25.279\hat{k}$); Δv_2 = 167.368 m/s ($\Delta\vec{v}_2 = 165.656\hat{i} + 0.525\hat{j} - 23.874\hat{k}$).

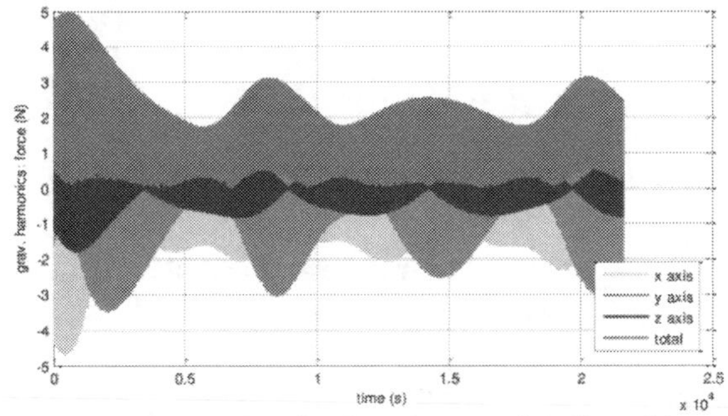

Figure 1. Mars gravitational potential disturbing applied during the maneuver

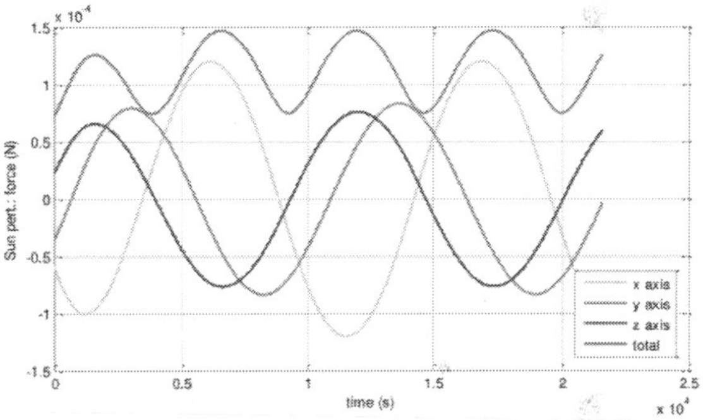

Figure 2. Sun gravitational disturbing applied during the maneuver

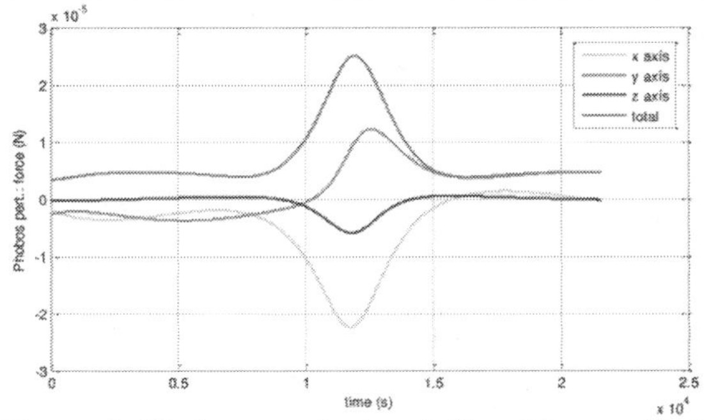

Figure 3. Phobos gravitational disturbing applied during the maneuver

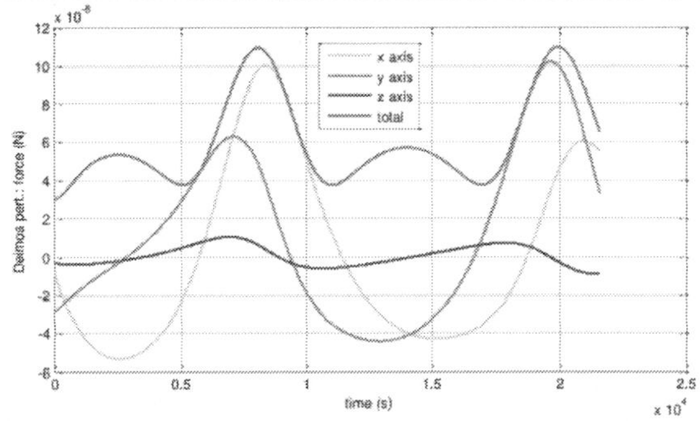

Figure 4. Deimos gravitational disturbing applied during the maneuver

The gravitational disturbing forces acting on the spacecraft during the maneuver are shown in Figs. 1 to 4. It can be verified that the major gravitational disturbing force is produced by the Mars gravitational potential. The influence of the Phobos and Deimos is not so relevant due to the small mass of the Mars's satellites. Although the influence of Phobos could be relevant if the spacecraft trajectory become too close of Phobos, since the orbit of Phobos is not too high.

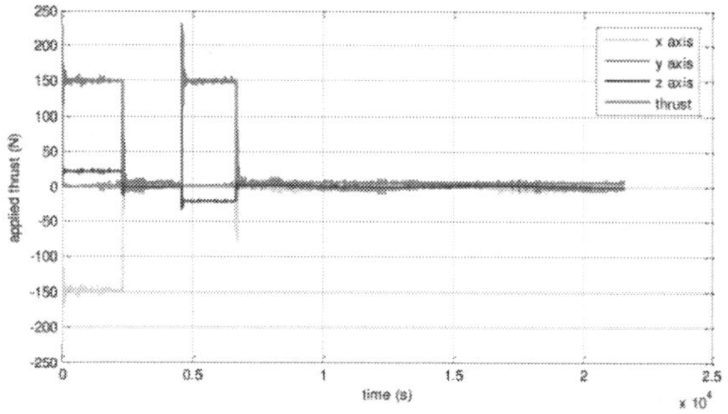

Figure 5. Case 2 (150N): applied thrust with automatic correction

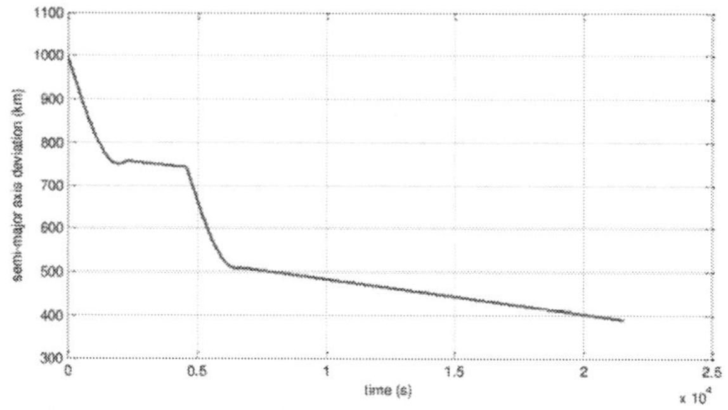

Figure 6. Case 2 (150N): semi-major axis error with automatic correction

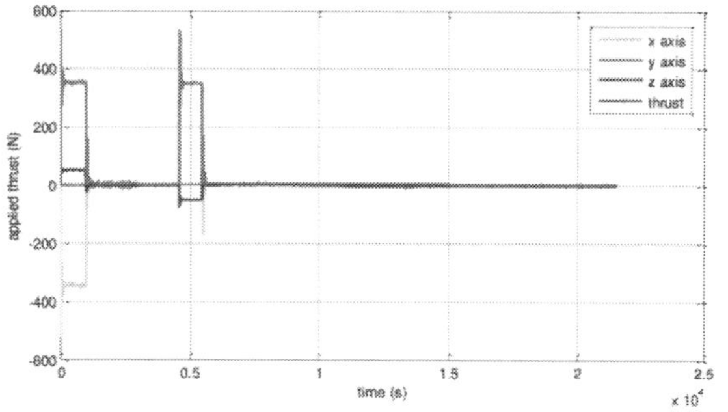

Figure 7. Case 6 (350N): applied thrust with automatic correction

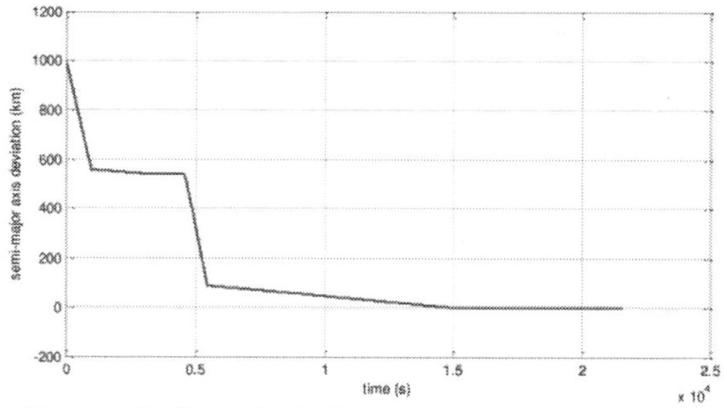

Figure 8. Case 6 (350N): semi-major axis error with automatic correction

The cases 1 to 6, taken into account several thrust capacities, some of them were shown in the previous figures. The error and the velocity increment for all cases are summarized in Table 1.

Table 1. Semi-major axis error and total applied velocity increment

	Semi-Major Axis Error (km)		Total Velocity Increment (m/s)	
Thrust Capacity	without automatic correction	with automatic correction	without automatic correction	with automatic correction
Case 1: Impulsive	0.00597855	0.00368516	353.516082	353.294809
Case 2: 150 N	526.228288	389.124696	359.442515	395.694899
Case 3: 200 N	316.722898	165.359838	359.130541	398.331927
Case 4: 250 N	207.462009	47.949564	361.249271	400.715093
Case 5: 300 N	145.138706	0.0001561	361.021206	397.495559
Case 6: 350 N	106.691771	0.0024326	363.167535	388.512041

When considered the application of the impulses, in case 1, all the velocity increment and all change in the semi-major axis are performed instantly. Therefore the applied thrust is in fact a pulse, however the continuous thrust was applied to counteract the disturbances. This continuous thrust was applied because the STRS was adjusted to control the trajectory, reducing the effect produced by the disturbances, but the main orbital maneuver in case 1 was performed in an impulsive way. From Table 1 can be verified that this case produces a small error in the final semi-major axis of the orbit. Both results, considering automatic correction and without automatic correction, presented very small errors, around 5 meters. Cases 2 to 4 in general presented results with bigger error. In these cases the change in the semi-major axis occurs gradually. When using the automatic correction the error was continually reduced after the application of the main thrust, nevertheless the amount of time available to correct the trajectory was not enough, since the thrust capacity adjusted for the automatic correction was 5N. With more time or with higher thrust capacity the results obtained for cases 2 to 4 could be improved. Thus, in a mission analysis and in the dimensioning of a spacecraft should be considered all possible combinations of the main thrust and the thrust capacity used in the automatic correction, to choose the best solution for the mission. In this study the best result was obtained in case 5 considering the main capacity of 300N. Without the automatic correction, the semi-major axis error was 145.13 km. However with the automatic correction the error was reduced to 15.61 cm. This result is much better than the results obtained in the other cases, even considering the impulsive approach. The performance of the automatic correction was able to correct the errors generated by the fact that the main thrust was distributed in a propulsive arc instead of an impulse applied instantly. With this new procedure was possible to eliminate almost all deviations related to the reference trajectory, reducing the errors to the same scale of the errors generated by the impulsive approach.

4. Conclusion

This work represents an effort in the design of the propulsion system of space vehicles. It was confirmed that the effect of a propulsive arc is not exactly equivalent to the application of an impulse. The difference produces a deviation of the final orbit relative to the reference orbit. This deviation depends on the magnitude of the impulse required for the maneuver, the maximum capacity of the propulsion system and the characteristics of the trajectory control system. Thus, the evaluation of the trajectory deviations is relevant to the analysis of a space mission and in the design of the trajectory control system if a more realistic model is considered instead of the ideal case represented by the impulsive approach. Beyond this, the STRS with the automatic correction of the orbital elements, which utilizes an innovative technology to reduce the errors in the spacecraft path, controlling the continuous thrust in closed loop, was tested with success in the presence of some perturbation forces.

5. References

[1] Battin R H 1999 *An Introduction to the Mathematics and Methods of astrodynamics, Rev. ed.* (Reston: AIAA Educational Series)

[2] Bond V R, Allman M C 1996 *Modern Astrodynamics: fundamentals and perturbation methods* (New Jersey: Princeton University Press)

[3] Rocco E M 2008 Perturbed Orbital Motion with a PID Control System for the Trajectory *XIV Colóquio Brasileiro de Dinâmica Orbital* Águas de Lindóia, November 17 - 21

[4] Rocco E M, Marcelino E W, Prado A F B A, Kuga H K 2010 Closed Loop Control System Applied to Earth-Moon Transfer Maneuvers Using Continuous Thrust *38th Scientific Assembly of the Committee on Space Research* Bremen, Germany, July 18-25

[5] Rocco E. M 2013 Automatic correction of orbital elements using continuous thrust controlled in closed loop *Journal of Physics: Conference Series* **465** doi:10.1088/1742-6596/465/1/012007

[6] Chobotov V A 1991 *Orbital Mechanics* (Washington: American Institute of Aeronautics and Astronautics, Inc.)

[7] Kaula W M 1966 *Theory of Satellite Geodesy Applications of Satellites to Geodesyc* (New York: Blaisdell Pub. Co.)

[8] Lemoine F G, Smith D E, Rowlands D D, Zuber M T, Neumann G A, Chinn D S 2001 An improved solution of the gravity field of Mars (GMM-2B) from Mars Global Surveyor *J. Geophys. Res.* **106** 23359-76

[9] Zuber M T, Lemoine F G, Smith D E, Konopliv A S, Smrekar S E, Asmar S W 2007 Mars Reconnaissance Orbiter Radio Science Gravity Investigation, *J. Geophys. Res.* **112** E05S07, doi:10.1029/2006JE002833

[10] Szebehely V 1967 *Theory of orbits* (New York: Academic Press Inc.)

[11] Edelbanum T N 1961 Propulsion Requirements for Controllable Satellites *ARS Journal* **31** 1079-89

[12] Rocco E M, Souza M L O, Prado A F B A. 2013 Station Keeping of Constellations Using Multiobjective Strategies *Mathematical Problems in Engineering* doi:10.1155/2013/476451

Journal of Physics: Conference Series **641** (2015) 012028 doi:10.1088/1742-6596/641/1/012028

Analysis of the influence of orbital disturbances applied to an artificial lunar satellite

L D Gonçalves[1], E M Rocco[1] and R V de Moraes[1, 2]

[1]Instituto Nacional de Pesquisas Espaciais, INPE, São José dos Campos, Brazil
[2]Universidade Federal de São Paulo, UNIFESP,São José dos Campos, Brazil

E-mail: lianadgon@gmail.com

Abstract. This paper analyzes the influence of the orbital disturbance forces in the trajectory of lunar satellites. The following gravitational and non-gravitational orbital disturbances are considered: the non-homogeneity of the lunar gravitational field; the gravitational attraction due to the third body, considering the Earth and the Sun; the lunar albedo; the solar radiation pressure. Numerical models were developed and implemented in an orbital trajectory simulator aiming to understand the dynamics of the orbital motion of an artificial satellite in lunar orbit when considering the simultaneous effect of all disturbances. Different orbits were simulated in order to characterize the major and the minor influence of each disturbing force as function of the inclination and the right ascension of the ascending node. This study can be very useful in the space mission analysis and in the selection of orbits less affected by environmental disturbances.

1. Introduction

The existence of environmental disturbing forces acting on an artificial satellite tends to cause variations in the elements that characterize the vehicle's orbit. To achieve the mission objectives it is necessary to analyze and study such disturbance forces previously, which can contribute to the efficiency with respect to important variables, such as the mission time and the fuel consumption.

To study the main disturbances acting on the orbital motion of a lunar artificial satellite, the disturbances models due to the lunar gravitational potential, lunar albedo, gravitational attraction of the Earth and Sun and solar radiation pressure were inserted into the Spacecraft Trajectory Simulator (STRS),[1], to be made the analysis of the magnitude of such forces applied in predefined orbits.

The model of the lunar gravitational field is based on spherical harmonics, according to the model presented by Konopliv [2] allowing to consider the spherical harmonics up to degree and order 100. The gravitational attraction of the Sun and Earth was modeled from the Newton's law of universal gravitational attraction, making use of the movement model for all bodies involved (Sun, Earth, Moon and Satellite). The lunar albedo model is based on the reflectivity of the Moon surface; it is possible to divide the lunar surface up to 51840 cells, to analyze the behavior of light reflected by each cell. The solar radiation pressure is modeled from the lunar albedo model, but considering the solar radiation shining directly into the satellite surface. Different orbits are simulated in order to characterize the major and minor influence of each disturbing force.

Journal of Physics: Conference Series **641** (2015) 012028 doi:10.1088/1742-6596/641/1/012028

2. Methodology

Considering that the objective of the study is to analyze the magnitude of the main disturbing forces acting on the motion of an artificial satellite around the lunar surface, models of lunar orbital perturbations were developed and inserted in simulation environment STRS. The STRS was developed by Rocco [1] and works with some specific features: the trajectory can be controlled using a closed loop control system, seeking to minimize the error in the vehicle's path and the deviation of the final state; uses a continuous propulsion system making it possible to assume that the magnitude of the thrust is finite and is applied during a time interval different of zero. A simplified block diagram of the orbital trajectory control system, used by the STRS, is shown in Figure 1.

The process starts with the determination of a reference state (Xref) which may vary from each mission objectives. Such reference state should be a good estimative of the trajectory to be followed by the satellite. The current state of the satellite (Xdet) is measured by sensors and compared with the reference state. The difference resulting from this comparison is an error signal that acts as controller input parameter. The control signal generated by the controller, which uses the techniques PID (proportional, integral and derivative), is sent to the actuators, which are able to define the magnitude and direction of corrections to be applied. The actuator output signal is added to external disturbances, and inserted to the dynamics of the orbital movement to be determined the current position and velocity of the artificial satellite. Then the current position of the satellite (Xdet) is determined, by using the sensors, from the information obtained by adding the signal of orbital dynamics with the sign of the disturbances.(Xactual). Finally the current state is compared to the reference position (Xref), generating an error signal and restarting the cycle of the control system.

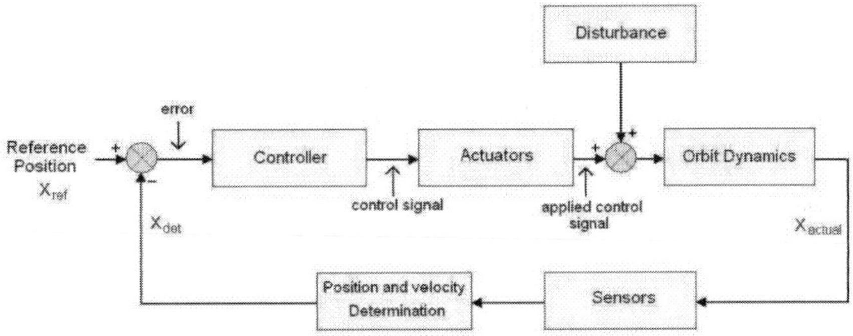

Figure 1. Trajectory closed loop control

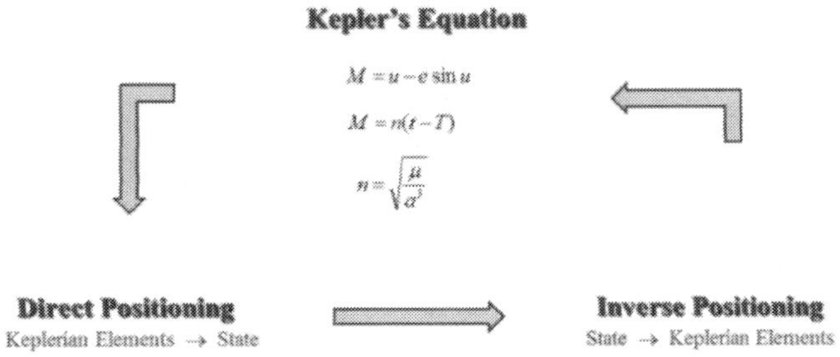

Figure 2. Dynamics of orbital motion

Solving the Kepler's equation ($M = u - e \sin u$ [3]) to each step defined as an input parameter in the simulator (STRS) the orbital motion can be determined, as shown in Figure 2. The Keplerian elements of the orbit can be determined from an initial state, considering a time interval, from the application of

Journal of Physics: Conference Series **641** (2015) 012028 doi:10.1088/1742-6596/641/1/012028

the inverse problem [3]. The propagated elements are obtained from the solution of Kepler's equation for the considered time interval. From the new Keplerian elements, the new state can be obtained solving the direct problem of satellite positioning [3].

3. Model for Perturbative Forces

The main disturbances forces capable of altering the orbit of an artificial lunar satellite are the lunar gravitational potential, the gravitational attraction of the Earth and Sun, the lunar albedo and the solar radiation pressure [3]. The models developed and used to study of the magnitude of these forces in predefined lunar orbits of an artificial satellite were based on theoretical study, briefly presented below.

3.1. Lunar Gravitational Potential

The Moon's gravitational potential is expressed by the expansion of normalized spherical harmonic coefficients, given by Equation (1) [2, 4]:

$$U(r,\lambda,\phi) = \frac{\mu}{r} + \frac{\mu}{r}\sum_{n=2}^{\infty}\sum_{m=0}^{n}\left(\frac{a_e}{r}\right)^n (\bar{C}_{nm}\cos m\lambda + \bar{S}_{nm}\,\mathrm{sen}\, m\lambda)\bar{P}_{nm}(\mathrm{sen}\,\phi) \tag{1}$$

where n is the degree, m is the order, μ is the gravitational constant and r is the lunar equatorial radius. $\bar{P}_{nm}$ are the fully normalized associated Legendre polynomials. a_e is the reference Moon radius, ϕ is the latitude, and λ is the longitude.

3.2. Gravitational Attraction of the Earth and the Sun

Considering the presence of a third body the function of the gravitational potential is given by Equation (2) [3], where μ' is the gravitational constant multiplied by the mass of the third body, r' is the module of the position vector of the third body; ψ is the angle between the position vector of the satellite relative to the Moon and the position vector of the satellite relative to the third body; r is the module of the satellite position vector with respect to Moon.

$$F' = \left(\frac{\mu'}{r'}\right)\left[1 + \sum_{n=2}^{\infty}\left(\frac{r}{r'}\right)^n P_n\cos\psi\right] \tag{2}$$

The general problem of three bodies presented by [5], and used by [6], provides the Equations (3) to calculate the disturbing accelerations due to the gravitational attraction of the bodies, where: $\vec{r}_1$, $\vec{r}_2$ e $\vec{r}_3$ are the positions of bodies, m_1, m_2 e m_3 are the masses of bodies and G is the gravitational constant.

$$\begin{aligned}
\ddot{\vec{r}}_1 &= -Gm_2\frac{\vec{r}_1 - \vec{r}_2}{|\vec{r}_1 - \vec{r}_2|^3} + Gm_3\frac{\vec{r}_3 - \vec{r}_1}{|\vec{r}_3 - \vec{r}_1|^3}\\
\ddot{\vec{r}}_2 &= -Gm_3\frac{\vec{r}_2 - \vec{r}_3}{|\vec{r}_2 - \vec{r}_3|^3} + Gm_1\frac{\vec{r}_1 - \vec{r}_2}{|\vec{r}_1 - \vec{r}_2|^3}\\
\ddot{\vec{r}}_3 &= -Gm_1\frac{\vec{r}_3 - \vec{r}_1}{|\vec{r}_3 - \vec{r}_1|^3} + Gm_2\frac{\vec{r}_2 - \vec{r}_3}{|\vec{r}_2 - \vec{r}_3|^3}
\end{aligned} \tag{3}$$

3.3. Lunar Albedo

The lunar albedo is defined as the fraction of solar energy reflected diffusely from the lunar surface into space, Equation (4), measured from the reflectivity of the Moon surface [1, 7].

$$albedo = \frac{radiation\ reflected\ back\ to\ space}{incident\ radiation} \tag{4}$$

Journal of Physics: Conference Series **641** (2015) 012028 doi:10.1088/1742-6596/641/1/012028

The lunar albedo model considers the Moon's reflectivity splitting the lunar surface up to 51840 cells. This innovative model was adapted from the Earth albedo model [8]. A detail description of the model and an analysis of the effect of the number of cells in the accuracy of results can be found in [9].

3.4. Solar Radiation Pressure

The solar radiation pressure is a non-gravitational force able to disturb the translational motion of a lunar artificial satellite due to the momentum of the photons that strike the satellite surface, causing a variation in the orbital elements.

In this work the model of the disturbance due to solar radiation pressure was developed based on the albedo model. Considering that is required to model the Sun movement, as well as its emitted radiation, and the Moon and the satellite's movement, for the albedo model, these same subroutines are also used to calculate the solar radiation pressure. However, it should be considered that the solar radiation directly reaches the satellite surface, instead of considering that solar radiation was reflected by the surface of the Moon before reaching the satellite, as in the albedo case.

4. Simulations and Results

With the objective to analyze the magnitude of each disturbance force were simulated some orbits, varying the following three orbital parameters: satellite altitude (15 km, 115 km, 215 km, 315 km, 415 km and 515 km); right ascension of the ascending node (0°, 90°, 180° and 270°); and inclination (0°, 90°, 180° and 270°).

The results for the integral of the perturbative force, for each disturbance adopted (lunar gravitational potential, attraction of the Earth and the Sun, solar radiation pressure, or lunar albedo), during one orbit around the Moon, are shown in Figures 3 to 7. Each figure presents a study for each orbital disturbance. The Figure 3 shows the study in the case of disturbance caused by the lunar gravitational potential, the Figure 4 for the Earth gravitational attraction, the Figure 5 for the Sun gravitational attraction, the Figure 6 for the solar radiation pressure and the Figure 7 for the lunar albedo.

In each situation were varied the values of altitude, inclination and right ascension of the ascending node. The colors change according to the variation of the values of the right ascension of ascending node and, consequently, the orbital plane. Blue for 0°, red for 90°, green for 180° and yellow for 270°. The altitudes are differentiated by lines connecting the dots in the graphs. For the cases of disturbance due to the gravitational potential and lunar albedo the top line shows the value of lower altitude (15 km) and the last line below shows the value of higher altitude (515 km). For cases of the Earth and Sun gravitational attraction and the solar radiation pressure, the top line shows the value of higher altitude (515 km) and the last line below shows the value of lower altitude (15 km). In the Figures 3 to 7, the altitudes are specified by numbers. Number 1 for altitude of 15 km, 2 for 115 km, 3 for 215 km, 4 for 315 km, 5 for 415 km and 6 for 515 km.

It is worth noting that for the cases studied separately for each perturbation, all other perturbations are also considered. If they were ignored, we would be considering the satellite at a different point in its actual position, which does not represent the real influence of disturbances. It can be observed an expected prevalence of lunar gravitational potential, as well as a significant variation of the disturbance forces integral as a function of the right ascension of the ascending node and the inclination. A detailed analysis of the gravitational potential influence in the orbit of an artificial lunar satellite, which were not considered other disturbances, and the performance of a control system to minimize such perturbative effects, can be found in [10, 11].

In view of the predominance of disturbance due to the lunar gravitational potential, was taken, for the case of inclination and right ascension of the ascending node, a quantitative analysis of all the perturbative forces, excluding the lunar gravitational potential. The results are shown in Table 1.

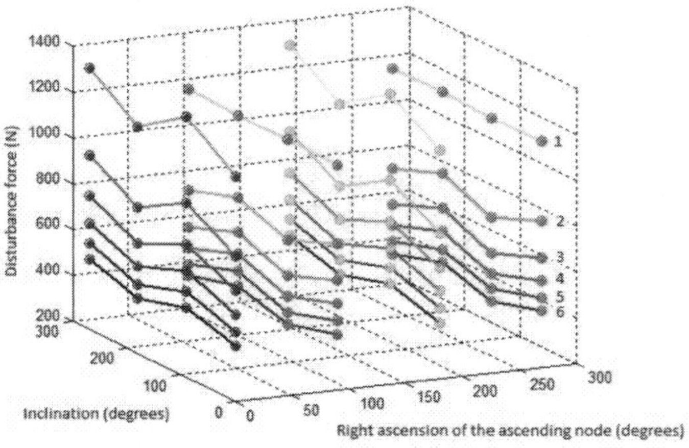

Figure 3. Integral of the disturbance force due to the lunar gravitational potential.

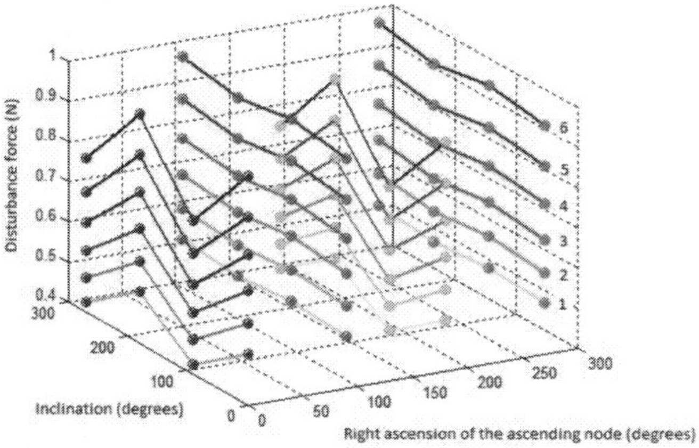

Figure 4. Integral of the disturbance force due Earth gravitational attraction.

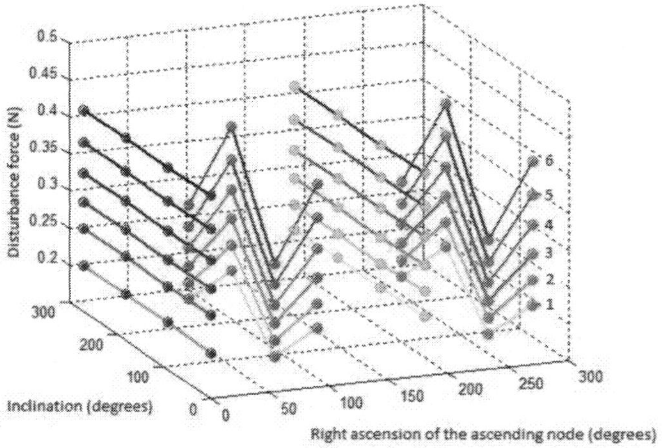

Figure 5. Sum of the perturbation force due Sun gravitational attraction.

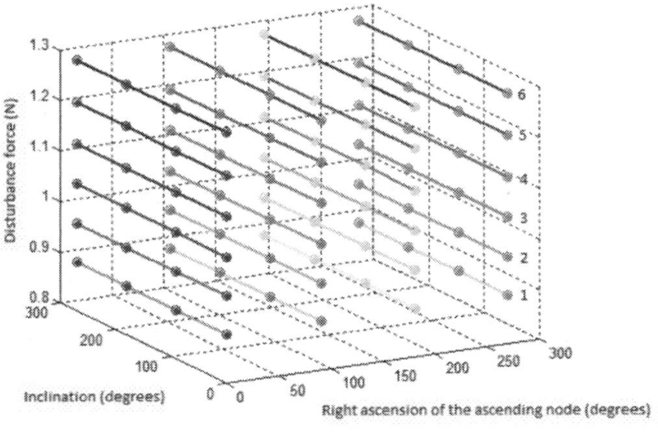

Figure 6. Sum of the perturbation force due to solar radiation pressure.

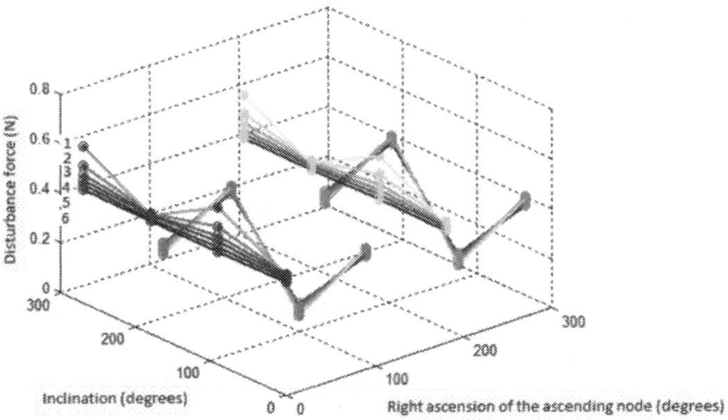

Figure 7. Integral of the disturbance force due to the lunar albedo.

Table 1. Disturbances Percentages

	Earth	Sun	Radiation	Albedo
15 km	19%	11%	41%	29%
115 km	21%	12%	43%	24%
215 km	23%	12%	43%	21%
315 km	24%	13%	44%	19%
415 km	25%	14%	44%	17%
515 km	26%	14%	44%	15%

It is seen a predominance of disturbance due to solar radiation pressure even when varying the altitude of the satellite. For altitudes closer to the lunar surface there is a higher relevance of albedo. This situation changes as the satellite reaches higher altitudes, where the influence of the Sun and Earth's gravitational attraction become more significant when compared to the lunar albedo. Note that the values of the disturbance changes as function of the altitude of the satellite. The data in Table 1 show the percentages of such disturbance forces.

Journal of Physics: Conference Series **641** (2015) 012028 doi:10.1088/1742-6596/641/1/012028

5. Conclusions

This paper presents the problem of quantize the magnitude of the forces that are able to disturb the orbit of a lunar satellite. For this, the influence of each force was considered and studied separately, but simultaneously with all disturbance forces. In other words, the contributions of each disturbance were obtained and then added to be obtained the total disturbance applied on the satellite. This kind of study can be very useful in the space mission analysis and in the selection of orbits less affected by environmental disturbances. For all the analyzed orbits, the disturbance due to the lunar gravitational potential was higher than the sum of all other disturbances considered. When the disturbance due to the lunar gravitational potential is excluded from the analysis was verified that the magnitude of non-gravitational forces is significant. This leads us to conclude that such disturbances should not be neglected in the studies of the orbital motion of a satellite around the lunar surface if it is necessary highly accurate results in the simulation of the trajectory, as is the case of a mission analysis of an artificial satellite.

6. Acknowledgements

The authors acknowledge the support provided by grants2012/21023-6 from São Paulo Research Foundation (FAPESP); by grants3085922/2014-0 from the National Council for Scientific and Technological Development (CNPq); and the financial support from the National Council for Scientific and Technological Development (CNPq).

References

[1] Rocco E M 2008 Perturbed orbital motion with a PID control system for the trajectory *XIV Colóquio Brasileiro de Dinâmica Orbital*, Águas de Lindóia, Brasil

[2] Konopliv A S, Asmar, S W, Carranza, E, Sjogren W L and Yuan, D N 2001 Recent gravity models as a result of the lunar prospector mission *Icarus* **150** 1-18

[3] Chobotov, V A 1991 *Orbital Mechanics* (Washington, DC: American Institute of Aeronautics and Astronautics) p 365

[4] Kuga H K, Carrara V and Kondapalli R R Artificial 2011 Satellites - Orbital Motion (in Portuguese) INPE Available in: http://urlib.net/8JMKD3MGP7W/3ARJ3NH

[5] Szebehely, V 1967 *Theory of orbits* (New York: Academic Press Inc)

[6] Prado, A F B A, Kuga, H K 2001. Space Technology Fundamentals (in Portuguese) Instituto Nacional de Pesquisas Espaciais, São José dos Campos ISBN: 85-17-00004-8

[7] Rocco E M 2008 Analysis of the deviations of the trajectory due to the terrestrial albedo applied to some scientific missions *International Conference on Mathematical Problems in Engineering, Aerospace and Sciences*, Genova, Italy

[8] Rocco, E M 2010 Evaluation of the Terrestrial Albedo Applied to Some Scientific Missions *Space Science Reviews* **151** 135-147

[9] Gonçalves, L D 2013 Evaluationofthetrajectorydeviationof a spacecraftduetothe lunar albedo *Congresso Brasileiro de Engenharia Mecânica*, Ribeirão Preto, Brasil

[10] Gonçalves L D, Rocco E M and de Moraes R V 2013 Orbital Disturbance Analysis due to the Lunar Gravitational Potential and Deviation Minimization through the Trajectory Control in Closed Loop *Journal of Physics: Conference Series* **465** 012008

[11] Gonçalves, L D, Rocco, E M, de Moraes, R V and Prado, A F B A 2013. Study of the potential effects on Lunar Spacecraft Trajectories (in Portuguese) *XI Conferência Brasileira de Dinâmica, Controle e Aplicações*, Fortaleza, Brasil

Journal of Physics: Conference Series **641** (2015) 012029 doi:10.1088/1742-6596/641/1/012029

Kalman filter application to mitigate the errors in the trajectory simulations due to the lunar gravitational model uncertainty

L D Gonçalves[1], E M Rocco[1], R V de Moraes[1, 2] and H K Kuga[1]

[1]Instituto Nacional de Pesquisas Espaciais, INPE, São José dos Campos, Brazil
[2] Universidade Federal de São Paulo, UNIFESP, São José dos Campos, Brazil

E-mail: lianadgon@gmail.com

Abstract. This paper aims to simulate part of the orbital trajectory of Lunar Prospector mission to analyze the relevance of using a Kalman filter to estimate the trajectory. For this study it is considered the disturbance due to the lunar gravitational potential using one of the most recent models, the LP100K model, which is based on spherical harmonics, and considers the maximum degree and order up to the value 100. In order to simplify the expression of the gravitational potential and, consequently, to reduce the computational effort required in the simulation, in some cases, lower values for degree and order are used. Following this aim, it is made an analysis of the inserted error in the simulations when using such values of degree and order to propagate the spacecraft trajectory and control. This analysis was done using the standard deviation that characterizes the uncertainty for each one of the values of the degree and order used in LP100K model for the satellite orbit. With knowledge of the uncertainty of the gravity model adopted, lunar orbital trajectory simulations may be accomplished considering these values of uncertainty. Furthermore, it was also used a Kalman filter, where is considered the sensor's uncertainty that defines the satellite position at each step of the simulation and the uncertainty of the model, by means of the characteristic variance of the truncated gravity model. Thus, this procedure represents an effort to approximate the results obtained using lower values for the degree and order of the spherical harmonics, to the results that would be attained if the maximum accuracy of the model LP100K were adopted. Also a comparison is made between the error in the satellite position in the situation in which the Kalman filter is used and the situation in which the filter is not used. The data for the comparison were obtained from the standard deviation in the velocity increment of the space vehicle.

1. Introduction

An artificial satellite around the Moon's surface is disturbed mainly by the non-homogeneity of the lunar gravitational field. This perturbation tends to cause deviations in the satellite trajectory, and, consequently, variations in the orbital elements that characterize the satellite orbit. It is common to use spherical harmonics to expand the expression of the lunar gravitational potential, as done in LP100K model, presented by [1]. The expansion of spherical harmonics, up to degree and order 100, was done to obtain the maximum precision of the model. However, the accuracy of the gravitational disturbing model can be varied according to the degree and order adopted. Studies and analysis of the influence of the accuracy of the model, as well as simulations considering the perturbation due to the lunar gravitational potential in the orbit of an artificial satellite can be found in [2, 3].

The LP100K model and the Kalman filter equations have been implemented in a simulator called Spacecraft Trajectory Simulator - STRS [4]. The orbital trajectory is calculated through the Kepler's equation solution for each simulation step, defined as simulator input parameter. This simulator uses continuous propulsion and trajectory control in closed loop to perform correction and transfer maneuvers, so that deviations and errors in the state variables are minimized. A more detailed description of the orbital trajectory controller can be found in [5].

The Kalman filter is recursive, thus it is an estimator with real-time characteristics, which provides estimates for the moment when the measure is processed. Therefore it is not necessary to have all measures to analyze them later with the estimator [6]. To do this, an estimate of the satellite position is made trying to approximate the results obtained using lower values for the degree and the order of the spherical harmonics, to the results that would be obtained if the maximum accuracy of the model LP100K were adopted.

2. Equations and models

The gravitational potential of the Moon is expressed by the coefficients of the normalized spherical harmonics, given by Equation (1) [1, 7]:

$$U(r,\lambda,\phi) = \frac{\mu}{r} + \frac{\mu}{r}\sum_{n=2}^{\infty}\sum_{m=0}^{n}\left(\frac{a_e}{r}\right)^n (\bar{C}_{nm}\cos m\lambda + \bar{S}_{nm}\,\mathrm{sen}\, m\lambda)\bar{P}_{nm}(\mathrm{sen}\,\phi) \tag{1}$$

where n is the degree, m is the order, μ is the gravitational constant and r is the orbit lunar equatorial radius. $\bar{P}_{nm}$are the fully normalized associated Legendre polynomials; a_e is the reference radius of the Moon, ϕ is the latitude, and λ is the longitude.

The LP100K model uses data obtained by the Lunar Prospector mission (1998-1999), which is the third mission of the Discovery NASA's exploration program. Its output provides the components x, y and z of the lunar gravitational acceleration, considering an inertial reference system centered on the Moon, at each instant of time along the simulation of the orbit of an artificial satellite, from the Equation 1. The gravity acceleration provided by LP100K model is compared with the gravity acceleration from a central field to obtain a disturber velocity increment applied in the satellite. Through the inverse problem it is possible to obtain the Keplerian elements that characterize the orbit of the artificial satellite.

Because it is a non-linear model, we used the extended Kalman filter.The extended filter updates the reference trajectory around the most current available estimate, wherein the filtering process consists of two stages: propagation phase and update phase.

The Equation 2 shows the propagation phase of the state and covariance, respectively.

$$\begin{gathered}\dot{\bar{w}} = f(\bar{w}) \\ \dot{\bar{P}} = F\bar{P} + \bar{P}F^T + GQG^T\end{gathered} \tag{2}$$

where$F$ is the jacobian matrix of f with respect to w, and$w$ represents one of the position coordinatesof the satellite uma das coordenadascartesianas da posição do satélitethe satellite position; P is the covariance matrix and $\bar{w}$ is the propagated state vector.

The Equation 3 shows the update phase.

$$\begin{gathered}K = \bar{P}H^T\left(H\bar{P}H^T + R\right)^{-1} \\ \hat{P} = (1 - KH)\bar{P} \\ \hat{w} = \bar{w} + K[y - h(\bar{w})]\end{gathered} \tag{3}$$

where K is the Kalman gain, H is the jacobian matrix of $h(w)$ with respect to w and $\hat{w}$ is the estimated state vector [6, 8].

3. Simulations and results

One of the objectives of the work is to vary the relation between the sensor and the uncertainty of the model, to thereby, choose a sensor which the uncertainty of the measures allows to generate an estimative that approximates the maximum precision of the LP100K model, when considered the use of the Kalman filter. In this way, the work begins with a study of the error in the trajectory of a lunar artificial satellite when one adopted values to degree and order less than 100. The Figure 1 shows this error through the sum of the standard deviation related to the values of the degree and order of the spherical harmonics. That is, we want to know the error that would be committed considering, for example, up to degree and order 10 (all terms added 1 to 10).

This analysis is made for all values of degree and order from 1 to 99, always compared with the value 100. This value is used as a reference basis, for being the maximum accuracy of the model, and for being considered the best way of describing the non-uniform distribution of Moon's mass. Thus, in this work, the value 100 to degree and order was regarded as a value without error, and therefore with a null standard deviation. This study was limited to the initial conditions of the last phase in Lunar Prospector mission: periapsis 17 km, apoapsis 43 km, inclination 90° and eccentricity 0.00735.

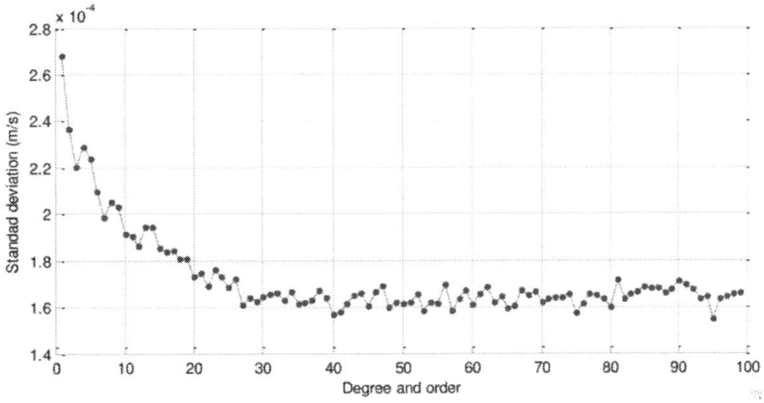

Figure 1. Standard deviation average for the values of degree and order from 1 until 99

In [3, 9] it was analyzed and quantified the influence of each term of the lunar gravitational potential between the values 1 to100 to degree and order. An expected prevalence of the first term of the gravitational potential (degree and order 2) was presented. It has been shown the existence of a significant contribution of the first 10 terms, however it was noted that the other terms are also relevants.

The results obtained in this work are consistent with previous published studies. Since it is a summation of values, the error decreases as higher values for degree and order are adopted. This decrease does not show a monotonic behavior due to the variability of the lunar gravitational potential. We cannot say that the latter term exerts less influence than the previous term, as shown in [3, 9]. Other studies have been conducted with the objective of evaluate errors in the gravitational potential due to the omission of higher terms of spherical harmonics, according to the covariance, especially for the geopotential case, as found in [10], [11], [12], [13] and [14].

The Figures 2, 4 and 6 show the average absolute deviation in the satellite position and the Figures 3, 5 and 7 show a study of the error in the component x of the satellite position vector to the value 2 of degree and order, value commonly used in simulations of lunar satellites trajectories. The results showed that the standard deviation for this case was 2.365×10^{-4} m/s.

The study below was made considering the situation with less accuracy as possible, wherein components x, y and z have the same uncertainty value. The case where the sensor and the model have the same value for the uncertainty (case 1), the case where the sensor uncertainty is 10 times

Journal of Physics: Conference Series **641** (2015) 012029 doi:10.1088/1742-6596/641/1/012029

greater than the model uncertainty (case 2), and, finally, the case where the sensor uncertainty is 100 times greater than the model uncertainty (case 3) were analyzed.

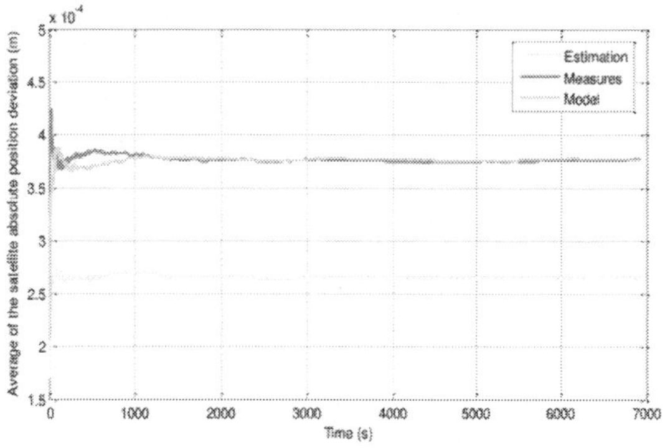

Figure 2. Average of the satellite absolute position deviation (case 1).

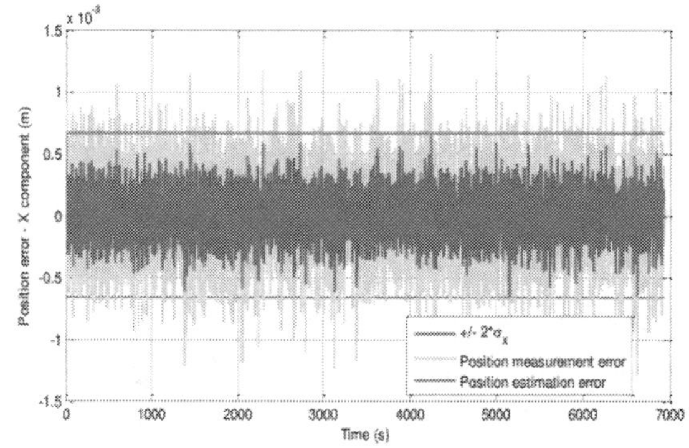

Figure 3. Position error (case 1).

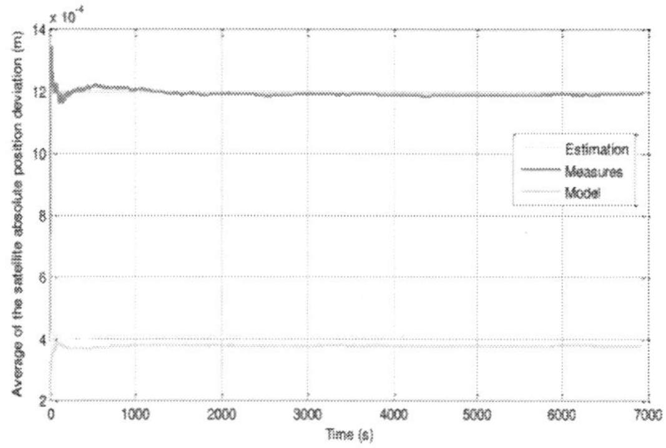

Figure 4. Average of the satellite absolute position deviation (case 2).

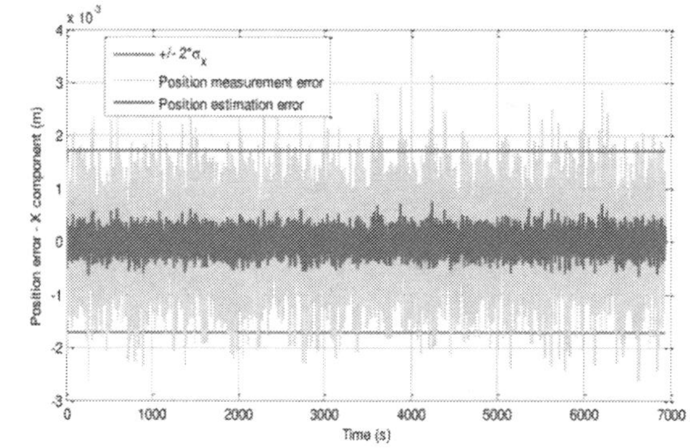

Figure 5. Position error (case 2).

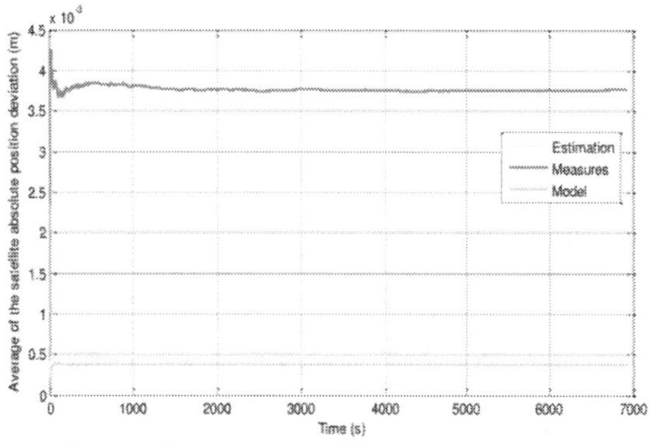

Figure 6. Average of the satellite absolute position deviation (case 3).

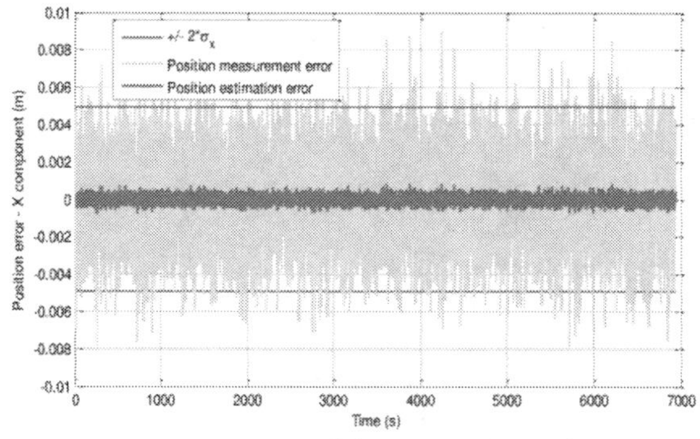

Figure 7. Position error (case 3).

In the Figures 2, 4 and 6 the blue line represents the mean error in the propagated position generated by the model, the pink line represents the mean error of the measured position, and the yellow line represents the mean error of the position estimated by the Kalman filter. In the Figures 3, 5 and 7 the blue line represents the error between the measured position and the propagated position, the red line represents the error between the propagated position and the position estimated by the Kalman filter and the pink line represents 2 standard deviations.

The results show that increasing the sensor uncertainty causes the model accuracy to become increasingly significant and the uncertainty in the estimated position approaches the estimated uncertainty of the model. Other studies comparing the uncertainty in the component x of the spacecraft position vector, for different cases, can be found in [15] and [16].

4. Conclusions

The work begins with a study of the inserted error in simulations due to the use of low values to degree and order for the expansion of spherical harmonics. This study was done by determining the standard deviation of the velocity increment, as function of the accuracy of the disturbance model adopted. The Kalman filter model has been successfully implemented and tested, for the use with lower values of degree and order of the expansion of spherical harmonics, allowing an estimate of the satellite position, as good as the position determined for the case when the maximum accuracy of the gravitational model is used. Finally, the study performed in order to find a sensor which the uncertainty in the measures, when using the Kalman filter, generate an estimate similar to the maximum accuracy of LP100K model was successfully completed. The approach adopted in this work to choose the accuracy of the sensor could be used in a real mission analysis to specify the sensor for the satellite.

5. Acknowledgements

The authors acknowledge the support providedby grants2012/21023-6 from São Paulo Research Foundation (FAPESP); by grants3085922/2014-0 from the National Council for Scientific and Technological Development (CNPq); and the financial support from the National Council for Scientific and Technological Development (CNPq).

References

[1] Konopliv A S, Asmar, S W, Carranza, E, Sjogren W L and Yuan, D N 2001 Recent gravity models as a result of the lunar prospector mission *Icarus* **150** 1-18

[2] Gonçalves LD 2013 Orbital Manuevers of Lunar Artificial Satellites with Continuous Propulsion Application (in Portuguese). Master thesis, Instituto Nacional de Pesquisas Espaciais (INPE)

[3] Gonçalves LD, Rocco E M and de Moraes RV 2013 Orbital Disturbance Analysis due to the Lunar Gravitational Potential and Deviation Minimization through the Trajectory Control in Closed Loop*Journal of Physics: Conference Series***465**012008

[4] Rocco E M 2008 Analysis of the deviations of the trajectory due to the terrestrial albedo applied to some scientific missions. *International Conference on Mathematical Problems in Engineering, Aerospace and Sciences*,Genova, Italy

[5] Rocco E M 2012 Trajectory Control with Continuous Propulsion for drag-free missions (in Portuguese) *Congresso Nacional de Engenharia Mecânica*, São Luís, Brasil

[6] Maybeck PS 1979 *Stochastic models, estimation and control* (New York: Academic Press)

[7] Kuga HK, Carrara V and Kondapalli R R Artificial 2011 Satellites - Orbital Motion (in Portuguese) INPE Available in:http://urlib.net/8JMKD3MGP7W/3ARJ3NH

[8] Kuga H K 2005 Practice Notions of Estimation Techniques (in Portuguese) Class noteson Optimization of Dynamical Systems

[9] Gonçalves L D, Rocco EM, de Moraes, RV, Prado and AFBA 2013 Evaluation of the Influence of zonal and sectorial harmonics in the orbit of a lunar satellite *Division on Dynamical Astronomy of the American Astronomical Society*, Paraty, Brasil

[10] Wright, J R 1981 Sequential Orbit Determination with auto-correlated Graviting Modeling Erros *Journal of Guidance and Control*

[11] Giacaglia, G E O, Velez, C E1986 Orbit Estimation with auto-correlated force field errors In:

Journal of Physics: Conference Series **641** (2015) 012029 doi:10.1088/1742-6596/641/1/012029

Advances in the Astronautical Science, San Diego

[12] Kondapalli, R R 1989 Influence Evaluation of Errors Involved in the Orbit Propagation of Artificial Satellites (in Portuguese), Doctorate thesis, Instituto Nacional de Pesquisas Espaciais (INPE)

[13] Yee, C, Kelbel, D, Lee,T, Samii, M, Mistreta and G, Hart, R1991 Study of geopotential error models used in orbit determination error analysis.Proceedings of Flight Mechanics/Estimation Theory Symposium 229-248 Paper N92-14070

[14] Wright, JR, Woodburn, J, Truong, S and Chuba, W Sample2008 Orbit Covariance Function and Filter-Smoother Consistency Tests *Advance in the Astronautical Sciences***130** Paper AAS-08-157

[15] Oliveira, T C, Rocco, EM and Kuga,HK 2011. Orbital Maneuvers Analysis Using of Low Thrust using Kalman filter (in Portuguese) *X Conferência Brasileira de Dinâmica, Controle e Aplicações*, Águas de Lindóia, Brasil

[16] Santos, W G, Kuga, H K and Rocco, EM 2012 Kalman Filter Application for State Estimation of Aeroassited Maneuver *VII Congresso Nacional de Engenharia Mecânica*, São Luis do Maranhão, Brasil

H infinity controller design to a rigid-flexible satellite with two vibration modes

A G de Souza[1], **L C G de Souza**[2]

[1] National Institute for Space Research (INPE), Av. dos Astronautas 1758, 12227-010 Sao Jose dos Campos, SP, Brazil.

[2] Brasilia University, Campus Gama, Area Especial de Industria-A-UnB, 72444-240 Brasilia, DF, Brazil.

E-mail: alain.souza@inpe.br

Abstract. The satellite attitude control system (ACS) design becomes more complex when the satellite structure has components like, flexible solar panels, antennas and mechanical manipulators. These flexible structures can interact with the satellite rigid parts during translational and/or rotational manoeuvre damaging the ACS pointing accuracy. Although, a well-designed controller can suppress such disturbances quickly, the controller error pointing may be limited by the minimum time necessary to suppress such disturbances thus affecting the satellite attitude acquisition. This paper deals with the rigid-flexible satellite ACS design using the H infinity method. The rigid-flexible satellite is represented by a beam connected to a central rigid hub at one end and free at the other one. The equations of motions are obtained considering small flexible deformations and the Euler-Bernoulli hypothesis. The results of the simulations have shown that the H-infinity controller was able to control the rigid motion and suppress the vibrations.

1. Introduction

There are several methodologies [1] to design satellite ACS, depending on the control system complexity; computer simulation cannot be the more appropriate one. Experimental platforms [2] have the important advantage of allowing the satellite dynamics representation in laboratory to accomplish experiments and simulations to evaluate satellites ACS. Experimental test has also the possibility of introducing more realism than the simulation; but it has the difficulty of reproducing zero gravity and torque free space condition. Examples of simulator dynamics and control system experimental investigations can be found in [3, 4]. A classic case of a phenomenon that was not investigated experimentally before launch, was the dissipation energy effect that has altered the satellite Explorer I rotation [5]. Several institutions and universities [6, 7] are investigating and testing the ACS performance by experimental prototypes. In [8,9] it was showed that the influence of the non-linearities introduced by the slosh motion, the panels flexibility and the system parameters variation can degrade the control system performance, indicating the necessity of new robust control technique. Examples of multi-objectives methods to design controllers space system can be found in [10, 11]. In this paper one investigates the rigid-flexible satellite ACS design using the H infinity method [12, 13]. The rigid-flexible satellite model is represented by the Rotary flexible link module which consists of a rigid central hub

Journal of Physics: Conference Series **641** (2015) 012030 doi:10.1088/1742-6596/641/1/012030

connected to a flexible appendage [14]. The model used can be interpreted as a solar panel or a flexible antenna coupled to a rigid satellite.

2. Rigid-flexible Satellite Model

The figure 1 shows the representations of the rigid-flexible satellite model; which consist of a rigid central part and by a flexible beam with the mass in the end. The actuator is located in the central part, and it is responsible for the control of the horizontal planar movement of the rigid structure and by the elastic displacement, where (X_0, Y_0) and (X, Y) are the coordinates system before and after deformation, respectively. The rigid central hub has radius R, the actuator rotor has viscous friction bm, and inertia J_{rotor} and develops a torque $\tau(t)$; the flexible beam has uniform linear mass density ρ, uniform bending stiffness EI and length L; the flexible beam deformation is $w(x, t)$; and the tip mass is m_{tip} and inertia J_{tip}; $\alpha(t)$ represent the tip angle and $\theta(t)$ is the rigid angle displacement; and $s(x)$ represent the distance between the referential axis to an element of mass in the link.

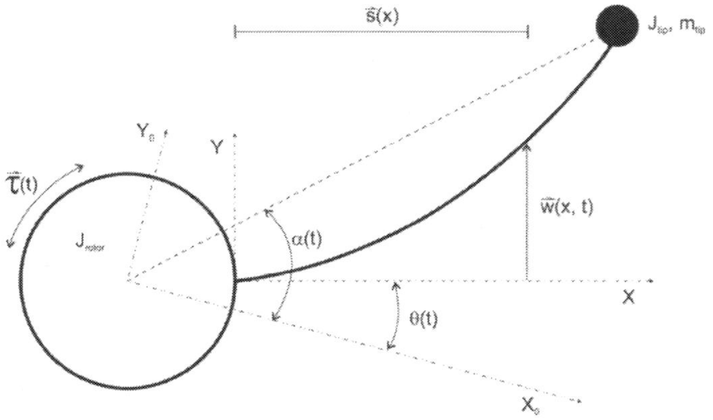

Figure 1. Rigid-flexible satellite parameters

The equations of motion of the entire structure are derived using the Lagragian approach [5] combined with the assumed modes method. The flexible link is modeled as an Euler-Bernoulli beam. The free vibration of the link is described by the partial differential equation

$$EI\frac{\partial^4 w(x,\,t)}{\partial x^4}+\rho\frac{\partial^2 w(x,\,t)}{\partial t^2}=0 \tag{1}$$

with boundary condition:

$$w(0,\,t)=0;\quad \frac{\partial w(0,\,t)}{\partial x}=0;\quad \frac{\partial^2 w\,(L,\,t)}{\partial x^2}=0;\quad \frac{m_{tip}}{EI}\frac{\partial^2 w(L,t)}{\partial t^2}=\frac{\partial^3 w(L,t)}{\partial x^3} \tag{2}$$

Solving this partial differential equation (PDE) of motion one obtains:

$$Y\,(x)=C_1\left(\cosh\left(\beta_i x\right)-\cos\left(\beta_i x\right)-\left(\frac{\left(\cosh\left(\beta_i L\right)+\cos\left(\beta_i L\right)\right)}{\left(senh\left(\beta_i L\right)+sen\left(\beta_i L\right)\right)}\right)\left(senh\left(\beta_i x\right)-sen\left(\beta_i x\right)\right)\right) \tag{3}$$

where: $\omega_i^2/C_2 = \beta_i^2$; $\omega_i^2 = C^2\beta_i^4$; $\omega_i = \beta_i^2\sqrt{(EI/\rho)}$; and C_1 is a normalization constrain and β_i is the real positive squares of the transcendental equation:

$$\cos\left(\beta_i L\right)\cosh\left(\beta_i L\right)-K\left(\cos\left(\beta_i L\right)\sinh\left(\beta_i L\right)-\cosh\left(\beta_i L\right)\sin\left(\beta_i L\right)\right)=1 \tag{4}$$

Journal of Physics: Conference Series **641** (2015) 012030 doi:10.1088/1742-6596/641/1/012030

with, $K = m_{tip}/\rho$.

The total kinetic energy [7] is given by the sum of the kinetic energies of the rotor, tip mass and a mass element in the link ($T = T_{rotor} + T_{dm} + T_{tip}$), which is given by

$$\begin{aligned} T = \tfrac{1}{2}\dot{\theta}^2 \left[J_{rotor} + \rho\left[\int_0^L w^2 dx + \tfrac{1}{3}(R+L)^3 - \tfrac{1}{3}R^3\right] + m_{tip}\left[w_L^2 + (R+L)^2\right] + J_{tip}\right] + \\ +\tfrac{1}{2}\dot{\theta}\left[2\rho\int_0^L \dot{w}(R+s)dx + 2m_{tip}\dot{w}_L + J_{tip}\left(\tfrac{\partial \dot{w}}{\partial x}\right)\right] + \tfrac{1}{2}\left[\rho\int_0^L \dot{w}^2 dx + m_{tip}\left(\tfrac{\partial \dot{w}}{\partial x}\right)\right] \end{aligned} \tag{5}$$

and the potential energy is

$$V = \frac{1}{2}EI\int_o^L \left(\frac{\partial^2 w}{\partial x^2}\right)^2 dx \tag{6}$$

One admits that the dissipation energy caused by the viscous damping of the rotor and by the deformation of the beam is given by

$$R = \frac{1}{2}b_m\dot{\theta}^2 + \frac{1}{2}K_e EI\int_0^L \left(\frac{\partial^2 \dot{w}}{\partial x^2}\right)^2 dx \tag{7}$$

According with the assumed mode method, one can write the elastic deflection as a function of the modal form ($\varphi_i(x)$) and a generalized coordinates of the deflection ($\eta_i(t)$).

$$w(x,\,t) = \sum_{i=1}^{n} \varphi_i(x)\,\eta_i(t) \tag{8}$$

The form function normalization rules are given by

$$\rho\int_o^L \varphi_i(x)\,\varphi_j(x)\,dx + m_{tip}\varphi_i(L)\,\varphi_j(L) = 0 \quad i \neq j; \qquad \rho\int_o^L \varphi_i^2(x)\,dx + m_{tip}\varphi_i^2 = 1 \quad i = j \tag{9}$$

As a result, the solution of the PDE [7] for the modal forms is:

$$\varphi_i(x) = C_1\left(\cosh(\beta_i x) - \cos(\beta_i x) - \left(\frac{(\cosh(\beta_i L) + \cos(\beta_i L))}{(senh(\beta_i L) + sen(\beta_i L))}\right)(senh(\beta_i x) - sen(\beta_i x))\right) \tag{10}$$

where the constant C_1 is defined by the normalizations rules. The angular frequency is $\omega_i = \beta_i^2\sqrt{(EI/\rho)}$.

For this paper one assumes two modes of vibration, where rigid angular displacement (θ) and the deflection variables (η_i) are used as generalized coordinates in the derivation of the equation of motion. For convenience one uses the matrix form, and then the vectors η_i and ϕ_i are represented by the matrices q and ϕ.

Using the eq. (5) e eq. (6) one obtains the Lagragian

$$\begin{aligned} \mathcal{L} = \tfrac{1}{2}\dot{\theta}^2\left[J_{rotor} + q^T\rho\int_0^L \phi\phi^T dx q + \rho\tfrac{1}{3}(R+L)^3 + m_{tip}\left(q^T\phi_L\phi_L^T q + (R+L)\right)^2 + J_{tip}\right] + \\ +\tfrac{1}{2}\dot{\theta}\left[2\rho\dot{q}^T\int_0^L \phi\,(R+s)\,dx + 2m_{tip}\dot{q}^T\phi_L\,(R+L) + J_{tip}\dot{q}^T\phi'\right] + \\ +\tfrac{1}{2}\left[\rho\dot{q}^T\int_0^L \phi\phi^T dx\dot{q} + m_{tip}\dot{q}^T\left(\phi_L\phi_L^T\right)\dot{q} + J_{tip}\dot{q}\left(\phi'\phi^{T'}\right)\dot{q}\right] - \tfrac{1}{2}q^T EI\int_0^L \phi''\phi^{T''}dx q \end{aligned} \tag{11}$$

Journal of Physics: Conference Series **641** (2015) 012030 doi:10.1088/1742-6596/641/1/012030

Applying the Lagragian methodology [7], the Lagrange equation is

$$\frac{d}{dt}\left[\frac{\partial}{\partial p_i}\mathcal{L}\right] - \frac{\partial}{\partial p_i}\mathcal{L} + \frac{\partial}{\partial p_i}R = Q_i \tag{12}$$

where $p_i = \{\theta,\ q\}$ e $Q = \{\tau, 0\}$, with that one have the following nonlinear dynamic, represented by a matrix form, is obtained.

$$\begin{bmatrix} I_t + q^T C_{rr} q & M_{rf}^T \\ M_{rf} & M_{ff} \end{bmatrix}\begin{bmatrix} \ddot{\theta} \\ \ddot{q} \end{bmatrix} + \begin{bmatrix} b_m & 0 \\ 0 & B_{ff} \end{bmatrix}\begin{bmatrix} \dot{\theta} \\ \dot{q} \end{bmatrix} + \begin{bmatrix} 0 & 0 \\ 0 & K_{ff} \end{bmatrix}\begin{bmatrix} \theta \\ q \end{bmatrix} + \begin{bmatrix} \dot{\theta} q^T C_{rr} \dot{q} \\ -\dot{\theta}^2 C_{rr} q \end{bmatrix} = \begin{bmatrix} \tau \\ 0 \end{bmatrix} \tag{13}$$

where,

$$\begin{gathered} \eta_i = \mathbf{q};\quad \varphi(x,\,t) = \phi;\quad \varphi(L,\,t) = \phi_L;\quad \mathbf{K}_{ff} = EI\int_0^L \phi''\phi''^T dx;\quad \mathbf{B}_{ff} = K_e EI\int_0^L \phi''\phi''^T dx \\ \mathbf{C}_{rr} = \rho\int_0^L \phi\phi^T dx + m_{tip}\phi_L{\phi_L}^T;\quad \mathbf{M}_{ff} = \rho\int_0^L \phi\phi^T dx + m_{tip}\phi_L\phi_L^T + \tfrac{1}{2}J_{tip}{\phi_L}'{\phi_L}'; \end{gathered} \tag{14}$$

To obtain a linear model the nonlinear terms $q^T C_{rr} q$, $q^T C_{rr} q$ e $-\dot{\theta}^2 C_{rr} q$ are neglected due to the fact that they are relatively very small [4]. So writing eq.(

$$\begin{bmatrix} \dot{X}_1 \\ \dot{X}_2 \end{bmatrix} = \begin{bmatrix} 0 & I \\ -M^{-1}K & -M^{-1}D \end{bmatrix}\begin{bmatrix} X_1 \\ X_2 \end{bmatrix} + \begin{bmatrix} 0 \\ M^{-1}Q \end{bmatrix} u \tag{15}$$

with $X = \begin{bmatrix} \theta & \eta_1 & \eta_2 & \dot{\theta} & \dot{\eta}_1 & \dot{\eta}_2 \end{bmatrix}$ is the states vectors and the control $u = K_{H_\infty} X$, where K_{H_∞} is the gain calculate by the H infinity method [15].

3. H infinity control method

One great advantage of the H_∞ control method is the possibility of designing robust controllers with respect to structured uncertainty at the same time that is possible to obtain good performance with respect to unstructured uncertainty. As a result, the H_∞ controller design is a multi-objective control problem where performance improvement and good robustness requirement are functions of the weighting functions selection as it will be explained in the sequel [15].

The norm H_∞ of the system transfer function $G(s)$ can be expressed by the major value of the singular value (σ) in frequency domain given by:

$$\|G(s)\|_\infty = \max_\omega \bar{\sigma}\left(G(j\omega)\right) \tag{16}$$

The H_∞ norm is usually computed numerically from a state-space realization as the smallest value of γ such that the Hamiltonian matrix eq. (17) has no eigenvalues on the imaginary axis [6],

$$\mathrm{H} = \begin{bmatrix} A + BR^{-1}D^T C & BR^{-1}B^T \\ -C^T\left(I + DR^{-1}D^T\right)C & -\left(A + BR^{-1}D^T C\right)^T \end{bmatrix} \tag{17}$$

where A, B, C e D are the space state matrices and $R = \gamma^2 I - D^T D$. This is an interactive procedure; where one may start with a large value of γ and reduce it until imaginary eigenvalues for H appear. Basically the H infinity control method consists of calculating a gain K that minimizing the H_∞ norm of the transfer function of the closed loop [15].

The closed loop transfer function is given by:

$$\left[\begin{array}{c} z \\ v \end{array}\right] = P(s)\left[\begin{array}{c} \omega \\ u \end{array}\right] = \left[\begin{array}{cc} P_{11}(s) & P_{12}(s) \\ P_{21}(s) & P_{22}(s) \end{array}\right]\left[\begin{array}{c} \omega \\ u \end{array}\right] \tag{18}$$

where, $u = K(s)v$. In the other hand one can write as:

$$F_l(P,K)\,\omega = P_{11} + P_{12}K(I - P_{22}K)^{-1}P_{21} \tag{19}$$

Minimizing the H norm:

$$\|F_l(P,K)\|_\infty = \max_\omega \bar{\sigma}\left(F_l(P,K)(j\omega)\right) \quad \|F_l(P,K)\|_\infty < \gamma \tag{20}$$

3.1. A weighting functions

The selection of weighting functions for a specific design problem often involves many interactions and fine-tuning. It is hard to give a general formula for the weighting functions that will work in every case [15]. Therefore in the literatures [16] is possible to find some suggestions, and/or starting point for analysis and iterations to find a starter weighting function. Then like was suggesting by [15] one uses the follow functions:

$$W_S = \left(\frac{\frac{s}{\sqrt[k]{M}} + \omega_b}{s + \sqrt[k]{A}\omega_b}\right)^k, \; W_T = \left(\frac{s + \frac{\omega_{bc}}{\sqrt[k]{M}}}{\sqrt[k]{A}s + \omega_{bc}}\right)^k, \; W_{KS} = cte \tag{21}$$

For the study case, the constants are: k- adjust the roll off ($k = 1$); ω_b- equal the value of the bandwidth of the plant ($_b = 0,03rad/s$; as showed in the figure 2); ω_{bf} - proportional to ($\omega_{bf} = 100 \cdot \omega_b$); A - limited the steady-error ($A = 10^{-3}$);M - limited the Overshoot ($M = 10$).

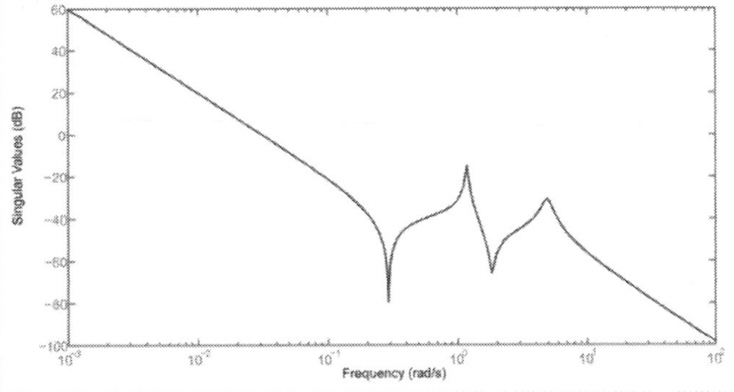

Figure 2. Singular values of the plant.

4. Results of the simulations

For the simulation the constant parameter are: $R = 0.05m$, $bm = 0.15m^2/s$, $J_{rotor} = 0.3kgm^2$, $L = 1m$, $\rho = 2700kg/m$, $ke = 0.03$, $EI = 18.4Nm^2$, $m_{tip} = 0.25kg$ and $J_{tip} = 0.04kgm^2$. The objective of the control is stabilize the tip mass angle (α) of the flexible link in a neutral position (0°). The angle α given by:

Appling the H infinity control method in the plant model showed in the eq. (15), one obtains the fallow results for the initial response (the link is in the position). In the figure 3 are presented the results the simulation for the states: the rigid angular displacement, the first vibration mode and second vibration mode, and its variation with respect the time, respectively. Figure 3 shows that the H_∞ control law stabilizes rigid angular displacement (θ) around in 60 seconds, the first

Journal of Physics: Conference Series **641** (2015) 012030 doi:10.1088/1742-6596/641/1/012030

vibration mode (η_1) stabilizes around 80 seconds and the second vibration mode (η_2) stabilizes around 60 seconds. Figure 4 shows that the H_∞ control law stabilizes the derivate of the states ($d\theta/dt$) and ($d\eta_1/dt$) around in 80 seconds, while ($d\eta_2/dt$) stabilizes around 40 seconds.

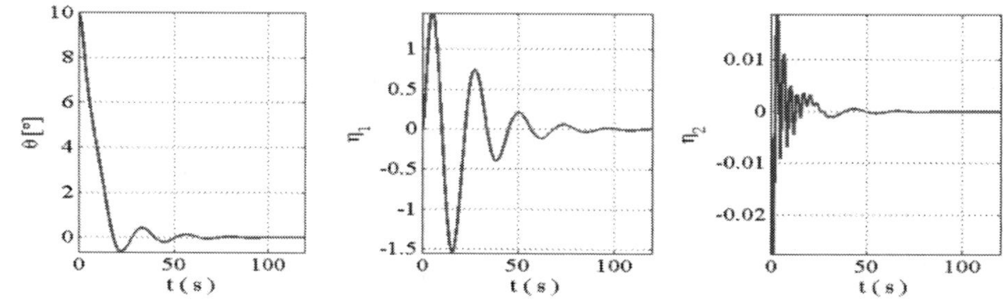

Figure 3. H_∞ control of the attitude and flexibility.

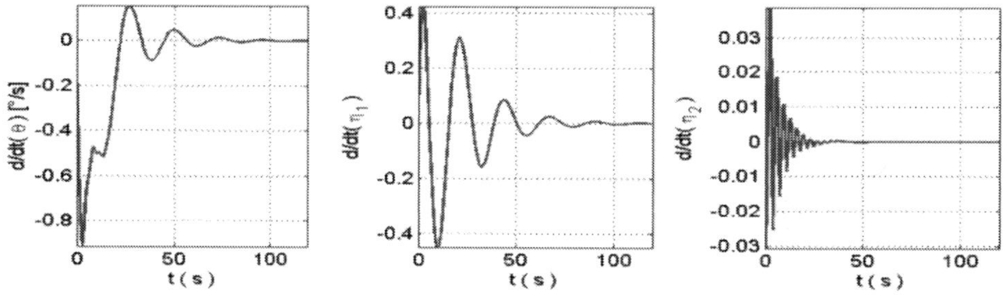

Figure 4. H_∞ control of the attitude and flexibility rates

Figure 5 shows the behavior of the tip mass angle (α) ; starting in an initial position 10° and finishing in a position 0°. The H_∞ control performance concludes in 100 seconds. Figure 6 shows the H_∞ control effort (the torque generate by the rotor) has the maximum about $3.8Nm$ and $-4.0Nm$.

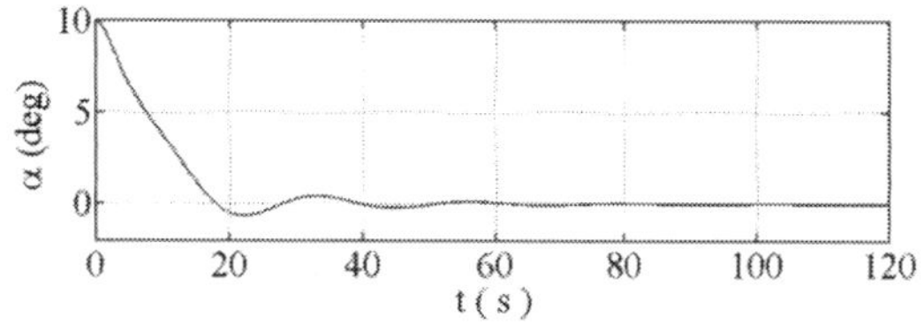

Figure 5. H_∞ control of the tip-mass.

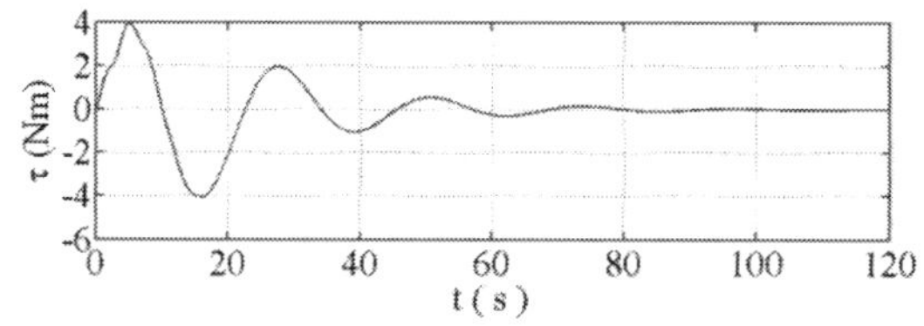

Figure 6. H_∞ control effort.

5. Conclusion

In this paper one applies the H-infinity method to design the attitude control system (ACS) for a rigid-flexible satellite. The satellite model is represented by a flexible beam connected to a rigid central hub with a tip mass in its free end. The Lagrange approach is used to obtain equations of motions considering small flexible deformations. The simulations results have showed that the H-infinity controller was able to control the rigid motion and suppress the vibrations. That good controller performance was obtained by selection of weighting functions parameters for that specific problem. The results has also showed that the model is representative of rigid-flexible satellite with one panel and that H_∞ controller has achieves the objectives controlling the rigid angular displacement and suppressing the vibrations modes, bringing the tip mass angle to a neutral position with acceptable control effort in a short time interval. Finally, one observes that the H_∞ control performance is dependent of weighting functions, so better performance can be archived if more investigation is spend in the selection of the weighting functions.

References

[1] Hall, C.D. , Tsiotras and Shen, H. Tracking Rigid Body Motion Using Thrusters and Momentum Wheels, *Journal of the Astronautical Sciences*, 2002.

[2] Schwartz, J. L.; Peck, M. A. and Hall, C. D. Historical review of air-bearing spacecraft simulators. *Journal of Guidance, Control and Dynamics*, 26, 4, (2003),513-522.

[3] Conti G.T. and Souza, L.C.G. Satellite attitude control system simulator. *Journal of Shock and Vibration.* 15, 3-4, (2008), 395-402.

[4] Souza, L.C.G. ; Gonzles,R.G.Application of the State-dependent Riccati Equation and Kalman Filter Techniques to Design of a Satellite Control System. *Shock and Vibration*, v. 19, p. 22-28, 2012.

[5] Kaplan M. H, *Modern Spacecraft Dynamics and Control*, John Wiley Sons, NY, 1976.

[6] Tsiotras, P. A Spacecraft Simulator for Research and Education, Georgia Institute of Technology, EUA, 2007.

[7] Souza, A. G. ; Souza, L. C. G. Design of Satellite Attitude Control System Considering the Interaction between Fuel Slosh and Flexible Dynamics during the System Parameters Estimation. *Applied Mechanics and Materials*, v. 706, p. 14, 2015.

[8] Bigot, P., and Souza, L. C. (2014). Investigation of the State Dependent Riccati Equation (SDRE) adaptive control advantages for controlling non-linear systems as a flexible rotatory beam. *International journal of systems applications, engineering and development.*Vol.8,pp 92-99.

[9] Mainenti-Lopes, I ; Souza, L. C. G. ; Souza, F.L . Design of a nonlinear controller for a rigid-flexible satellite using multi-objective Generalized Extremal Optimization with real codification. *Shock and Vibration*, v. 19, p. 1-10, 2012.

[10] Pinheiro, E. R.; Souza, L. C. G.. Design of the Microsatellite Attitude Control System Using the Mixed Method via LMI Optimization. *Mathematical Problems in Engineering (Print)*, v. 2013, p. 1-8, 2013.

[11] Souza, L. C. G. ; Control of a Rigid-Flexible Satellite Solar Panel with Non-collocated Sensor/Actuator. Journal of Mechanics Engineering and Automation, v. 5, p. 173-179, 2015.

[12] Souza, L C G ; Souza, A. G. . Satellite Attitude Control System Design considering the Fuel Slosh Dynamics. *Shock and Vibration*, v. 2014, p. 1-8, 2014.

[13] Quanser homepage. at: http://www.quanser.com/.

[14] Skogestad, S.; Postlethwaite, I. *Multivariable Feedback Control Analysis and Design.* 2. ed. Chichester: John Wiley Sons, 2001

[15] Zhou, K. ; Doyle, J. C. *Essentials of Robust Control.* New Jersey: Prentice Hall, 1998.

Journal of Physics: Conference Series **641** (2015) 012031 doi:10.1088/1742-6596/641/1/012031

Navigation and control of an UAV quadrotor in search and surveillance missions

L A Frederico[1], **L S Martins-Filho**[1] **and A L Da Silva**[2]
[1]Universidade Federal do ABC, Centro de Engenharia, Modelagem e Ciências Sociais Aplicadas, Santo André, São Paulo, Brazil
[2]Universidade Federal de Santa Maria, Cachoeira do Sul, Rio Grande do Sul, Brazil

E-mail: leo_de_avellar@hotmail.com

Abstract. This study addresses the problem of exploration of areas for surveillance and searching using an Unmanned Aerial Vehicle quadrotor (UAV). Their agility provides them operational flexibility, as required by such applications. Other desirable characteristics are unpredictable motion path (from the point of view of intruders) and fast scan of the area. Random trajectories have been studied in these cases. Here, we study trajectories based on a specific random motions, known as Levy flights. In addition, this work concerns the study of the flight dynamics and the control to achieve the mission objectives. The backstepping control developed was implemented in simulation model, and tested together the Levy flight path generator. A procedure to determine the control gains in order to satisfy the Levy flight requirements and quadrotor constraints was proposed. Results show this control is adequate for the execution of the Levy flight respecting the operational and dynamic constraints.

1. Introduction

This work deals with the application of Lévy flight in an UAV for searching and surveillance missions. The Lévy flight has been observed in some animal species as an optimal solution to the problem of finding food, shelter or other target [1-2]. Fast area scan and unpredictability of the trajectory are desirable characteristics, becoming the Lévy flights an interesting solution. The Lévy function has the following form [1-2]:

$$p(l) \approx l^{-\mu}, \qquad l \geq l_0 \tag{1}$$

In equation (1), l is the trajectory length in each step of the mission, μ is an exponent and l_0 is the minimum trajectory length possible. For each value of the trajectory length will be there a probability value of that trajectory length be chosen. The approximation in equation (1) shows it depends on a maximum value; so that its distribution is equal the unit between these values. Levy flight is a super diffusive motion, so that tends to stay in the same direction any longer. The exponent μ must be: $1 < \mu < 3$. In this gap the motion is super diffusive, or, in other words, a Lévy flight motion type.

1.1 Model of Mission

The task model is given as follows: the flight is defined by waypoints. These waypoints are determined by the length of trajectory (randomly, according to Lévy function) and the yaw angle (according to a uniform density function). In this work, $\mu = 1.125$. The operational altitude is 60 m, and the operating speed is 20 km/h. Besides, the maximum range of the radio link and weather conditions are disregarded. The monitoring is accomplished by imaging, with the camera's viewing

Journal of Physics: Conference Series **641** (2015) 012031 doi:10.1088/1742-6596/641/1/012031

range $\alpha = 40°$. The UAV autonomy is 20 min and the overflown region is square, 180 m long. The surveillance area size, the camera's viewing angle and autonomy of quadrotor UAV are related: the area is divided into square subareas whose side length is 30 m long [5]. Each area is a square subscribed in the circumference of specified radius. Once the UAV has imaged all subareas, the search is considered complete. It was observed that the mean total length for the mission is approximately 5.8 km, which, given the values of autonomy and maximum speed considered valid [5]. The steps of the mission are: 1-Takeoff; 2-Choice of the random variables (trajectory length and yaw angle); 3-Implementation of yaw motion; 4-Implementation of horizontal motion. If the full scan of the area has not been reached, the process returns to step 2, otherwise the UAV comes back to the launch point. If the next waypoint is specified outside the surveillance area, the UAV moves to the edge of the area in the predetermined direction, covering the remaining distance at a mirrored angle.

1.2. Dynamic Model of Quadrotor Flight

The adopted definitions are basically the same as other studies [3-7]. The following assumptions are adopted: the vehicle is considered to be a rigid body; its structure is symmetrical; the upper quadrotor is one where the propellers are located; the four propellers are contained in the same plane (propellers plan); the distance between the center of mass and the plane of the propellers is negligible; the engines are numbered 1-4, clockwise, looking up from top to bottom; engines 1 and 3 rotate in a counterclockwise direction; the motors 2 and 4, clockwise; the propellers cause tensile forces parallel and perpendicular to the plane of propellers; the four motors are identical; the drag force is negligible. Two coordinate systems are used: an inertial system and a mobile system. Figure 1 shows them:

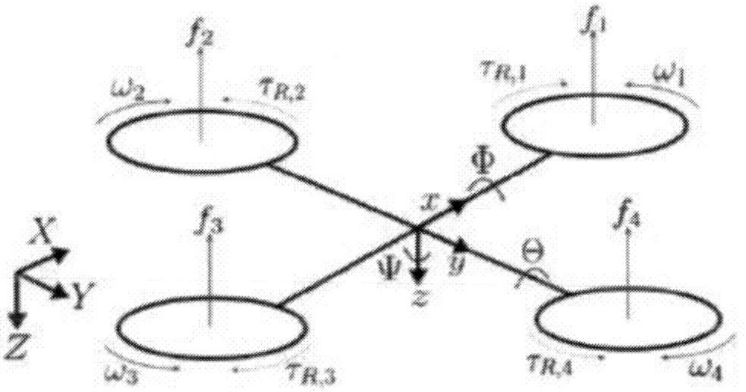

Figure 1. Inertial coordinate and UAV body systems. Adapted from [4].

The mobile system is fixed on quadrotor mass center. Such as [3], [5] and [6], for quadrotor system, the Euler-Lagrange formalism states the generalized coordinates are $\mathbf{q} = [x_n\ y_e\ z_d\ \phi\ \theta\ \psi]^T$. The origin of the inertial system is taken to be the place of the take-off. The variables, x_n, y_e and z_d are the translations in the X, Y and Z, respectively. The variables ϕ, θ and ψ represent the orientation in relation to the inertial system through successive rotations about the axes x, y and z. The sequence of rotations that takes the orientation of the body system to the orientation of the inertial one is [3-7]:

$$C_0{}^B = \begin{pmatrix} c_\theta c_\psi & s_\phi s_\theta c_\psi - c_\phi s_\psi & c_\phi s_\theta c_\psi + s_\phi s_\psi \\ c_\theta s_\psi & s_\phi s_\psi s_\theta + c_\phi c_\psi & c_\phi s_\psi s_\theta - s_\phi c_\psi \\ -s_\theta & s_\phi c_\theta & c_\phi c_\theta \end{pmatrix} \tag{2}$$

And the Lagrangian function Λ of the quadrotor system is:

$$\Lambda = (1/2)[m(\dot{x}^2 + \dot{y}^2 + \dot{z}^2) + I_{xx}(\dot{\phi} - \dot{\psi}s_\theta)^2 + I_{yy}(\dot{\theta}c_\phi + \dot{\psi}c_\theta s_\phi)^2 + I_{zz}(\dot{\psi}c_\phi c_\theta - \dot{\theta}s_\phi)^2] + mgz \tag{3}$$

From equation (3), considering the first three variables of the vector $\mathbf{q}$, the forces related to these coordinates are respectively:

$$F = \begin{vmatrix} F_x \\ F_y \\ F_z \end{vmatrix} = \begin{vmatrix} m\ddot{x} \\ m\ddot{y} \\ m\ddot{z} - mg \end{vmatrix} \tag{4}$$

For the rotational case it is considered the Euler angles are near zero. This simplifies the control, wherein each angle: sin(α) ≈ α ≈ 0 and cos(α) ≈ 1 (where α is ϕ, θ). Such approach is also considered valid in [3-7]. From equation (3) again, and after considering α ≈ 0 [3-7]:

$$\tau = \begin{vmatrix} \tau_\phi \\ \tau_\theta \\ \tau_\psi \end{vmatrix} = \begin{vmatrix} I_{xx}\ddot{\phi} + (I_{zz} - I_{yy})\dot{\theta}\dot{\psi} \\ I_{yy}\ddot{\theta} + (I_{xx} - I_{zz})\dot{\phi}\dot{\psi} \\ I_{zz}\ddot{\psi} + (I_{yy} - I_{xx})\dot{\phi}\dot{\theta} \end{vmatrix} \tag{5}$$

Equation (5) is generic. The forces generated by the propellers make the equations particular. The four efforts those governing the motion are: the lift force and the three moments associated with the angular motions. The relationship between effort and rotations is given by [3-7]:

$$\begin{vmatrix} \Omega_1^2 \\ \Omega_2^2 \\ \Omega_3^2 \\ \Omega_4^2 \end{vmatrix} = \begin{vmatrix} \frac{1}{4k_f} & 0 & \frac{1}{2k_f l_c} & \frac{1}{4k_m} \\ \frac{1}{4k_f} & \frac{-1}{2k_f l_c} & 0 & \frac{-1}{4k_m} \\ \frac{1}{4k_f} & 0 & \frac{-1}{2k_f l_c} & \frac{1}{4k_m} \\ \frac{1}{4k_f} & \frac{1}{2k_f l_c} & 0 & \frac{-1}{4k_m} \end{vmatrix} \begin{vmatrix} T \\ L \\ M \\ N \end{vmatrix} \tag{6}$$

In which k_f is the thrust factor, k_m is the drag factor and l_c is the quadrotor arm length. Each of the motors will produce a force along the normal direction of the propellers plane, so that:

$$m\begin{vmatrix} \dot{v}_n \\ \dot{v}_e \\ \dot{v}_d \end{vmatrix} = mg\begin{vmatrix} 0 \\ 0 \\ 1 \end{vmatrix} - C_0^B \begin{vmatrix} 0 \\ 0 \\ k_f(\Omega_1^2 + \Omega_2^2 + \Omega_3^2 + \Omega_4^2) \end{vmatrix} \tag{7}$$

Once ϕ ≈ 0 e θ ≈ 0, it implies $\tau_\phi \approx L$, $\tau_\theta \approx M$ e $\tau_\psi \approx N$. So:

$$\ddot{\phi} = \frac{L + (I_{yy} - I_{zz})\dot{\theta}\dot{\psi}}{I_{xx}} \qquad \ddot{\theta} = \frac{M + (I_{zz} - I_{xx})\dot{\phi}\dot{\psi}}{I_{yy}} \qquad \ddot{\psi} = \frac{N + (I_{xx} - I_{yy})\dot{\phi}\dot{\theta}}{I_{zz}} \tag{8}$$

1.3. Flight Control

Given the nonlinear nature of quadrotor dynamics, a nonlinear control technique is implemented: the backstepping control. The use of this control is widespread in literature [3-7]. The state vector is:

Journal of Physics: Conference Series **641** (2015) 012031 doi:10.1088/1742-6596/641/1/012031

$$X = |\phi \quad \dot{\phi} \quad \theta \quad \dot{\theta} \quad \psi \quad \dot{\psi} \quad z \quad \dot{z} \quad x \quad \dot{x} \quad y \quad \dot{y}|^T$$

$$X = |x_1 \quad x_2 \quad x_3 \quad x_4 \quad x_5 \quad x_6 \quad x_7 \quad x_8 \quad x_9 \quad x_{10} \quad x_{11} \quad x_{12}|^T$$

$$\dot{X} = |\dot{x}_1 \quad \dot{x}_2 \quad \dot{x}_3 \quad \dot{x}_4 \quad \dot{x}_5 \quad \dot{x}_6 \quad \dot{x}_7 \quad \dot{x}_8 \quad \dot{x}_9 \quad \dot{x}_{10} \quad \dot{x}_{11} \quad \dot{x}_{12}|^T$$

$$\dot{X} = \begin{vmatrix} x_2 \\ x_4 x_6 a_1 + (L/I_{xx}) \\ x_4 \\ x_2 x_6 a_2 + (M/I_{yy}) \\ x_6 \\ x_2 x_4 a_3 + (N/I_{zz}) \\ x_8 \\ g - (1/m) c_{x1} c_{x3} T \\ x_{10} \\ -(1/m)(u_x)T \\ x_{12} \\ -(1/m)(u_y)T \end{vmatrix} \tag{9}$$

Wherein $c_x = \cos(x)$ e $s_x = \sin(x)$. And:

$$a_1 = \frac{(I_{yy} - I_{zz})\dot{\theta}\dot{\psi}}{I_{xx}} \qquad a_2 = \frac{(I_{zz} - I_{xx})\dot{\phi}\dot{\psi}}{I_{yy}} \qquad a_3 = \frac{(I_{xx} - I_{yy})\dot{\phi}\dot{\theta}}{I_{zz}} \tag{10}$$

$$u_x = c_\phi s_\theta c_\psi + s_\phi s_\psi \qquad u_y = c_\phi s_\psi s_\theta - s_\phi c_\psi \tag{11}$$

For x_{2k-1} (k = 1,2,3,4,5,6), the following variable change are adopted:

$$z_{2k-1} = x_{2k-1}^{ref} - x_{2k-1} \qquad z_{2k} = x_{2k} - \dot{x}_{2k-1}^{ref} - \alpha_{2k-1} z_{2k-1} \tag{12}$$

The candidate Lyapunov function that stabilizes the system is:

$$V(z_{2k-1}, z_{2k}) = \frac{(z_{2k-1})^2 + (z_{2k})^2}{2} \tag{13}$$

From its derivative (considering the derivatives of x_{2k-1}^{ref}, in time, is zero) the solution is [3, 5, 6]:

$$\dot{x}_{2k} = z_{2k-1} - \alpha_{2k} z_{2k} - \alpha_{2k-1}(z_{2k} + \alpha_{2k-1} z_{2k-1}) \tag{14}$$

For each of k values, it is possible to write an expression for each of efforts. The intention is that the control of the motion does not ask for input values that exceed the capacity of the engines. During each stage, only the motion in question is executed. For the mission it is considered that the motions are critically damped. In this way, from the last assumptions and from equation (14) is possible write:

$$\begin{gathered} x_{2k-1} = [x_{2k-1}^{ref} - x_{2k-1}(0)]\{1 - e^{-\omega_n t}[1 + \omega_n t]\} + x_{2k-1}(0) \\ x_{2k} = [x_{2k-1}^{ref} - x_{2k-1}(0)]\omega_n^2 t e^{-\omega_n t} \end{gathered} \tag{15}$$

From equation (15) and the assumption during a motion that one will be the unique, efforts will be:

$$\begin{vmatrix} T \\ L \\ M \\ N \\ u_x \\ u_y \end{vmatrix} = \begin{vmatrix} m\{g + [x_7^{ref} - x_7(0)]\omega_n{}^2 e^{-\omega_n t}(\omega_n t - 1)\} \\ I_{xx}[(x_1^{ref} - x_1(0))\omega_n{}^2 e^{-\omega_n t}(1 - \omega_n t)] \\ I_{yy}[(x_3^{ref} - x_3(0))\omega_n{}^2 e^{-\omega_n t}(1 - \omega_n t)] \\ I_{zz}[(x_5^{ref} - x_5(0))\omega_n{}^2 e^{-\omega_n t}(1 - \omega_n t)] \\ \frac{c_{x3}}{g}[(\alpha_9 + \alpha_{10})(x_{10}) + (1 + \alpha_9\alpha_{10})(x_9 - x_9^{ref})] \\ \frac{c_{x3}}{g}[(\alpha_{11} + \alpha_{12})(x_{12}) + (1 + \alpha_{11}\alpha_{12})(x_{11} - x_{11}^{ref})] \end{vmatrix} \tag{16}$$

Wherein:

$$(\alpha_{2k-1} - \alpha_{2k}) = \pm 2 \tag{17}$$

$$\omega_n = \frac{(\alpha_{2k-1} + \alpha_{2k})}{2} \tag{18}$$

One criterion adopted here is: at no time, mission control must request values for the efforts that exceed the limits of the actuators. So, the following inequality must be satisfied:

$$0 \le \Omega_i^2 \le \Omega_{max}^2 \tag{19}$$

1.3.1. Roll and Pitch Control

The first four values of gains, α_1 to α_4, are related. The pair α_1 and α_3 follow the same gap, as well as the pair α_2 and α_4. It occurs because the conditions under the pitch motion that are extended to roll motion, due to the symmetry. From equations (6) and (19), the constraints to the pitch motion are:

$$\Omega_{max}^2 \ge \frac{T}{4k_f} + \frac{M}{2k_f l_c} \qquad 0 \le \frac{T}{4k_f} + \frac{M}{2k_f l_c} \qquad \Omega_{max}^2 \ge \frac{T}{4k_f} - \frac{M}{2k_f l_c} \qquad 0 \le \frac{T}{4k_f} - \frac{M}{2k_f l_c} \tag{20}$$

Now, from the equation (6) is possible explicit the frequency associate with this motion, or in other words, the gains α_3 and α_4. In this way, more one consideration is made: $|\theta| \le 45°$. For calculation of these gains, a limit case motion is considered:

$$\theta(t = 0) = \pm 45° \qquad \theta^{ref} = \mp 45° \qquad \dot{\theta} = 0 \tag{21}$$

The limit case would be a state where the quadrotor would be in a position in $\theta = +45°$ at the initial instant of the pitch motion, trying to reach the opposite state. Now two instants are considered: the initial instant of the motion and the instant after $2/\omega_n$ seconds. The first is the instant in which the effort's modulus is maximum, and the second one, the minimum modulus [5]. So, considering an instant in that $\theta(t=0) = -45°$ and $\theta^{ref} = +45°$, taking the instant $t = 0$ for the first and the last equations from (20), and the instant $t = 2/\omega_n$ for the other two ones, the most restrictive gap is given by the last equation of (20):

$$1 < \alpha_1 \le 8.153 \qquad -1 < \alpha_2 \le 6.153 \qquad 1 < \alpha_3 \le 8.153 \qquad -1 < \alpha_4 \le 6.153 \tag{22}$$

In fact, the consideration of the instant $t = 2/\omega_n$ have not taken to restrict conditions [5]. The taking into account of this instant in these equations was made only to ensure that the constraints would be respected yet.

1.3.2. Yaw Control

Again, four constraints are considered to ensure satisfactory values of the efforts:

$$\Omega^2_{máx} \geq \frac{T}{4k_f} + \frac{N}{4k_m} \qquad 0 \leq \frac{T}{4k_f} + \frac{N}{4k_m} \qquad \Omega^2_{máx} \geq \frac{T}{4k_f} - \frac{N}{4k_m} \qquad 0 \leq \frac{T}{4k_f} - \frac{N}{4k_m} \tag{23}$$

It occurs not only because equation (6), but because the duality of motion of the propeller pairs. When a pair rotates in clockwise another one rotates in counterclockwise. So, the considerations for only one case ensure the validity of the motion. The maximum amplitude considered for the yaw motion is $x_5^{ref} - x_5(0) = \pi$, a half total rotation. Now, taking equation for N, the instant $t = 0$ for the first and the last equations from equation (23), and the instant $t = 2/\omega_n$ (as in the previous case) for the other two ones, the most restrictive gap is given by the last equation of (23) [5]:

$$1 < \alpha_5 \leq 2.382 \qquad -1 < \alpha_6 \leq 0.382 \tag{24}$$

1.3.3. Altitude Control

In this case, the constraints from the equation (19) were supplanted for other two conditions: during takeoff and landing, the values of the rates of climb and descent should not exceed certain values. The values of maximum rate of climb and descent, respectively, are 7 m/s and 4 m/s, from the platform considered [5]:

$$-7 \leq \left[x_7^{ref} - x_7(0)\right]\omega_n{}^2 te^{-\omega_n t} \qquad 4 \geq [x_7^{ref} - x_7(0)]\omega_n{}^2 te^{-\omega_n t} \tag{25}$$

And:

$$\begin{matrix} x_7^{ref} = -60 & x_7(0) = 0 \\ x_7^{ref} = 0 & x_7(0) = -60 \end{matrix} \tag{26}$$

The first line of (26) represents the takeoff. The second one does the landing. The maximum amplitude for both occurs at $t = 1/\omega_n$ [5]. So, from these, the second equation from (25) gives the following gap:

$$1 < \omega_7 \leq 1.181 \qquad -1 < \omega_8 \leq -0.819 \tag{27}$$

1.3.4. Horizontal Motion Control

For the horizontal motion, is considered the average horizontal velocity. The average horizontal velocity is considered approximately constant, being the same in each step:

$$\bar{x}_{2k} = \frac{1}{(\frac{5}{\omega_n})}\int_0^{\frac{5}{\omega_n}} x_{2k}dt \qquad \bar{x}_{2k} \approx 0.192\omega_n[x_{2k-1}^{ref} - x_{2k-1}(0)] \tag{28}$$

In this case, is considered the same gain pair ($\alpha_9 = \alpha_{11}$, $\alpha_{10} = \alpha_{12}$). In these cases, k = 5 e k = 6. So:

$$\bar{v}_h = \sqrt{x_{10}^2 + x_{12}^2} = 0.192\omega_n l \tag{29}$$

$$\alpha_9 = \alpha_{11} = 1 + \frac{10.2}{l} \qquad \alpha_{10} = \alpha_{12} = -1 + \frac{10.2}{l} \tag{30}$$

The average horizontal velocity has supplanted other considered conditions, showing itself the best one [5]. The four last values of the gains are dependent on the trajectory length, as shows the equation (29). The average speed adopted for the steps is 2m/s [5]. The control diagram is shown in Figure 2:

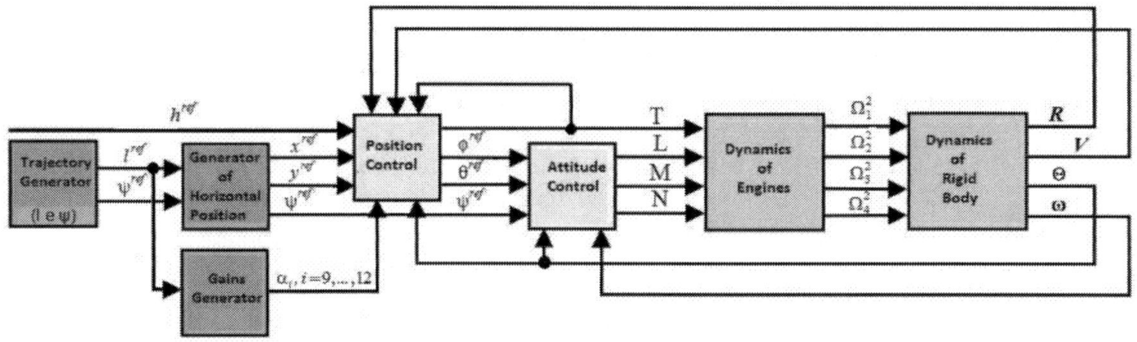

Figure 2. Control Diagram.

2. Methodology

The methodology is divided in two parts. The first one is about Levy flight simulations, around the mean total length needed to the full scan of the area. The simulations considered a squared area, 180 m long. During the mission, the quadrotor motion should be kept inside the area. During the implementation of the horizontal motion, if the quadrotor tried to leave the area, a change of direction (a reflexive motion in the area border) is made. The area was divided into squared subareas, 30 m long. In this way, a mission was considered completed when all subareas were visited by the quadrotor. 200 simulations about the vigilance mission were done. The second simulations cases concerned the proposed control laws. Firstly, each control law was tested separately, and after all of them were tested together.

3. Results and Discussions

Figures 3 to 7 show the UAV flight for trajectories generated from the design based on Levy flights. The tests were conducted using numerical simulations, from the simulation model developed in [6] and adapted in this work. The gains are: $\alpha_1 = \alpha_3 = 8$, $\alpha_2 = \alpha_4 = 6$, $\alpha_5 = 2.3$, $\alpha_6 = 0.3$, $\alpha_7 = 1.18$ and $\alpha_8 = -0.82$. The other gains depend on the length path step. The values of the gains α_{2k} can assume negative values. This is true, since the stability is given by ω_n, ($\omega_n > 0$).

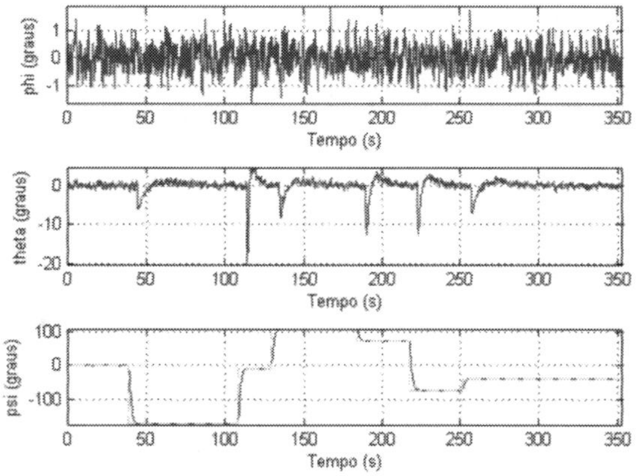

Figure 3. Euler angles.

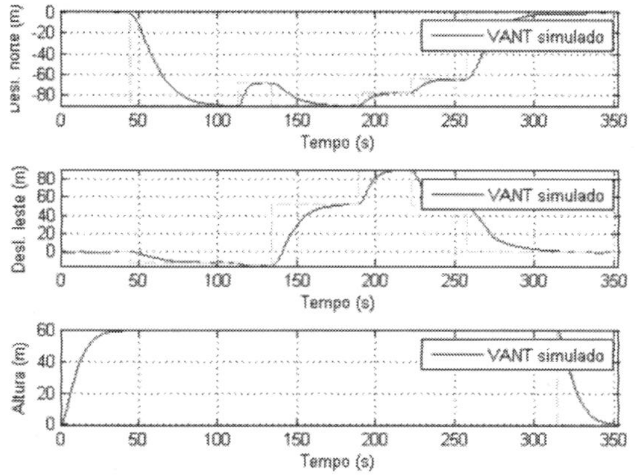

Figure 4. Horizontal and vertical displacements.

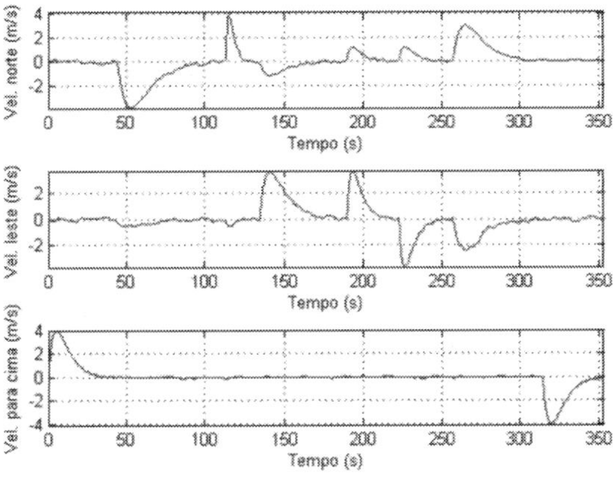

Figure 5. Horizontal and ascent speeds.

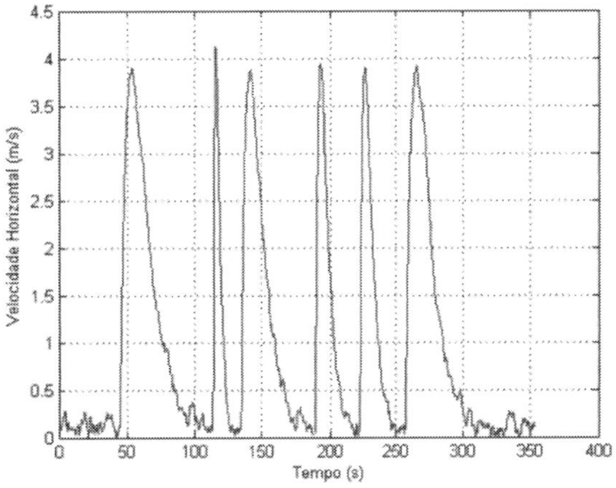

Figure 6. Horizontal speed.

Figures 3 to 6 show the performance of the UAV. Throughout the mission, the quadrotor managed to reach the points determined by the Lévy flight. Figure 3 shows the yaw angle and the other two ones. As can be seen, at no instant roll and pitch modules are bigger than the value proposed. Yaw angle reach the chosen values determined. Green lines represent the references. Figure 4 shows the horizontal and vertical displacements. As can be seen, each waypoint is reached. The maximum rates of climb and descent values did not exceed 4 m/s -4 m/s respectively, as can be seen in Figure 5. In Figure 6, at no instant the horizontal speed exceeded the value of 20 km/h (approximately 5.56 m/s). The success of the mission shows the constraint represented by equation (19) was satisfied.

4. Conclusion

In this study, we addressed the control of a vehicle quadrotor for search and surveillance missions, through a random trajectories generator biologically inspired by the Lévy flight. For this, it agreed on a mission model and a backstteping to control the motion of the UAV. Given the control law, were imposed certain restrictions on the motion of quadrotor, leading to gains value ranges. Thus, it was possible to implement the Lévy flight for a search mission and surveillance restricted to a certain area.

References

[1] Viswanathan G M, Afanasyev V, Buldyrev S V, Murphy E J, Prince P A and Stanley, H E 1996 Lévy flight search patterns of wandering albatrosses. Nature 381, p. 413–415

[2] Viswanathan G M, Raposo E P and Da Luz M G E 2008 Lévy flights and superdiffusion in the context of biological encounters and random searches. PLR, 5, p. 133–150

[3] Bouabdallah S 2007. Design and Control of Quadrotors with Application to Autonomous Flying. PhD Thesis - École Polytechnique Fédérale de Lausanne. 154

[4] Silva A L 2012. Voo Autônomo de Veículo Aéreo Não Tripulado Tipo quadrotor. Relatório Final de Pós Doutorado – Universidade de São Paulo. 71

[5] Frederico L A 2015. Planejamento de Trajetórias e Controle de Voo em Missões de Busca Vigilância de um Veículo Aéreo Não Tripulado. Dissertação – Universidade Federal do ABC. 203

[6] Santana P H R Q A and Borges G A 2009. Modelagem e controle de quadrotores. Brasília. In: SBA (Ed.). IX Simpósio Brasileiro de Automação Inteligente - SBAI 2009.

[7] Mellinger D, Michael N and Kumar V 2010. Trajectory generation and control for precise aggressive maneuvers. The International Journal of Robotics Research, p. 11